Exploring Fundamental Issues in Nuclear Physics

Nuclear Clusters — Superheavy, Superneutronic, Superstrange, of Anti-Matter

Exploring Fundamental Issues in Nuclear Physics

Nuclear Clusters — Superheavy, Superneutronic, Superstrange, of Anti-Matter

Proceedings of the Symposium on Advances in Nuclear Physics in Our Time

Goa, India   28 November – 2 December 2010

Editor

Debades Bandyopadhyay

Saha Institute of Nuclear Physics, India

World Scientific

NEW JERSEY • LONDON • SINGAPORE • BEIJING • SHANGHAI • HONG KONG • TAIPEI • CHENNAI

Published by

World Scientific Publishing Co. Pte. Ltd.
5 Toh Tuck Link, Singapore 596224
USA office: 27 Warren Street, Suite 401-402, Hackensack, NJ 07601
UK office: 57 Shelton Street, Covent Garden, London WC2H 9HE

British Library Cataloguing-in-Publication Data
A catalogue record for this book is available from the British Library.

EXPLORING FUNDAMENTAL ISSUES IN NUCLEAR PHYSICS
Nuclear Clusters — Superheavy, Superneutronic, Superstrange, of Anti-Matter
Proceedings of the Symposium on Advances in Nuclear Physics in Our Time

ISBN-13 978-981-4355-72-8
ISBN-10 981-4355-72-0

Printed in Singapore by Mainland Press Pte Ltd.

PREFACE

It is a great pleasure for me to edit the proceedings of the symposium held in Goa, India from 28 November - 2 December, 2010. This event was jointly organised by Frankfurt Institute for Advanced Studies (FIAS) and Saha Institute of Nuclear Physics (SINP). The inspiration and driving force behind this symposium in picturesque Goa was Professor (Dr.) Walter Greiner. The symposium was held just after the 75th birthday of Professor (Dr.) Walter Greiner. We were happy to dedicate this symposium in honour of Professor Walter Greiner on the occasion of his 75th birthday.

This symposium focused on recent advances in superheavy and exotic nuclei, relativistic heavy ion collisions, DUBNA and FAIR facilities, novel phases of dense matter in neutron stars, black holes, gravitational waves, cosmology and other important issues like applications of heavy ions in cancer therapy and renewable energy. We are celebrating the 100th year of the publication of Rutherford's seminal paper on atomic nucleus in 2011. In this context, this symposium was very timely. Experts of the above mentioned fields and young researchers gathered to discuss their work in this symposium. Talks and discussions reached a great height and provided up-to-date accounts of various fields.

This symposium was shadowed by an unfortunate event. We can not forget the sudden demise of Mrs. Irina Oganessian, wife of Professor Yuri Oganessian on the very first day of the symposium. Mrs. Oganessian was a renowned violinist. We pay our homage to Mrs. Irina Oganessian; we convey our heart felt condolences to Professor Yuri Oganessian.

We thank all speakers and participants of the Goa symposium for making it a successful event. Last but not the least, we thank FIAS, SINP and Alexander von Humboldt Foundation for their support.

Debades Bandyopadhyay
Saha Institute of Nuclear Physics
Kolkata, India
15 April, 2011

CONTENTS

Three giants

Last supper with Irina Oganessian

Dinner in the garden

Keep attention!

An important point

Mikhail Itkis lectures

Attention everywhere

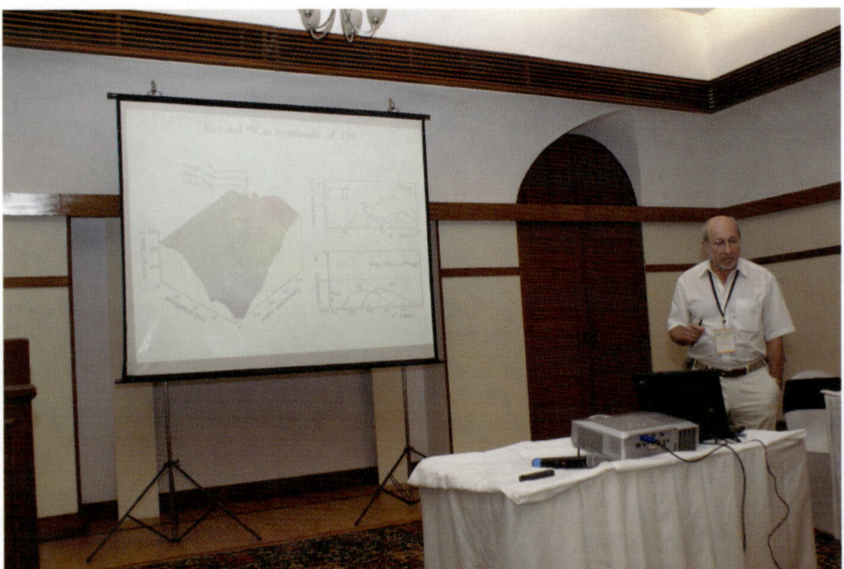

Valery Zagrebaev lectures

Yuri Oganessian and the superheavies

Martin Greiner explains

Natalya and Mikhail Itkis

Larry McLerran lectures

Jörg Aichelin objects

Neutron rich nuclei in bomb explosions

The beach

Jan Vaagen explains

Laura Quist takes care of everything

Yuri Oganessian lectures

Konstantin Gridnev and superneutronic nuclei

Andrey Popeko: Das Leben ist gelungen!

Mark Gorenstein objects

Natalya Itkis and Walter Greiner

Igor Mishustin – Robert Eichler – Valery Zagrebaev

Chhanda Samanta and Walter Greiner – Party time

Valery Zagrebaev – Igor Mishustin – Andrey Popeko

Christine and Sigurd Hofmann

Debades Bandyopadhyay

Walter Greiner – Final Remarks and Good Bye

COMPETITION BETWEEN FUSION-FISSION AND QUASIFISSION IN THE REACTIONS WITH HEAVY IONS

M. G. ITKIS*, I.M. ITKIS, G.N. KNYAZHEVA and E. M. KOZULIN

Flerov Laboratory of Nuclear Reactions, Joint Institute for Nuclear Research, Dubna, 141980 /Moscow region, Russia
**E-mail: itkis@jinr.ru*

The main aim of the present study is to evaluate the fusion probabilities and investigate competing quasifission process in the reactions with heavy ions leading to the formation of superheavy composite systems. The mass-energy distributions, as well as capture cross-sections of fission-like fragments for the reactions of ^{22}Ne, ^{26}Mg, ^{36}S, ^{48}Ca, ^{58}Fe and ^{64}Ni ions with actinides leading to the formation of superheavy compound systems with Z=102-120 at energies near the Coulomb barrier have been measured. The relative contribution of quasifission to the capture cross section becomes dominant for superheavy composite systems. Fusion-fission cross sections were estimated from the analysis of mass and total kinetic energy distributions.

Keywords: Superheavy nuclei, reaction dynamics, fusion probability, quasifission.

1. Introduction

In reactions with heavy ions complete fusion and quasifission (QF) are competing processes.[1,2] The relative contribution of QF to the capture cross section becomes dominant for superheavy composite systems, and compound nucleus (CN) formation is hindered by the QF process. The balance between the two processes strongly depends on the entrance channel properties, such as mass asymmetry, deformation of interacting nuclei, collision energy, and the Coulomb factor Z_1Z_2.

A large success was achieved in the synthesis of superheavy elements with Z$= 108 - 118$ in the reactions of "cold" and "warm" fusion. Reactions with Pb/Bi targets, leading to lowly excited compound nucleus (CN), are used for "cold" fusion,[13] whereas the reactions with actinide targets, leading to excited CN, are used for "warm" fusion.[4] In the "cold" fusion reactions the excitation energy of CN is 10-20 MeV near the reaction threshold, the

CN formation cross section is suppressed strongly by quasifission (QF) and deep-inelastic reactions.

The excitation energy of a formed CN is about 30-40 MeV in "warm" fusion reactions and the de-excitation of CN to the ground state is due to the emission of three or four neutrons and several γ-rays. Nevertheless in the "warm" fusion reactions the formed CN is more neutron rich and nearer to the closed neutron shell at N= 184 than in the case of "cold" fusion reactions. Consequently, the "warm" fusion reactions are more preferable for synthesis of the SHE. At FLNR double magic ^{48}Ca ions were used for the productions of elements with Z=112-118.[4] The cross sections of SHE in these reactions do not change practically with increasing atomic number of CN and maintain the level of a few picobarn.

Nuclei with Z $>$ 118 cannot be synthesized in ^{48}Ca induced reactions since ^{249}Cf is the heaviest target material available for these purposes. One of the possible ways for further progress in the field of the SHE synthesis is to use the complete fusion reactions of ^{238}U, ^{244}Pu and ^{248}Cm nuclei with heavier projectiles, such as ^{58}Fe or ^{64}Ni, leading to the formation of even heavier elements with Z= $120 - 124$ and N= $179 - 183$. However, the increase of projectile charge leads to the increasing of the QF contribution.

This paper presents the results of the experimental investigations of the influence of the entrance channel properties on the competition between fusion-fission (FF) and QF in "warm" fusion reactions. The properties of binary reaction products obtained in the reactions of ^{22}Ne, ^{26}Mg, ^{36}S, ^{48}Ca, ^{58}Fe and ^{64}Ni with actinides have been measured at energies near the Coulomb barrier. Binary reaction products were detected by the two-arm time-of-flight spectrometer CORSET.[5]

Previously in the experimental investigations the symmetric fragment region with mass $A_{CN} \pm 20\mu$ was often attributed with CN-fission. However, a realistic description of the mass, energy and angular distributions of the reaction fragments formed in deep inelastic scattering, QF and CN-fission processes in low energy heavy ion collisions show[6] that the potential energy surface for these systems is strongly modulated by shell effects and leads to the appearance of deep valleys corresponding to the formation of well bound magic nuclei. In accordance with these calculations, at least three paths leading to the formation of fission-like fragments can be distinguished: asymmetric QF caused by the influence of proton shells with Z= $28, 82$ and neutron shells with N= $50, 126$; symmetric QF determined by the shells with Z $= 50$ and N= 82; CN-fission leading to the formation of symmetric fragments. Therefore, in present analysis we supposed that the mass region

$A_{CN} \pm 20\mu$ is populated as CN-fission fragments, as well as QF fragments and paid a special attention to the properties of symmetric fragments in order to assign probabilities of CN-fission and QF processes.

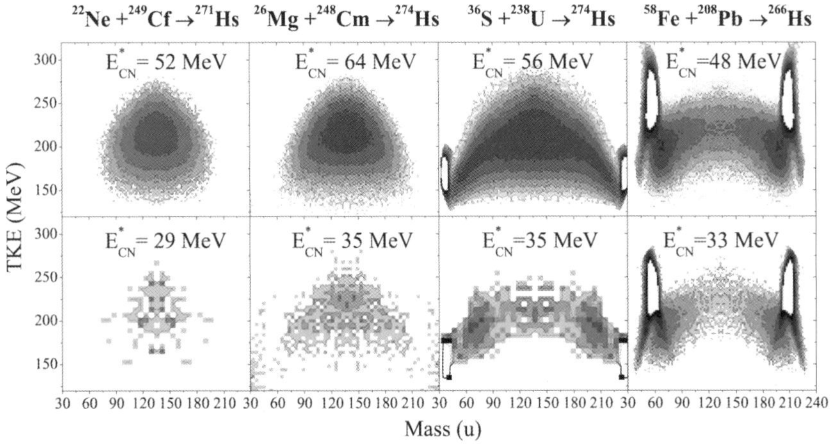

Fig. 1. Mass-energy distributions of binary reaction fragments for the reactions ^{22}Ne+^{249}Cf, ^{26}Mg+^{248}Cm, ^{36}S+^{238}U, ^{58}Fe+^{208}Pb leading to the formation of elements with Z= 108 at energies above (top panel) and below (bottom panel) the Coulomb barrier.

2. CN-Fission and QF in the Reactions Leading to the Hs Composite Systems

Mass and Total Kinetic Energy (TKE) distributions of binary reaction products obtained in the reactions ^{22}Ne+^{249}Cf, ^{26}Mg+^{248}Cm, ^{36}S+^{238}U, ^{58}Fe+^{208}Pb have been measured. All reactions lead to Hs isotopes. In the chosen reactions the parameter of mass-asymmetry in the entrance channel η varies strongly: for the reaction ^{58}Fe+^{208}Pb: $\eta = 0.571$, for ^{36}S+^{238}U: $\eta = 0.737$, for ^{26}Mg+^{248}Cm: $\eta = 0.810$, and for ^{22}Ne+^{249}Cf: $\eta = 0.838$. As demonstrated in Fig. 1 the TKE-mass distributions change with decreasing asymmetry η in the entrance channel from symmetric for incoming ^{22}Ne-ions to strongly asymmetric for incoming ^{58}Fe-ions. These changes are understood as reflecting the relative contributions of CN-fission and QF to the fission process excited Hs* isotopes as depending on the reaction studied.

It is clearly seen that even at similar CN excitation energies the TKE-mass distributions are vastly different for these reactions. In the case of the reactions ^{22}Ne+^{249}Cf and ^{26}Mg+^{248}Cm the mass distributions have a near Gaussian shape with no evidence for asymmetric fission. The reactions are considered to be mainly attributed to the CN-fission process. For the ^{36}S+^{238}U reaction the mass distributions of the fission-like fragments change markedly. This difference in mass distribution for the ^{26}Mg+^{248}Cm and ^{36}S+^{238}U reactions is connected with an increasing contribution of the QF process for the ^{36}S-induced reaction. At low excitation energies QF is the dominant process for the reaction ^{36}S+^{238}U. At higher excitation energies the mass distribution becomes symmetric and similar to the reaction ^{26}Mg+^{248}Cm though - due to a remaining trace of QF - slightly wider.

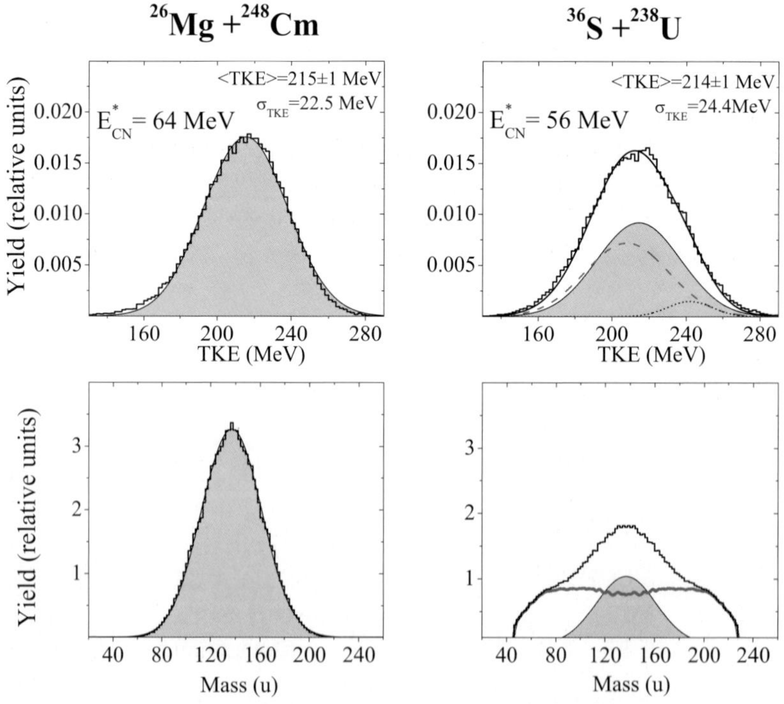

Fig. 2. Top panel: TKE distributions of fragments with masses $A_{CN}/2 \pm 20$ u for the reactions ^{26}Mg+^{248}Cm and ^{36}S+^{238}U. Bottom panel: Mass distributions for the same reactions decomposed into CN-fission (filled region) and QF (solid line) components.

For the reactions ^{26}Mg+^{248}Cm at excitation energy of 64 MeV of the CN, the mass distribution is well described by a single Gaussian, and the average TKE and its variance are close to the prediction of the Liquid Drop Model (LDM) at high excitation energies of compound nuclei and may hence be attributed to CN-fission process (see Fig. 2 left-hand panel). As opposed to the ^{26}Mg+^{248}Cm reaction where CN-fission is the main process in the case of the reaction ^{36}S+^{238}U both CN-fission and QF contribute significantly. The mean value and dispersion of TKE increase with increasing energy of ^{36}S ions and exceed the LDM predictions at the highest energy. This is clearly seen in Fig. 2. At an excitation energy of 56 MeV the TKE distribution seems to be a simple Gaussian. But the standard deviation is higher than that from the LDM, whereas in the reactions with Mg ions at the high excitation energies the TKE distributions are well described by single Gaussian with the parameters coming from the LDM. One may speculate about the presence of other processes together with the CN-fission in the symmetric mass region for the ^{36}S+^{238}U reaction. We assume that the mass symmetric fragments may be formed by three different modes: CN-fission, symmetric QF, and a tail of the asymmetric QF process. To evaluate the contribution of the CN-fission process in the symmetric mass region, the TKE distributions are decomposed as a sum of three Gaussians. One of them is associated with the CN-fission process (filled region on the top panel in Fig. 2). The yield of CN-fission fragments in the mass distribution being described by a Gaussian (filled region on the bottom panel in Fig. 2) was weighted using this estimation. More detailed analysis of mass and TKE distributions for these reactions at different excitation energies are presented in Ref 7.

The TKE-mass distribution for the ^{58}Fe+^{208}Pb reaction has a wide U shape even at excitation energy of 48 MeV (well above the Bass barrier). For this reaction the QF process dominates at energies below and above the Bass barrier. This reaction takes a special place due to the fact that one of the partners is double-magic lead. It is known that for heavy-ion induced reaction the formation of QF fragments is determined by the strong influence of the nuclear shell at Z= 82 and N= 126 (double magic lead). Therefore, for this reaction the QF fragments overlap strongly with quasi-elastic and deep-inelastic events. It is clearly seen on the TKE-mass matrices in the Fig. 1. The QF fragments for the reactions with ^{36}S and ^{58}Fe ions are mainly formed in the mass region near the closed shells with Z= 28 and N= 50 for the bump at low masses and with Z = 82 and N= 126 for the bump at large masses.

For the reactions $^{26}Mg+^{248}Cm$ and $^{36}S+^{238}U$ the cross section of the evaporation residues (ER) amounts to a few nanobarns[8,9] and contributes insignificantly to the fusion cross section. Thus we can estimate the fusion probability using the measured TKE-mass distributions as the ratio between the number of events attributed to fusion-fission in the frame of the LDM and all fission-like fragments. The so-obtained fusion probabilities as a function of the energy above the barrier are presented as open symbols in Fig. 3 for these reactions.

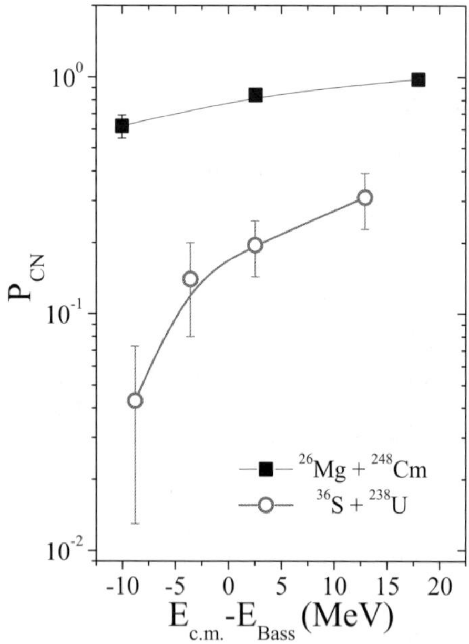

Fig. 3. The fusion probabilities in the reactions $^{26}Mg+^{248}Cm$ and $^{36}S+^{238}U$ obtained from the present analysis of mass-energy distributions of fissionlike reaction products.

3. Reactions with an ^{238}U Target

Figure 4 shows the mass energy distributions of binary fragments obtained in the reactions of ^{36}S, ^{48}Ca, ^{64}Ni ions with an uranium target. The Coulomb factors are 1472, 1840 and 2576 for the $^{36}S+^{238}U$, $^{48}Ca+^{238}U$ and $^{64}Ni+^{238}U$, respectively. Some noteworthy features of the asymmetric QF component of fragment mass distributions for the studied reactions can be

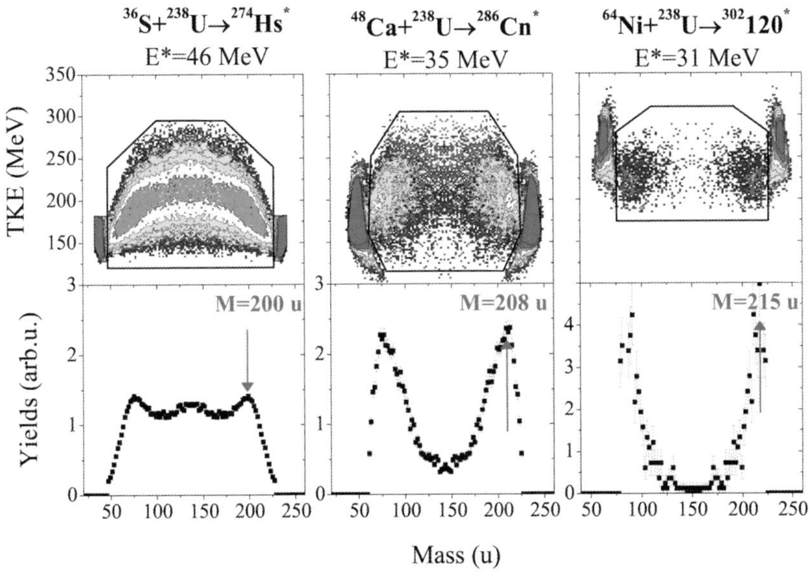

Fig. 4. Mass-energy distributions of binary fragments for the reactions ^{36}S, ^{48}Ca, ^{64}Ni+^{238}U at energies close to the Coulomb barrier [from top to bottom: the (TKE, Mass) matrices for binary products; the mass yields for fissionlike fragments inside the outlined contour on the (TKE, Mass) matrices].

highlighted at this point. Generally, in heavy-ion-induced reactions the formation of asymmetric QF fragments is connected with the strong influence of the nuclear shell at Z = 82 and N = 126 (doubly magic lead). In fact, for the ^{48}Ca+^{238}U reaction the maximum yield corresponds to fragments with masses 208 μ. However, in reactions with lighter projectiles on a uranium target, the asymmetric QF peak shifts toward more symmetric masses. In the reaction with ^{36}S-ions the asymmetric QF peaks around 200 μ. By contrast, for the heavier projectile ^{64}Ni, the maximum yield of asymmetric QF fragments corresponds to the heavy mass 215 μ. This trend is illustrated in Table 1, where the positions of heavy QF fragments for these reactions are presented. But, in the formation of the asymmetric QF component, also the closed shell in the light fragment at Z = 28 and N= 50 could be effective, together with the shells Z= 82 and N= 126, and could lead to the shift of the asymmetric QF peak. Based on the simple assumption of an N/Z equilibration, the masses of the light and heavy fragments corresponding to these closed shells were calculated. In Table 1 M_H^{Shell} is a heavy fragment mass averaged over all these shells. The obtained values of M_H^{Shell} are in good agreement with the experimental ones, except for the more asymmetric $^{30}Si+^{238}U$ reaction. For this reaction the Coulomb repulsion is expected to be smaller. This may lead to longer reaction times before separation for asymmetric QF and thus allow for larger numbers nucleons to be exchanged. For the other more symmetric reactions with heavier projectiles, the major part of the asymmetric QF peak fits into the region of the Z= 82, N = 126 and Z= 28, N = 50 shells. The maximum yield of the asymmetric QF component is a mixing between all these shells.

Table 1. Positions of heavy peaks in the primary mass distributions of asymmetric QF fragments in reactions with heavy ions.

Reaction	Z_1Z_2	M_H	M_H^{Shell}	Reference
$^{30}Si+^{238}U$	1288	178	199.3	Ref. 10
$^{36}S+^{238}U$	1472	200	202.5	Ref. 7,10
$^{40}Ar+^{238}U$	1656	204	204.5	Ref. 10
$^{48}Ca+^{238}U$	1840	208	208.5	Ref. 11
$^{64}Ni+^{238}U$	2576	215	216.5	Ref. 11

Unfortunately, the transition to heavier projectile results in the large increase in the Coulomb factor Z_1Z_2 that is crucial for the competition between CN-fission and QF. In figure 5 the TKE distributions of fissionlike fragments in the mass region $A_{CN}/2 \pm 20$ u for the reactions $^{48}Ca+^{238}U$, $^{58}Fe+^{244}Pu$ and $^{64}Ni+^{238}U$ are presented. It is seen that the TKE distributions have a complex structure for all reactions. The TKE distributions were described by the sum of the three Gaussians for all the reactions studied.

The high energy part may arise instead from the symmetric mode of the QF. Furthermore, we note that the mean TKE of this mode is about 40 MeV higher that the mean TKE for the CN-fission process. Considering that both processes give rise to symmetric mass fragments, the difference in mean TKE can be taken as an evidence that in the QF process a complete dissipation of the entrance channel energy does not occur. As a consequence, the symmetric fragments with high TKE do not originate from complete fusion because the final fragments retain part of the entrance channel total kinetic energy. In contrast to $^{48}Ca+^{238}U$, for the reaction $^{64}Ni+^{238}U$ the TKE distribution has more pronounced low and high energy components (see Fig. 5c), while the component with average value of 252 MeV (corresponding to the Viola systematics) is highly hindered. Because of the low

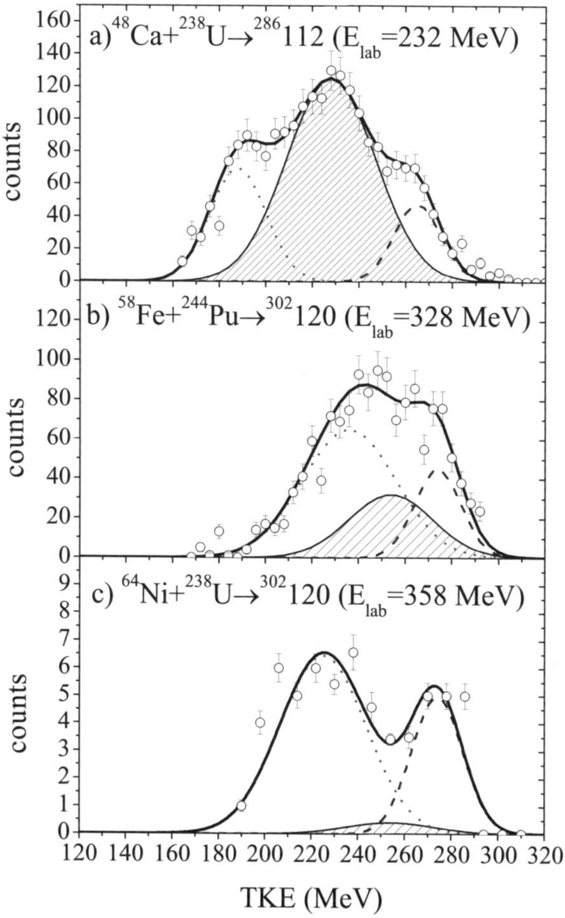

Fig. 5. TKE distributions of fragments with masses $A_{CN}/2 \pm 20$ u for the reaction ^{48}Ca+^{238}U, ^{58}Fe+^{244}Pu and ^{64}Ni+^{238}U.The hatched region corresponds to fusion-fission with energy taken from the Viola systematics.

statistics, only an upper value for the relative yield of the CN-fission component can be reasonably given. Table 2 gives the relative contribution of all symmetric fragments in the mass range $A_{CN}/2 \pm 20$ u and symmetric fragments with TKE corresponding to the Viola systematics. The obtained capture cross sections are about a few hundred millibarns for Ca and Ni induced reactions, whereas the formation of symmetric fragments is one order of magnitude less for the reaction ^{64}Ni+^{238}U. Yet, in the case of the Ca+U at the energies above the barrier, approximately 70% of the events have the TKE expected for the CN-fission process, whereas in the case of the Ni reaction only a few percents of the symmetric fragments have the TKE compatible with the Viola prediction for the ^{302}120 CN-fission. While the ^{64}Ni+^{238}U reaction has lower excitation energy at the reaction threshold energy, the CN-fission cross section is suppressed by stronger QF process.

Table 2. Relative contributions of all symmetric fragments ($\sigma_{(A_{CN}/2\pm20)}$) and symmetric fragments with TKE corresponding to the older Viola systematics (σ_{CNF}) to the capture cross-section (σ_{cap}) for the ^{48}Ca+^{238}U, ^{64}Ni+^{238}U and ^{58}Fe+^{244}Pu reactions at CN excitation energies of ∼45 MeV.

Reaction	$\frac{\sigma_{(A_{CN}/2\pm20)}}{\sigma_{cap}}$ (%)	$\frac{\sigma_{CNF}}{\sigma_{(A_{CN}/2\pm20)}}$ (%)	$\frac{\sigma_{CNF}}{\sigma_{cap}}$ (%)
^{48}Ca+^{238}U	12±2	68±3	8±4
^{64}Ni+^{238}U	4±1	≤ 5	≤ 0.2
^{58}Fe+^{244}Pu	8±3	≤ 25	≤ 2

4. Conclusions

Previously for separation of QF and CN-fission the angular and mass distributions of fissionlike fragments were used. The present analysis of the TKE distributions of fragments with mass $A_{CN}/2 \pm 20$ u for different reactions studied shows that the variance of the TKE distribution is sensitive to the presence of the QF process. At excitation of CN larger than 40 MeV (when the shell in CN-fission is practically disappeared) broadening of the TKE distribution point to the presence of the both QF and CN-fission processes. Narrowing of the TKE distribution indicate that QF is a dominant process.

The fusion probabilities for different reactions were estimated from the analysis of the TKE distribution in addition to mass distributions. The CN-fission cross-section in the ^{64}Ni+^{238}U→^{302}120 case drops three orders of magnitude with respect to the ^{48}Ca+^{238}U →^{286}112 case. This is unfortunately a limiting factor. The relative contribution of the CN-fission from ^{64}Ni+^{238}U is much lower than in the case of ^{58}Fe+^{244}Pu →^{302}120. Recently the experiments aimed at the synthesis of isotopes of element Z=120 have been performed using the ^{244}Pu(^{58}Fe,xn)$^{302-x}$120 reaction[12] and ^{238}U(^{64}Ni,xn)$^{302-x}$120 reaction.[13] A cross-section limit of 0.4 pb at

E* = 44.7 MeV for the former reaction and 0.09 pb at E* = 36.4 MeV for the latter reaction were obtained. In the case of ^{48}Ca+^{238}U reaction the evaporation residue cross-section for 3n,4n channels is about a few pb. Thereby in the transition from Ca to Fe and Ni ions, the evaporation residue cross-section drops down at least one and two orders of magnitude, respectively.

Hereby, a further progress in the field of synthesis of superheavy nuclei can be achieved using hot fusion reactions between actinide nuclei and ^{48}Ca ions as well as actinide nuclei and ^{50}Ti, ^{54}Cr, ^{58}Fe ions. At the transition from Ca to Ni projectiles the contribution of the QF process rises sharply and Ni ions is not suitable for the synthesis of element Z$\geq$120 in the complete fusion reactions.

Acknowledgments

The work was supported by the Alexander von Humboldt foundation.

References

1. R. Bock et al., *Nucl. Phys. A* **388**, 334 (1982).
2. M.G. Itkis et al., *Int. J. Mod. Phys. E* **16**, 957 (2007).
3. S. Hofmann and G. Mnzenberg, *Rev. Mod. Phys.* **72**, 733 (2000).
4. Yu. Ts. Oganessian,*J. Phys. G: Nucl. Part. Phys.* **34**, R165 (2007).
5. E.M. Kozulin et al., *Instrum. Exp. Tech.* **51**, 44 (2008).
6. V.I. Zagrebaev and W. Greiner, *J. Phys. G* **31**, 825 (2005).
7. I.M. Itkis et al., *Phys. Rev. C* **83**, 064613 (2011).
8. J. Dvorak et al., *Phys. Rev. Lett.* **100**, 132503 (2008).
9. R.Graeger et al., *Phys. Rev. C* **81**, 061601(R) (2010).
10. K.Nishio et al., *in Proceedings of the International Symposium on Exotic Nuclei*, edited by Yu.E. Penionzhkevich and S.M. Lukyanov, AIP Conf. Proc. No. 1224 (AIP Melville, NY, 2009), p. 301. 061601(R) (2010).
11. E.M. Kozulin et al., *Phys. Lett. B* **686**, 227 (2010).
12. Yu.Ts. Oganessian et al., *Phys. Rev. C***79**, 024603(2009).
13. S. Hofmann et al., *PGSI Scientific Reports*, 131 (2008).

EXPERIMENTS ON SUPERHEAVY ELEMENTS AT THE GSI SHIP

S. HOFMANN

GSI Helmholtzzentrum für Schwerionenforschung
64291 Darmstadt, Germany
and
Institut für Physik, Goethe-Universität
60438 Frankfurt, Germany
E-mail: S.Hofmann@gsi.de
www.gsi.de

An overview of present experimental investigation of superheavy elements is given. Using cold fusion reactions which are based on lead and bismuth targets, relatively neutron deficient isotopes of the elements from 107 to 112 were synthesized at GSI in Darmstadt, Germany, a neutron deficient isotope of element 113 at RIKEN in Wako, Japan. In hot fusion reactions of ^{48}Ca projectiles with actinide targets more neutron rich isotopes of element 112 and new elements up to 118 were produced at FLNR in Dubna, Russia. Recently, part of these data which represent the first identification of nuclei located on the predicted island of SHEs, was confirmed in independent experiments. The measured data combined with theoretical results were used for estimating cross-sections for production of element 120 isotopes. Also evaluated were their decay properties. An experiment for searching of isotopes of element 120 is planned at the GSI SHIP.

1. Introduction and Status of Experiments

For the synthesis of heavy and superheavy elements (SHE) fusion-evaporation reactions are used. Two approaches have been successfully employed. Firstly, reactions of a medium mass ion beam impinging on targets of stable Pb and Bi isotopes (cold fusion). These reactions have been successfully used to produce elements up to Z=112 at GSI [1] and to confirm the results of these experiments at RIKEN [2] and LBNL.[3] Recently, a number of neutron deficient odd element isotopes were produced in a combination with ^{208}Pb targets and odd element projectiles at LBNL.[4,5] Using a ^{209}Bi target the isotope ^{278}113 was synthesized at RIKEN.[6]

Secondly, reactions using lighter ions, particularly ^{48}Ca beams and radioactive actinide targets (hot fusion) were used for producing more neutron rich isotopes of element 112 and for synthesizing the new elements from 113 to 118 at FLNR.[7,8] Recently, the results of four of these reactions, $^{48}Ca + ^{242}Pu$,[9–11] $^{48}Ca + ^{238}U$,[12] $^{48}Ca + ^{244}Pu$,[13] and $^{48}Ca + ^{248}Cm$,[14] were confirmed in independent experiments. A new isotope of element 114, $^{285}114$ and its α decay daughters, was synthesized by evaporation of five neutrons in the reaction $^{48}Ca + ^{242}Pu$ at LBNL.[15] Figure 1 summarizes the data as they are presently known.

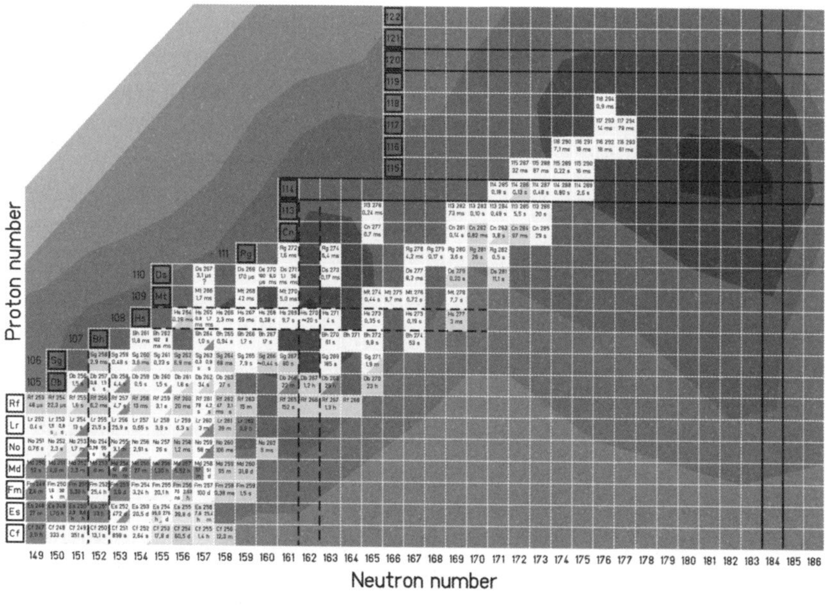

Fig. 1. Upper end of the chart of nuclei showing the presently (2010) known isotopes. For each known nucleus the element name, mass number, and half-life are given. The magic numbers for the protons at element 114 and 120 and for the neutrons at N=184 are emphasized. The bold dashed lines mark proton number 108 and neutron numbers 152 and 162. Nuclei with that number of protons or neutrons have increased stability, however, they are deformed contrary to the spherical superheavy nuclei. The crossing at Z=114 and N=162 reflects the uncertainty, whether nuclei in that region are deformed or spherical. The background structure in grey shows the calculated shell correction energy according to the macroscopic-microscopic model.[16,17]

Besides the insight that nuclei with such a high number of protons and resulting extremely high repulsive Coulomb forces are existing, two more important observations emerged. Firstly, the expectation that half-lives of the new isotopes should lengthen with increasing neutron number as one approaches the island of stability seems to be fulfilled. Secondly, the measured cross-sections for the relevant nuclear fusion processes reach values up to 5 pb, which is surprisingly high. Furthermore, the cross-sections seem to be correlated with the variation of shell-correction energies as predicted by macroscopic-microscopic calculations.[16–18]

2. Continuation of SHE Experiments using ^{248}Cm Targets

A comparison of various theoretical studies reveals that the location of the next closed proton shell beyond Z=82 is uncertain. The question is still open whether Z=120 is a closed proton shell or if strong shell closures exist at Z=114 or 126. In addition the possibility has to be considered that the island of superheavy nuclei is relatively flat and extents between sub-shells at 114, 120 and 126. Concerning the closed neutron shell, most theories agree with N=184 as a strong shell. Also experimental data, longer half-lives and decreasing negative shell-correction energies with increasing neutron number, are in accord with this finding.

As most important activity of our work on the synthesis and properties of superheavy elements in the future we proposed to systematically study hot fusion reactions based on ^{248}Cm targets. This target material has special properties which makes it favorable for the synthesis of heavy nuclei. It is one of the heaviest (Z=96) and most neutron rich targets available. Increased shell effects at its neutron number N=152 result in a relatively long half-life of 3.4×10^5 years and thus low specific activity. In combination with strongly bound projectile nuclei like ^{48}Ca or the neutron rich isotopes of the heavier elements up to nickel, relatively low excitation energies of the compound nuclei result, which are approximately 30 MeV at the fusion barrier. This advantageous property increases the probability for neutron emission instead of fission and thus results in relatively high fusion-evaporation cross-sections.

In the following we list a number of general arguments which have to be considered selecting the best reaction with respect to cross-sections for production of new elements beyond 118, in particular the new element 120.

Production cross-sections are strongly determined by fission barriers which again are built by shell effects in the region of SHEs. The rising up of cross-sections to several picobarns for elements 114 and 116 is due to increasing shell effects when N=184 is approached. This systematic suggests using the most neutron rich projectile and target nuclei available for synthesis of element 120.

Shell effects and thus fission barriers are considerably reduced with increasing excitation energy of the compound nucleus. Therefore, selection of a reaction resulting in minimum excitation energy is mandatory.

Cross-sections are further strongly influenced by Coulomb re-separation (quasi-elastic and quasi-fission processes) in the entrance channel of the reaction. In order to reduce this most unwanted effect, reaction partners have to be used resulting in lowest repulsive Coulomb forces. This can be achieved using reaction partners of high asymmetry.

Other phenomena which also influence the cross-sections, but are difficult to predict quantitatively and in detail, originate from isotopic effects. The number of neutrons determines the nuclear radius, relatively more for the relatively small projectiles, which influences the compactness of the system at the contact configuration. In the case of deformed nuclei of the actinides nuclear orientation is another property which strongly determines cross-section and beam energy. However, it is not possible to align the target nuclei in order to increase the cross-sections.

Finally, the reaction must be technically possible, i.e. projectiles and targets have to be available.

Heaviest isotopes which could be used as targets are ^{257}Fm ($T_{1/2}$ = 100 d) and ^{254}Es (276 d). However, the production of these isotopes is complex and only amounts of nanograms and micrograms, respectively, can be produced at high costs.

The next lighter isotopes available in principle are ^{252}Cf (2.6 y), ^{249}Cf (351 y), and ^{249}Bk (320 d). The isotope ^{252}Cf can be handled only with special radiation protection because of the high neutron flux being emitted from this fissioning material. The isotope ^{249}Bk has a relatively short half-life. It must be produced on demand and it is not available regularly. Due to the relatively short half-life, also the isotope ^{249}Cf has a high specific activity. In addition, the compound nucleus ^{299}120 which can be made in reactions with a ^{50}Ti beam, has three neutrons less then the compound nucleus ^{302}120 which can be produced with a ^{248}Cm target and a ^{54}Cr beam.

Considering all pros and cons we conclude that the reaction ^{54}Cr + ^{248}Cm $\rightarrow$ ^{302}120* is presently the most promising one being technically feasible to search for element 120.

What cross-sections and decay properties do we expect?

Several calculated cross-sections for synthesis of element 120 were published. We reproduce in Tab. 1 the most recent results without describing explicitly the various methods used. It should be mentioned that all of these theoretical studies were able to predict or reproduce reasonably well the existing data measured at FLNR.

Table 1. Calculated maximum cross-sections for synthesis of element 120 in various hot fusion reactions.

Reaction	σ(3n)/fb	σ(4n)/fb	Year	Ref.
$^{54}Cr + ^{248}Cm \rightarrow ^{302}120^*$	13	25	2008	19
	170	11	2010	20
	8000	800	2009	21
	–	0.07^a	2009	22
	–	54^b	2009	22
$^{50}Ti + ^{249}Cf \rightarrow ^{299}120^*$	40	43	2008	19
	800	30	2010	20
	–	0.45^a	2009	22
	–	89^b	2009	22
$^{58}Fe + ^{244}Pu \rightarrow ^{302}120^*$	1.2	5.3	2008	19
	7.0	1.0	2009	21
	–	0.015^a	2009	22
	–	32^b	2009	22
exp.	<400	<400	2007	23
$^{64}Ni + ^{238}U \rightarrow ^{302}120^*$	4.3	3.2	2008	19
	0.0003	0.0022	2009	21
	–	0.02^a	2009	22
	–	5^b	2009	22
exp.	<90	<90	2008	24

Note: [a] Using mass-predictions of.[25]
[b] Using mass-predictions of.[26]

Obviously, there exist significant differences. The reason is the extremely sensitive dependence of cross-sections from fusion barriers and resulting excitation energies at the barrier, from Coulomb re-separation and from fission barriers as outlined before. The model by Siwek-Wilczynska et al.[20] assumes a lower fusion barrier which results in an increase of the 3n cross-section. In the model by Nasirov et al.[21] the quasifission processes result in strong reduction of cross-sections with increasing symmetry, whereas this effect changes the cross-section only within a factor of ten in the model by Zagrebaev and Greiner.[19] Finally, in the paper by Adamian et al.[22] various mass formula and various damping parameters of the fission barrier at increasing excitation energy were compared. Two of the results are reproduced here, which predict cross-sections differing by two and three orders of magnitude.

Experimental limits were obtained for the reactions with ^{58}Fe and ^{64}Ni beams at FLNR[23] and at SHIP,[24] respectively. Although these limits are still high, they allow to reject unusual high fission barriers at element 120.

Using the rule of thump that a 1 MeV change of the fission barrier changes the cross-section by one order of magnitude at least,[20] we obtain experimental limits for the fission barrier of element 120 isotopes of <8.9 and <8.3 MeV, respectively. As a starting point in this estimate we used the calculation of Zagrebaev and Greiner,[19] who determined their cross-sections with a fission barrier of 7 MeV. At a fission barrier of 8.3 MeV their cross-section estimates would be a factor of 20 higher. This simple consideration and the very different predictions given in Tab. 1 show, that a sufficiently accurate estimate for the time necessary to produce element 120 cannot be made on the basis of the presently existing data and calculations.

Half-life and decay mode are nuclear properties which could hamper the identification, although isotopes of element 120 could be produced with high enough cross-sections. Theoretical calculations show that the heaviest elements decay by α emission. This result is proved by experimental data on elements up to $^{294}118$ which has a measured Q_α value of 11.81 MeV and decays with a half-life of 890 μs.[7] In the region of interest, β decay and spontaneous fission are predicted to have significantly longer half-lives. This result is in agreement with the measured α-decay chains which end by spontaneous fission only at copernicium or below.

Calculated Q_α values and deduced α half-lives for isotopes of element 120 are listed in Tab. 2. Q_α values slightly higher than 13 MeV are obtained by the models of Sobiczewski et al.[16,27] and Möller et al.[18] A different trend is obvious for the data by Liran et al.[28] The explanation is given by the different location of the closed proton shell used in the models, which is in the former case at Z=114 and in the latter one at Z=126. Nuclei above closed shells have high Q_α values, the opposite is the case for nuclei below closed shells. This effect is increased in the model by Liran et al. due to a strong shell closure at N=184, which further reduces the Q_α values when this shell is approached with increasing neutron number. It should be noted that in the case of $^{294}118$ also this model reproduces well the measured Q_α value of 11.81 MeV. The calculated value is 11.55 MeV.

Deduced half-lives are given in the right side of Tab. 2. In the upper part are listed the values originally given in the quoted papers. Hyphens mark cases where no half-life was given. In the lower part we calculated the half-lives using the most recent formula worked out by Parkhomenko et al.[29] for the same Q_α values as given in the upper half of the table. We notice that the recalculated half-lives using the new formula are longer by a factor of ten. The new values are in a window between 3 and 13 μs. Less modified are the half-lives given by Möller et al. We also note that the odd N

Table 2. Table 2. Calculated Q_α values and deduced partial α half-lives of isotopes of element 120. The isotopes expected to be produced in the reaction $^{54}Cr + ^{248}Cm \rightarrow ^{302}120^*$ are marked in bold. See text for details.

Q_α/MeV							$T_{1/2}/\mu s$							
294	295	296	297	**298**	**299**	300	294	295	296	297	**298**	**299**	300	A
174	175	176	177	**178**	**179**	180	174	175	176	177	**178**	**179**	180	N
														Ref.
13.24	–	13.23	–	**13.44**	–	13.11	0.67	–	0.70	–	**0.26**	–	1.2	27
13.42	–	13.40	–	**13.36**	–	13.10	0.28	–	0.31	–	**0.37**	–	1.3	16
13.29	13.36	13.69	13.54	**13.35**	**13.10**	13.40	1.5	13	0.26	5.9	**1.2**	**43**	0.93	18
12.68	12.39	12.11	11.79	**11.47**	**11.12**	10.78	–	–	–	–	–	–	–	28
13.24	–	13.23	–	**13.44**	–	13.11	6.9	–	7.2	–	**2.9**	–	12	29
13.42	–	13.40	–	**13.36**	–	13.10	3.1	–	3.4	–	**4.1**	–	13	29
13.29	13.36	13.69	13.54	**13.35**	**13.10**	13.40	5.5	8.6	0.99	3.9	**4.3**	**28**	3.4	29
12.68	12.39	12.11	11.79	**11.47**	**11.12**	10.78	88	820	1.4E3	18E3	**41E3**	**780E3**	2.1E6	29

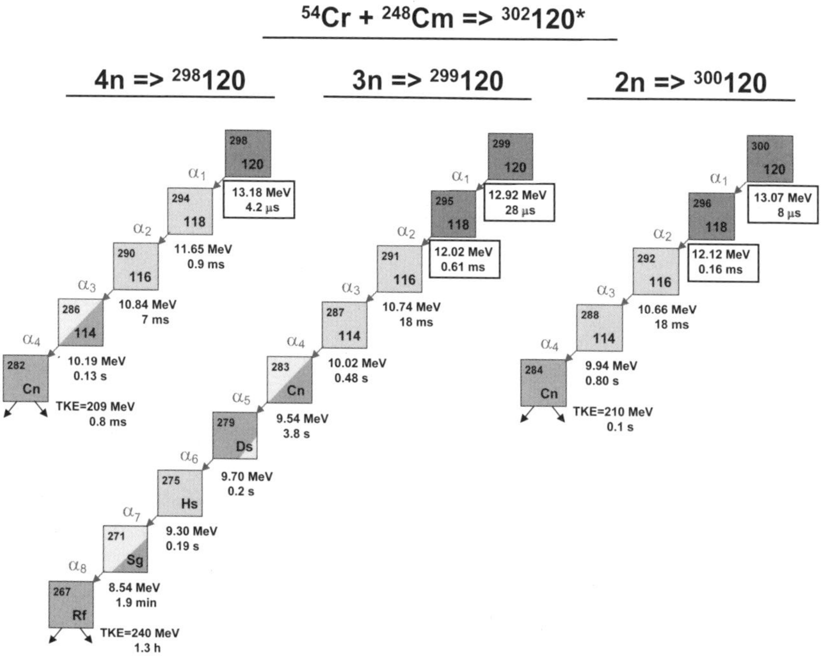

Fig. 2. Expected decay chains populated in the reaction ^{54}Cr + ^{248}Cm → ^{302}120*. Predicted decay data of so far unknown isotopes are given in frames, see Tab. 2.

nuclei, and here especially the isotope ^{299}120, have significantly longer half-lives than the neighboring even-even nuclei. Despite similar Q_α values the prolongation is due to an odd neutron hindrance factor and the probability that α decay may populate low lying excited states in the daughter nucleus. This latter effect was taken into account in the half-life formula given by Parkhomenko et al.[29] Unusual long half-lives were obtained for the isotopes having the relatively low Q_α values calculated by Liran et al.[28]

In conclusion we realize that half-lives of the isotopes of interest are predicted to be in the range from 1 to 30 μs, but could be significantly longer if the proton shell is at Z=126 or 120 and not at 114. In any case special technical preparations are needed for detection of short living isotopes of element 120.

A minimum lifetime of 2 μs is needed so that the residues can pass SHIP, otherwise they will decay inside the separator. In this case also the daughter nucleus after α decay will be lost with high probability due to the recoil momentum from the emitted α particle, an effect which reduces the transmission by a factor of ten.

The decay chains expected in the case of three and four neutron evaporation will populate isotopes of element 116, $^{290}116$ and $^{291}116$, which were measured previously at FLNR and which were confirmed indirectly by identification of the daughter nuclei $^{286}114$ and $^{287}114$ at LBNL recently.[11] In this case a well founded identification of element 120 by genetic correlation to known nuclei is given. The expected decay chains are shown in Fig. 2. In the case of an also possible two neutron channel, we would observe as a granddaughter the isotope $^{292}116$ and its daughter decays, which we observed in a confirmation experiment at SHIP recently using the reaction $^{48}\text{Ca} + {}^{248}\text{Cm} \rightarrow {}^{292}116 + 4\text{n}$.[14]

3. Conclusion and Outlook

The experimental work of the last three decades has shown that cross-sections for the synthesis of the heaviest elements do not decrease continuously as it was measured up to the production of element 113 using cold fusion reactions. Recent data on the synthesis of elements 112 to 118 in Dubna using hot fusion show that this trend is broken when the region of spherical SHEs is reached. Some of the results originally obtained in Dubna were confirmed in independent experiments and with different methods, including the use of chemical, element specific properties. We conclude that the region of the predicted spherical SHEs has finally been reached and the exploration of the 'island' has started and can be performed even on a relatively high cross-section level.

An opportunity for the continuation of experiments in the region of SHEs at low cross-sections afford, among others, further accelerator developments. High current beams and radioactive beams are options for the future. A wide range of half-lives encourages the application of a wide variety of experimental methods in the investigation of SHEs, from the safe identification of short lived isotopes by recoil-separation techniques to atomic physics experiments on trapped ions, and to the investigation of chemical properties of SHEs using long-lived isotopes.

References

1. S. Hofmann and G. Münzenberg, *Rev. Mod. Phys.* **72**, 733 (2000).
2. K. Morita et al., *J. Phys. Soc. Jpn.* **76**, 043201 (2007).
3. T.N. Ginter et al., *Phys. Rev. C* **67**, 064609 (2003).
4. C.M. Folden III et al., *Phys. Rev. Lett.* **93**, 212702 (2004).
5. C.M. Folden III et al., *Phys. Rev. C* **73**, 014611 (2006).
6. K. Morita et al., *J. Phys. Soc. Jpn.* **73**, 2593 (2004).

7. Yu.Ts. Oganessian, *J. Phys. G: Nucl. Part. Phys.* **34**, R165 (2007).
8. Yu.Ts. Oganessian *et al.*, *Phys. Rev. Lett* **104**, 142502 (2010).
9. R. Eichler et al., *Nucl. Phys. A* **787**, 373c (2007).
10. R. Eichler et al., *Nature* **447**, 72 (2007).
11. L. Stavsetra et al., *Phys. Rev. Lett.* **103**, 132502 (2009).
12. S. Hofmann et al., *Eur. Phys. J. A* **32**, 251 (2007).
13. Ch. E. Düllmann *et al., Phys. Rev. Lett.* **104**, 252701 (2010).
14. S. Hofmann et al., to be published.
15. P.A. Ellison et al., *Phys. Rev. Lett.* **105**, 182701 (2010).
16. R. Smolanczuk and A. Sobiczewski, *Proc. of the XV. Nucl. Phys. Div. Conf. on Low Energy Nuclear Dynamics*, eds. Yu.Ts. Oganessian et al. (World Scientific, Singapore, 1995), p. 313.
17. R. Smolanczuk et al., *Phys. Rev. C* **52**, 1871 (1995).
18. P. Möller et al., *At. Data Nucl. Data Tables* **66**, 131 (1997).
19. V. Zagrebaev and W. Greiner, *Phys. Rev. C* **78**, 034610 (2008).
20. K. Siwek-Wilczynska et al., *Int. J. Mod. Phys. E* **19**, 500 (2010).
21. A.K. Nasirov et al., *Phys. Rev. C* **79**, 024606 (2009).
22. G.G. Adamian et al., *Eur. Phys. J. A* **41**, 235 (2009).
23. Yu.Ts. Oganessian et al., *Phys. Rev. C* **79**, 024603 (2009).
24. S. Hofmann, *et al., GSI Scientific Report 2008, GSI Report* **2009-1**, 131 (2009).
25. P. Möller and J.R. Nix, *At. Data Nucl. Data Tables* **39**, 213 (1988).
26. W.D. Myers, *Droplet Model of Atomic Nuclei* (Plenum, New York, 1977).
27. I. Muntian et al., *Acta Phys. Pol. B* **32**, 691 (2001).
28. S. Liran et al., *Phys. Rev. C* **62**, 047301 (2000).
29. A. Parkhomenko and A. Sobiczewski, *Acta Phys. Pol. B* **36**, 3095 (2005).

NEW VISTAS OF THE STRUCTURE OF NEUTRON-RICH NUCLEI - SINGLE PARTICLE STATES AND g-FACTORS

A.V. RAMAYYA[1], J.H. HAMILTON[1], Y.X. LUO[1,2], S.J. ZHU[1,3], J.K. HWANG[1], J.O. RASMUSSEN[2], G.M. TER-AKOPIAN[4], A.V. DANIEL[4], N. BREWER[1], Y. OGANESSIAN[4]

[1]Physics Department, Vanderbilt University, Nashville, TN 37235, USA
[2]Lawrence Berkeley National Laboratory, Berkeley, CA 94720, USA
[3]Physics Department, Tsinghua University, Beijing 1100084, China
[4]Flerov Laboratory for Nuclear Reactions, JINR, Dubna, Russia

Prompt γ-γ-γ coincidence studies of neutron-rich nuclei populated in the spontaneous fission of ^{252}Cf led to the discoveries of many important processes like neutron-less binary and ternary fragmentation in spontaneous fission, binary and ternary hot fission modes, hyper-deformation, cluster radioactivity and quasi-molecular states in ternary fission. The high statistics data, 5.7 x 10^{11} triples and higher folds, opened up discoveries of new band structures and significant extensions of previously known bands. These data were used to measure the angular correlations of cascades of γ-rays from the excited states of neutron-rich fission fragments stopped in an unmagnetized iron foil. By using this γ–γ angular correlation technique, the single particle configurations of the known rotational bands were assigned and confirmed in the present work. In other words, the multipole mixing ratios of ΔI=1 transitions in the rotational bands were measured. These experimental mixing ratios are compred with the theoretical ones of particle and axial-rotor model. The configuration of the ground states in ^{109,111}Ru were, for the first time, assigned. And magnetic g-factors of excited states in even-even neutron-rich nuclei were meaured, too. Two sets of ΔI=1 alternating parity doublets with opposite parities were observed in Ba and Ce isotopes. The B(E1)/B(E2) branching ratios and D_0 values may indicate that the octupole correlations are strong in Ba and Ce isotopes because of the Z=56 and N=88 octupole shell gaps. High-spin, even-parity bands in neutron-rich ^{108,110,112}Ru nuclei indicate two-phonon quasi-gamma bands in ^{110,112}Ru.

1. Introduction

The single particle configurations for odd-A deformed nuclei are based on either Nilsson shell model calculations or by comparing the states with the states in neighboring nuclei where shell model calculations are performed. The multipole mixing ratios of ΔI = 1 transitions between levels in rotational bands built on single-particle states in odd neutron nuclei are dependent on the configurations of the states. In particular, the mixing ratio can be used to distinguish between several possible single-particle configurations if interpreted with PRM. The neutron single-particle configurations of ^{101}Zr, ^{103,105,107}Mo, and ^{109,111}Ru are determined by measuring the mixing ratios of ΔI = 1 transitions in their rotational bands. The mixing ratios are interpreted in terms of PRM. These nuclei are known to be highly deformed [1], with the $h_{11/2}$ neutron orbital playing a role in driving the deformation although the proton-neutron interaction may be also important. These nuclei are produced in spontaneous fission of ^{252}Cf. In this paper, we described the technique of measuring angular correlations with large gamma detector arrays and discussed the assignment of single particle configurations. The details on angular correlation techniques and the measurements can be found in references [2,3].

Nuclear g-factors give insight into the collective and single-particle structure of nuclear states. It is valuable to measure g-factors for isotopic chains to observe the behavior moving away from closed shells. Systematic trends in g-factors of 2^+ states or higher yrast states can indicate the presence of subshell closures or other structure changes. In this work, the systematics of g-factors in neutron rich, even-even Xe isotopes have been extended. The g-factors of 2^+ states in ^{146}Ba and ^{146,148}Ce have been measured and are compared to previous work. The Xe nuclei with Z=54 and N=86 (88) with two proton pairs above the Z=50 shell closure and two (four) neutron pairs above the N=82 shell closure are transitional between spherical and deformed shapes. Assuming vibrational collectivity is dominant for low energy excitations in ^{140,142}Xe, g-factors in these nuclei are compared with predictions of the IBM-2. Since ^{146}Ba and ^{146,148}Ce show moderate axial quadrupole deformation with $\beta_2 = 0.2$ [1], both the IBM-2 and rotation-vibration model with independent proton and neutron deformations [2] are considered in discussion of the results for these nuclei.

The first evidence for octupole deformation in an odd-*A* system in this region was discovered in ^{143}Ba. Both parity doublets were then reported in ^{143}Ba and confirmed in our work. Evidence for octupole correlations and deformation is also observed in ^{139}Xe, ^{140,141,142}Ba, and ^{144}Ce. The odd-*Z* ^{145,147}La also are reported to have strong octupole correlations with the evidence significantly extended in our work. Thus, an island of stable octupole deformation around the *Z*=56, *N*=88 region is established. Here we report the first evidence of octupole deformation in ^{145}Ba and ^{148}Ce. In the present work, we show and discuss the even-parity bands of neutron-rich ^{108,110,112}Ru nuclei. Bands with (4^+) bandhead and energies between two and three times the energy of the second 2^+ bandheads are found in ^{110,112}Ru, which we assign as a collective band associated with two-phonon quasi-gamma excitation. A two-phonon quasi-gamma bandhead (4^+) is assigned to ^{108}Ru.

2. Experimental Method

The data for this analysis were taken by using the Gammasphere detector array, which was located at Lawrence Berkeley National Laboratory. A ^{252}Cf spontaneous fission source with an α activity of 62 μCi was sandwiched between two iron foils. The foils were thick enough to stop the fission fragments, eliminating the need for a Doppler correction. Approximately 5.7×10^{11} three- and higher-fold γ coincidence events were recorded.

The coincidence intensity between a cascade of two γ rays as a function of angle is fitted to Eq. (1).

$$W(\theta) = 1 + G_2A_2(\delta)P_2(\cos\theta) + G_4A_4(\delta)P_4(\cos\theta) \tag{1}$$

In this equation the A_k s are calculable coefficients depending on the spin sequence of the levels involved and the multipolarities and mixing ratios of the transitions between them. The G_k coefficients relate to attenuations of the correlations and are equal to unity unless some interaction is present during the lifetime of the intermediate state of the cascade that can cause precession of the nuclear spin. For the cascades involved in this work, the nuclear hyperfine field in iron is less than about 50 T. Unless the life-time of the intermediate state is greater than ~0.7 ns, the G_ks are close to unity.

The mixing ratios of $\Delta I = 1$ transitions within a rotational band are given by

$$\delta\left(\frac{E2}{M1}\right) = \sqrt{\frac{5}{12}}\frac{1}{K}\frac{E_\gamma}{1.2\times10^3}\frac{Q_0}{(g_K-g_R)}\frac{<I_fK20|I_iK>}{<I_fK10|I_iK>}, \tag{2}$$

where Q_0 is the quadrupole moment; g_K and g_R are the intrinsic and rotational g factors, respectively; and E_γ is the $\Delta I=1$ transition energy in keV. For these calculations, $2g_R$ is taken to be equal to Z/A and the quadrupole moment Q_0 is calculated from the deformation parameter, β, using the relation

$$\beta = 91.7\frac{Q_0}{ZA^{2/3}} \tag{3}$$

with β assumed to take a common value, $\beta = 0.3$. The single-particle g_K factor depends on the Nilsson configuration of the particle and can be calculated by using the $M1$ matrix elements tabulated in Ref. [4] for various single-particle configurations by

$$g_K = g_l + \frac{g_s-g_l}{2K}\Sigma(\alpha^2_{l,K-\frac{1}{2}} + \alpha^2_{l,K+\frac{1}{2}}) \tag{4}$$

where the α's are Nilsson coefficients tabulated in Ref. [4] as a function of the deformation parameter β, and g_l, g_s are the orbital and spin g factors. In these calculations, which involve neutron states, the values $g_l=0$ and $g_s^{eff} = 0.6g_s^{free} = -2.296$ are used.

Gammasphere consists of 110 germanium detectors, corresponding to 5995 unique detector pairs. The detectors are placed at 17 different azimuthal angles and 60 various polar angles. For each azimuthal angle, there are 5–10 detectors placed symmetrically with respect to the polar angle. Due to the symmetries of Gammasphere, the angle between any two detectors will be one of only 64 possible values, and each of these 64 angle bins has many pairs of detectors. For our experiment, only 101 detectors were present, so the number of detectors in each bin is slightly less than if Gammasphere were at full capacity. The angular properties of Gammasphere and the process of determining angle bins are given in greater detail in [5]. For the angular correlation, we sorted our triple coincidence data into 64 two dimensional histograms corresponding to the 64 angle bins. The histograms were produced by reading the full data set event by event. For each γ–γ coincidence, the angle between the detectors which detected γ_1 and γ_2 was calculated and the event was added to the appropriate histogram. unattenuated angular correlations.

3. Single Particle States and g-Factors

One example of angular correlation measurements is shown in Fig. 1. In Fig 1, ^{138}Ba has the 2^+ level with a very short lifetime of 0.202(8) ps which indicates that the correlation should be unattenuated. Then, we obtained the expected value of $G_2 \cong 1$. The consistency of this result with the expected value shows the viability of our method for measuring angular correlations with Gammasphere.

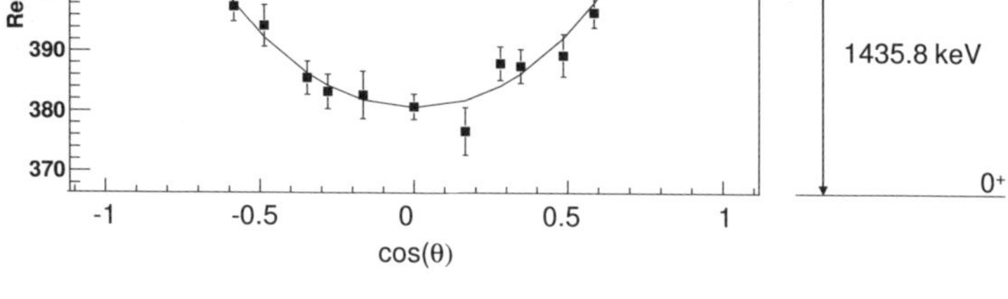

Fig. 1. Angular correlation of the $4^+ \rightarrow 2^+ \rightarrow 0^+$ cascade in ^{138}Ba. The levels are shown on the right.

The relevant partial level schemes of odd-A Mo and Ru nuclei can be seen in Ref. [3]. From the A_2 and A_4 coefficients, we extracted the δ values for the $\Delta I=1$ transitions within the rotational bands. To interpret the mixing ratios of the $\Delta I=1$ transitions within rotational bands, the experimental values were compared to calculations made with a particle plus axial rotor model for various single-particle states. Orlandi *et al.* [7] established the 3/2[411] assignment of the ground band of ^{101}Zr, which is also supported in this work. They found the ground band in ^{103}Mo to have the same configuration, which is also supported by this work. They further showed the negative parity $5/2^-$ bandhead at 347 keV in ^{103}Mo to have the 5/2[532] configuration and identified this configuration as the ground-state band in ^{105}Mo. The present results in ^{105}Mo support this assignment. In ^{107}Mo, the configuration of the ground state was assigned by Urban *et al.* [8] to be 5/2[413] based on the measurement of *E2/M*1 branching ratios in the rotational band. Two independent values of the mixing ratio of the 152.1-keV transition between the lowest levels of this band in ^{107}Mo, are reported in this work. They are both negative and, taking the stronger *M*1 limit from the experimental fit ranges, give a δ around −0.5, showing strong preference for the 5/2[402] assignment over the 5/2[413] for which the predicted δ value is positive and close to unity. It would seem that Urban *et al.* [8], who obtained δ, without sign, chose δ positive contrary to the present direct sign measurement. For the negative-parity band-head at 348.4 keV, the configuration was assigned 7/2[523] [8], which is supported by the present result on mixing in the 110.2-keV transition. In ^{109,111}Ru, Wu *et al.* [9] attempted to determine the single-particle configurations of the ground-state bands based on *E2/M*1 branching ratios but were not able to distinguish between the 5/2[413] and 5/2[402] configurations. In this work angular correlation data gives clear evidence that the ground states of ^{109,111}Ru both have a 5/2[402] configuration. Calculations for a 5/2[402] state yield $\delta = -0.26$ and $\delta = -0.21$, respectively, in excellent agreement with the measured values of −0.25(6) and −0.32(2), whereas calculations for the 5/2[413] configuration yield large positive values. The more details can be seen in Ref. [3].

Table 1. g-factors of the 2+ state measured in the even-even nuclei [10].

Fragment	t (ns)	B (Tesla)	g
^{100}Zr	0.78(3)	27.4(4)	0.32(5)
^{102}Zr	2.76(36)	27.4(4)	0.25(4)
^{104}Mo	1.040(59)	25.6(1)	0.28(6)
^{106}Mo	1.803(43)	25.6(1)	0.24(3)
^{108}Mo	0.72(43)	25.6(1)	0.3(3)
^{146}Ce	0.36(4)	41(2)	0.30(6)
^{148}Ce	1.46(9)	41(2)	0.37(5)

In our work, g-factors of the first 2$^+$ states in ^{140,142}Xe and ^{146}Ba and ^{146,148}Ce found by using new techniques developed for measuring angular correlations with large detector arrays are reported. Those g-factors are shown in Table 1 [10]. The g-factors of 2$^+$ states in ^{140,142}Xe are measured for the first time, while the g-factors in ^{146}Ba and ^{146,148}Ce are found to have good overall agreement with previously measured values, thus confirming the reliability of the experiment and analysis. The experimental results compare well with the predictions of the IBM-2, suggesting the presence of vibrational collectivity in the 2$^+$ states in question, although single particle components in the wave functions cannot be excluded. Using measured values of ^{146}Ba and ^{146,148}Ce, a method is outlined which potentially yields information on the ratio of neutron to proton deformation in deformed nuclei. The ratios obtained, although not very precise, are in a sensible agreement with results of microscopic calculation using HF+BCS model with a Skyrme interaction in these less deformed nuclei.

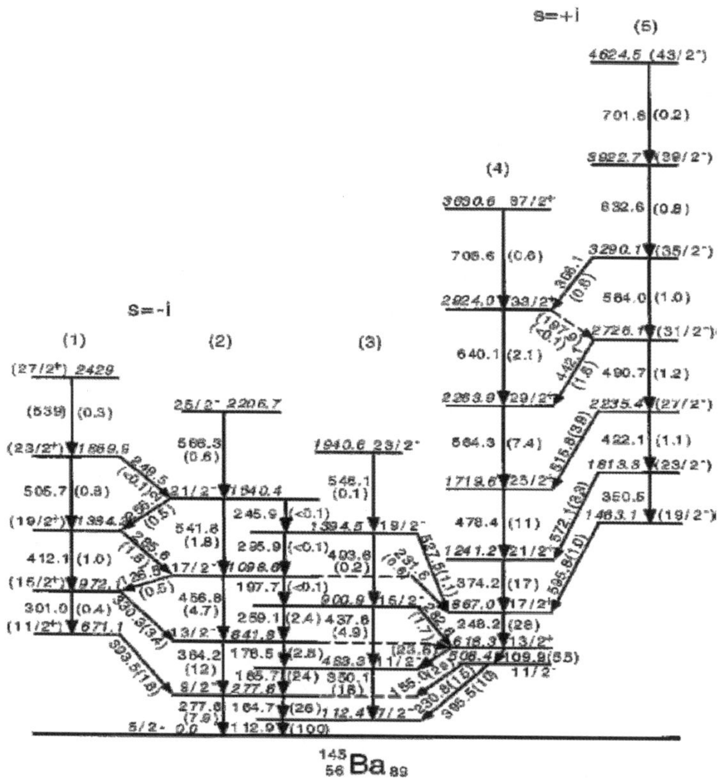

Fig. 2. Partial level scheme of ^{145}Ba [11].

4. Octupole Correlations in ^{143,145}B and ^{148}Ce

The new level schemes for ^{143,145}Ba [11] and ^{148}Ce [12] were established. All previously reported transitions were confirmed. Many new transitions and levels are observed. The level schemes of ^{145}Ba and ^{148}Ce are shown in Figs. 2 and 3. Spins and parities for each bands have been assigned based on systematics and some angular correlation, as well as internal conversion coefficient measurements. The similarity of $B(E1)/B(E2)$ values in our investigation supports the spin and parity assignments. These data indicate the two sets of parity doublets expected for octupole deformation at higher spins.

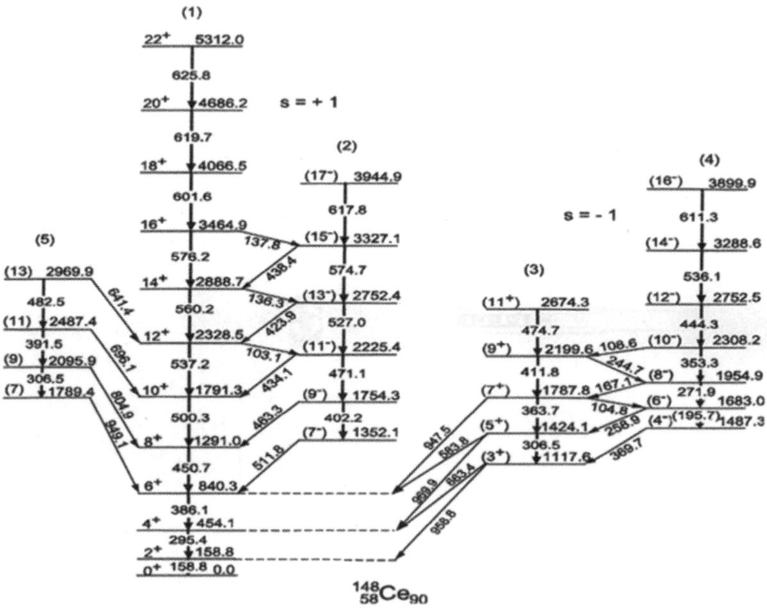

Fig. 3. Partial level scheme of ^{148}Ce.

New high spin states in ^{143,145}Ba and ^{148}Ce have been investigated. Stable octupole deformation or at least strong octupole correlations are observed in these nuclei. These new data confirm the long-standing theoretical prediction of stable octupole deformation in ^{145}Ba. A new band with rigid body moments of inertia in ^{145}Ba may be the first example in neutron rich nuclei of a pairing-free structure or of a type of superdeformed band. The strong-coupling ground band and octupole deformation structures in ^{145}Ba show competition and coexistence between symmetric and asymmetric shapes. Octupole bands in ^{148}Ce might be a first observation of such kind of octupole deformed structure in the even-even nuclei.

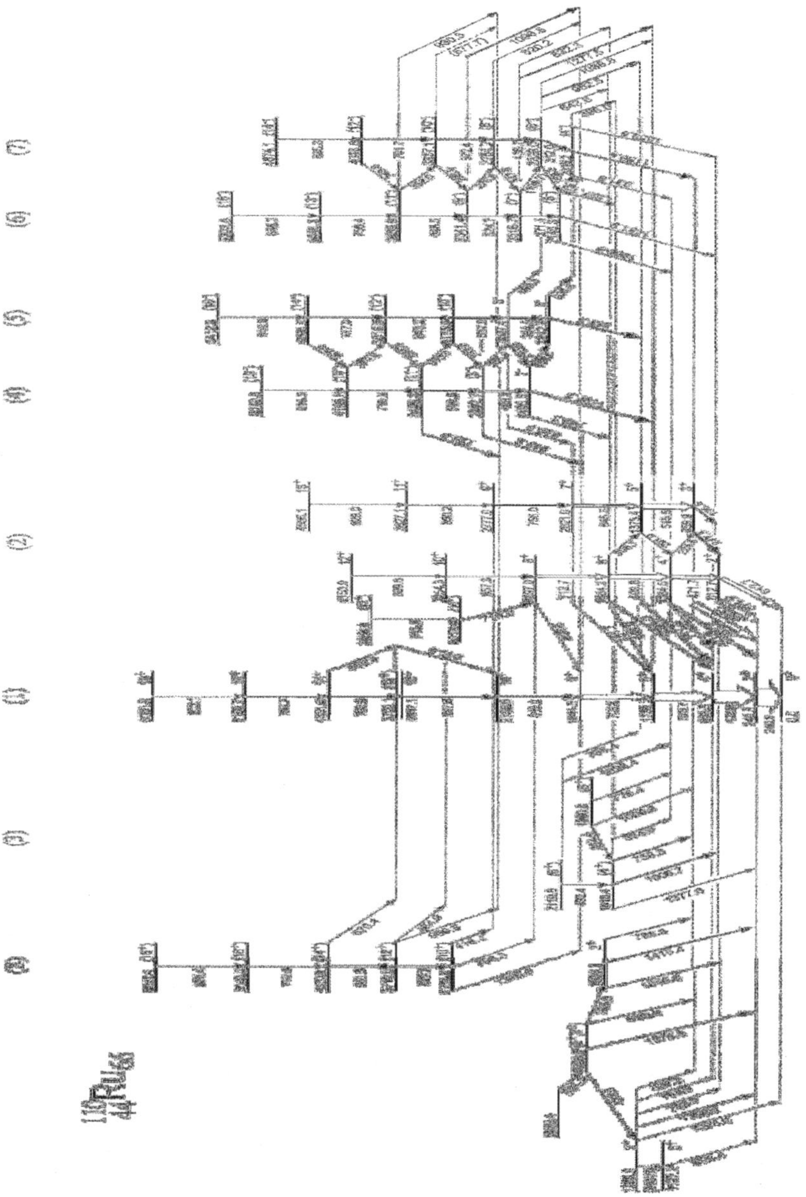

Fig. 4. Level scheme of ^{110}Ru [13].

5. Even Parity Bands of ^{108,110,112}Ru

In the present work [13], we show and discuss the even-parity bands of neutron-rich ^{108,110,112}Ru nuclei. The ground bands and one-phonon quasi-gamma bands were carefully remeasured with our high statistics data. Our higher statistics data from stopped fragments enabled us to identify weak bands and branching transitions in these bands, not reported before. Bands with (4^+) bandhead and energies between two and three times the energy of the second 2^+ bandheads are found in ^{110,112}Ru, which we assign as a collective band associated with two-phonon quasi-gamma excitation. The level scheme of ^{110}Ru is shown in Fig. 4.

A two-phonon quasi-gamma bandhead (4^+) is assigned to ^{108}Ru [13]. A weakly populated doubling of levels to both ground and quasi-gamma bands is observed in ^{110}Ru. The odd-even spin energy staggering are observed and discussed in ^{110,112}Ru. An excited 0^+ band is confirmed and extended in ^{108}Ru [13]. Our new TRS calculations indicate that ^{108,110,112}Ru have triaxial deformation. The band-crossings in the ground bands are interpreted by CSM calculations as $h_{11/2}$ neutron-pair alignments. Further evidence is provided for the two-phonon quasi-gamma bands. There are unusual doublings of the ground and one-phonon quasi-gamma band above the 8^+ level, where a neutron-pair alignment band-crossing starts in ^{110}Ru. Likewise, there are curious reversals of sign in signature splitting of the one-phonon quasi-gamma band in going through this series of nuclei.

Acknowledgments

The work at Vanderbilt University and Lawrence Berkeley National Laboratory are supported by US Department of Energy under Grant No. DE-FG05-88ER40407 and Contract No. W-7405-ENG48. The Joint Institute for Heavy Ion Research is supported by the University of Tennessee, Vanderbilt University, and the US DOE through Contract Nos. DE-FG05-87ER40311 and DE-FG02-96ER40983 with University of Tennessee is gratefully acknowledged.

References

1. J. Skalski, S. Mizutori, and W. Nazarewicz, 1997 *Nucl. Phys.* A **617,** 282 (1997).
2. A. V. Daniel et al., *Nucl. Inst. And Meth. in Phys. Res.* B **262,** 399 (2007).
3. C. Goodin et al., *Phys. Rev.* C **80**, 014318 (2009).
4. E. Browne and F. Femenia, *Nucl. Data Tables* **10,** 81 (1971) and S.G. Nilsson, *Nat. Fys. Medd. Dan. Vis. Selsk.* **29,** (1955).
5. D. Patel et al., *J. Phys.* G **28,** 649 (2002).
6. C. L. Lawson and R. J. Hanson, *Solving Least Square Problems* (NJ: Prentice Hall, Englewood Cliffs 1974).
7. R. Orlandi *et al., Phys. Rev.* C **73,** 054310 (2006).
8. W. Urban *et al., Phys. Rev.* C **72,** 027302 (2005).

9. C.Y. Wu *et al. Phys. Rev.* C **73,** 034312 (2006).
10. C. Goodin et al., *Phys. Rev.* C **79**, 034316 (2009).
11. S.J. Zhu et al., *Phys. Rev.* C **60**, 051304 (1999).
12. Y.J. Chen et al., *Phys. Rev.* C **73**, 054316 (2006).
13. S.J. Zhu et al., *Int. J. Mod. Phys.* E **18**, 1717 (2009).

SUPERHEAVIES: THEORETICAL INCITEMENTS AND PREDICTIONS

V.I. ZAGREBAEV, A.V. KARPOV

Flerov Laboratory of Nuclear Reactions, JINR, Dubna, Moscow Region, Russia

I.N. MISHUSTIN and WALTER GREINER

Frankfurt Institute for Advanced Studies, Frankfurt am Main, Germany

It is well known that in fusion reactions one may get only neutron deficient superheavy nuclei located far from the island of stability. The multi-nucleon transfer reactions allow one to produce more neutron enriched new heavy nuclei but the corresponding cross sections are rather low. Neutron capture process is considered here as alternative method for production of long-lived neutron rich superheavy nuclei. Strong neutron fluxes might be provided by nuclear reactors and nuclear explosions in laboratory frame and by supernova explosions in nature. All these cases are discussed in the paper.

Keywords: Superheavy nuclei, fusion and transfer reactions, neutron capture, nuclear explosions, pulsed reactors.

1. Motivation

A ten years epoch of ^{48}Ca irradiation of actinide targets for the synthesis of superheavy (SH) elements is over. The heaviest available targets of Berkelium ($Z = 97$) and Californium ($Z = 98$) have been used to produce the elements 117[1] and 118.[2] Note that the predicted cross sections and excitation functions[3,4] have been fully confirmed by the experiments performed in Dubna and later in Berkeley and GSI. These predictions are based on the shell corrections to the ground states of superheavy nuclei calculated by P. Möller and others[5] which we used to estimate survival probabilities of these nuclei against fission. It is the gradual increase of the fission barriers of the compound nuclei (CN) formed in these reactions which explains the almost constant value (of a few picobarns) of the cross sections for the production of SH elements with $Z = 112 \div 118$ in hot fusion reactions.

As mentioned above, ^{249}Cf ($T_{1/2} = 351$ yr) is the heaviest available target that can be used in experiment. Thus, to get SH elements with $Z > 118$

in fusion reactions, one should proceed to heavier than ^{48}Ca projectiles. The strong dependence of the calculated evaporation residue (EvR) cross sections for the production of element 120 on the mass asymmetry in the entrance channel[6] makes the nearest to ^{48}Ca projectile, ^{50}Ti, most promising for further synthesis of SH nuclei. Of course, the use of the titanium beam instead of ^{48}Ca also decreases the yield of SH nuclei mainly due to a worse fusion probability. The calculated excitation functions for the synthesis of SH elements 119 and 120 in the fusion reactions of ^{50}Ti with ^{249}Bk and ^{249}Cf targets are shown in Fig. 1 taken from Ref. 6.

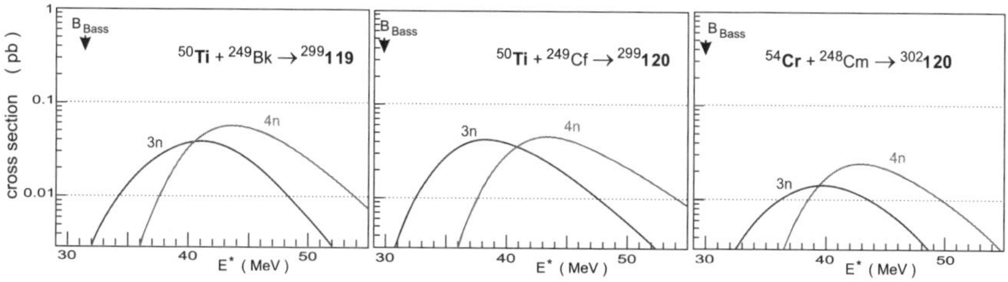

Fig. 1. Excitation functions for production of superheavy elements 119 and 120 in the 3n and 4n evaporation channels of the ^{50}Ti+^{249}Bk, ^{50}Ti+^{249}Cf and ^{54}Cr+^{248}Cm fusion reactions.

The orientation effects are known to play an important role in fusion reactions of statically deformed heavy nuclei.[3,4] The fusion probability (formation of CN) is strongly suppressed for more elongated nose-to-nose initial orientations. As a result, the preferable beam energies for the synthesis of SH elements in the hot fusion reactions are shifted to values that are several MeV higher than the corresponding Bass barriers (calculated for spherical nuclei). As can be seen from Fig. 1, the estimated EvR cross sections for the 119 and 120 SH elements synthesized in the ^{50}Ti induced fusion reactions are quite reachable at available experimental setups, though one needs much longer time of irradiation than for the ^{48}Ca fusion reactions.

The yield of superheavy nuclei (number of events per day) depends not only on the cross section but also on the beam intensity and target thickness. In this connection the other projectile–target combinations should be also considered. Most neutron-rich isotopes of element 120 may be synthesized in the three different fusion reactions ^{54}Cr+^{248}Cm, ^{58}Fe+^{244}Pu and ^{64}Ni+^{238}U leading to the same SH nucleus ^{302}120 with neutron number near to the predicted closed shell $N = 184$.[7] These three combinations are not of equal value. The estimated EvR cross sections for more symmetric

^{58}Fe+^{244}Pu and ^{64}Ni+^{238}U reactions are lower than those of the less symmetric ^{54}Cr+^{248}Cm combination,[6] which, in its turn, is quite comparable with the Ti-induced fusion reaction, see the right panel of Fig. 1. The advantage factor 2 or 3 for the ^{50}Ti+^{249}Bk and ^{50}Ti+^{249}Cf fusion reaction as compared with ^{54}Cr+^{248}Cm is definitely within the theoretical uncertainty for calculation of such small cross sections.

Such uncertainty originates from two main factors. First, it is connected with the complex dynamics of CN formation in competition with the dominating quasi-fission process. There are too few experimental data on low-energy collisions of very heavy ions to fix finally all the parameters (potential energy, friction forces, nucleon transfer rate) and to obtain an explicit value for the probability of CN formation for such heavy systems. The second source of uncertainty comes from the estimation of the survival probability of the excited CN, which is calculated within the statistical model. The basic quantities here are the fission barriers of SH nuclei appearing in the neutron evaporation chain. For example, in the 4n evaporation channel one needs to know the fission barriers and neutron separation energies for 5 subsequent isotopes of SH element.

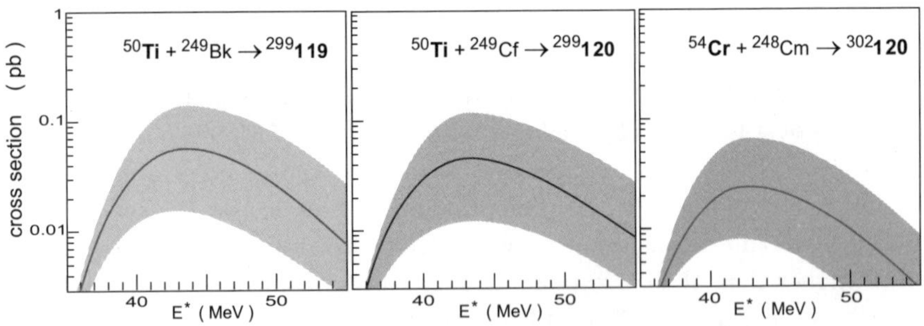

Fig. 2. Excitation functions for the production of superheavy elements 119 and 120 in the 4n evaporation channels of the ^{50}Ti+^{249}Bk, ^{50}Ti+^{249}Cf and ^{54}Cr+^{248}Cm fusion reactions. A factor 3 is assumed for the uncertainty of the survival probability caused by the uncertainty of the fission barriers of SH nuclei.

Different theoretical models give quite different values for the fission barrier for heavy nuclei. As mentioned above, we used the fission barriers of SH nuclei predicted by the macro–microscopic model,[5] which gives lower fission barriers for the isotopes of elements 119 and 120 than for elements $114 \div 116$ synthesized in ^{48}Ca fusion reactions. On the other hand, the full microscopic models based on self-consistent Hartree-Fock calculations usually predict much higher fission barriers for these isotopes (up to 10 MeV) if the Skyrme forces are used[8] (though these predictions are not unambiguous and depend strongly on chosen nucleon-nucleon forces). The exponential dependence of the survival probability on the values of the fission barriers leads to a change of it approximately by a factor 10 if the barriers are changed by 1 MeV.

Thus, we may conclude that the theoretical uncertainty in the predictions of the cross sections for formation of SH nuclei in hot fusion reactions with actinide targets is not less than a factor 3. If we consider such an uncertainty in our calculations, the excitation functions for synthesis of the new SH elements 119 and 120 in 4n evaporation channels of the ^{50}Ti+^{249}Bk, ^{50}Ti+^{249}Cf and ^{54}Cr+^{248}Cm fusion reactions will look as shown in Fig. 2. From this figure it is clear that all these reactions must be considered as quite promising for the synthesis of new elements. The final choice between them depends not so much on the difference in the cross sections as on other experimental conditions (availability of appropriate targets, beam intensities, etc.).

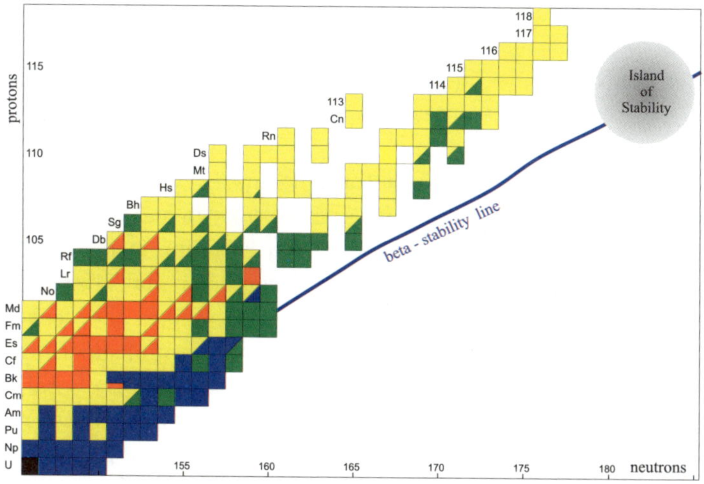

Fig. 3. Upper part of the nuclear map. The island of stability is shown schematically.

Note, however, that the present limits for neutron rich isotopes located at the upper part of the nuclear map ($Z > 60$) are very close to the line of stability while the unexplored area of heavy neutron-rich nuclides (to the east of the stability line) is extremely important for nuclear astrophysics investigations and, in particular, for the understanding of the r-process of astrophysical nucleogenesis. For nuclei with $Z > 100$ only neutron deficient isotopes (located to the left of the stability line) have been synthesized so far, see Fig. 3. Due to the bending of the stability line to the neutron axis, in fusion reactions of stable nuclei one may produce only proton rich isotopes of heavy elements (it's true also for the isotopes of 119 and 120 elements discussed above). This is the main reason for the impossibility of reaching the center of the "island of stability" in the superheavy mass region ($Z \sim 114 \div 120$ and $N \sim 184$) in fusion reactions with stable projectiles. In the ^{50}Ti+^{249}Cf and ^{54}Cr+^{248}Cm fusion reactions the 120-th element can be synthesized with cross sections of about 0.04 and 0.02 pb, correspondingly,[6] but again only the short-lived neutron deficient isotopes of this element may be obtained. The use of beams of radioactive nuclei hardly may solve this problem due to their low intensities.

Multi-nucleon transfer processes in low-energy collisions of actinide nuclei (like U+Cm) may really lead to the formation of neutron rich long-living superheavies. The shell effects (antisymmetrizing quasi-fission process with a preferable formation of nuclei close to the doubly magic lead isotope) might significantly enhance the corresponding cross sections.[6,9] In spite of difficulties of separation of the transfer reaction products, experiments of such kind are planned to be performed in nearest future.

The neutron capture process is an alternative (oldest and natural) method for the production of new heavy elements. Strong neutron fluxes might be provided by nuclear reactors and nuclear explosions under laboratory conditions and by supernova explosions in nature. However the "Fermium gap", consisting of the short lived Fermium isotopes $^{258-260}$Fm located at the beta stability line while have very short half-lives for spontaneous fission, impedes formation of nuclei with Z>100 by the weak neutron fluxes realized in existing nuclear reactors. In nuclear and supernova explosions (fast neutron capture) this gap may be bypassed, if the total neutron fluence is high enough. Theoretical models predict also another region of short lived nuclei located at Z=106÷108 and A∼270 (see Fig. 4).

In this paper we study the possibility of synthesizing heavy elements in multiple "soft" nuclear explosions and in pulsed reactors. We have found that in the first case the both gaps may be easily bypassed and, thus, a measurable amount of the neutron rich long-lived superheavy nuclei of the island of stability may be synthesized. For the second case we have formulated requirements for the pulsed reactors of the next generation which could be also used for the production of long-lived superheavy nuclei.

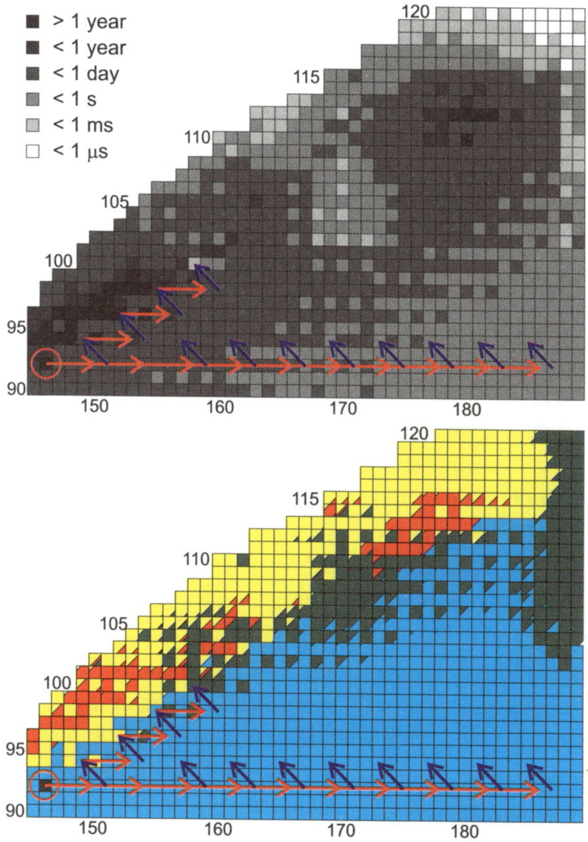

Fig. 4. Half-lives (up) and decay modes (bottom) of nuclei in the upper part of the nuclear map. Schematic view of slow (terminated at the short-lived fission Fermium isotopes) and fast neutron capture processes with subsequent beta-minus decays are shown by the arrows.

2. Nucleosynthesis by Neutron Capture

The synthesis of heavier nuclei in the reaction of neutron capture with subsequent beta-minus decay is a well studied process (see, e.g.,[10–12]). Relative yields of the isotopes formed in such a process may be found as a solution of the following set of differential equations (somewhat simplified here)

$$\frac{dN_{Z,A}}{dt} = N_{Z,A-1}n_0\sigma_{n\gamma}^{Z,A-1} - N_{Z,A}n_0\sigma_{n\gamma}^{Z,A} \tag{1}$$

$$-N_{Z,A}[\lambda_{Z,A}^{\beta-} + \lambda_{Z,A}^{fis} + \lambda_{Z,A}^{\alpha}] + N_{Z-1,A}\lambda_{Z-1,A}^{\beta-} + N_{Z+2,A+4}\lambda_{Z+2,A+4}^{\alpha},$$

where n_0 is the neutron flux (number of neutrons per square centimeter per second) and $\lambda^i_{Z,A} = ln2/T^i_{1/2}$ is the decay rate of the nucleus (Z, A) into the channel i (i.e., beta-minus and alpha decays and fission). For simplicity we ignore here the energy distribution of the neutrons and, thus, the energy dependence of the neutron capture cross section $\sigma^{Z,A}_{n\gamma}$. Neutrons generated by fission in nuclear reactors and in explosions are rather fast (far from the resonance region). In the interval of 0.1–1 MeV the neutron capture cross section is a smooth function of energy with the value of about 1 barn, which is used below for numerical estimations. Note that integration over the neutron energy may be performed very easily and does not change the obtained conclusions. The simplest version of the reaction chain, where only the neutron capture reactions are retained [first line in Eq.(1)], was considered recently in Ref. 13.

To solve Eq.(1) numerically one needs to know the decay properties of neutron rich nuclei which are not studied yet. That is the main problem, which significantly complicates the analysis of the multiple neutron capture processes. Theoretical estimations of half-lives for α-decay are rather reliable because they depend only on ground state masses which are very close in different theoretical models. For the half-lives of α-decays we used the well-known Viola-Seaborg formula with the coefficient proposed by Sobiczewski et al.[14] Half-lives of allowed β-decays depend also on the ground state masses of the nuclei and may be estimated by the empirical formula $\log_{10}\left[f_0 T^{\beta}_{1/2}(\text{sec})\right] = 5.7$, where the Fermi function f_0 is calculated by the standard formulas (see, e.g., Ref. 15).

The fission half-lives are the most uncertain quantities in Eqs.(1). Reliable analysis of the fission process requires knowledge of the multidimensional potential energy surface as well as of the collective inertia parameters. Such calculations can be performed only in very restricted areas of the nuclear map. We used the ground state masses and shell corrections for heavy and superheavy nuclei proposed by P. Möller et al.[5] and then we applied the empirical formula for the estimation of the fission half-lives. For this purpose we employed the relation by Swiatecki[16] based on the idea of the dominant role of the fission barrier for the fission probability

$$\log_{10} T^{fis}_{1/2}(\text{sec}) = 1146.44 - 75.3153 Z^2/A + 1.63792\left(Z^2/A\right)^2$$
$$-0.0119827\left(Z^2/A\right)^3 + B_f\left(7.23613 - 0.0947022 Z^2/A\right)$$
$$+\begin{cases} 0, & \text{Z and N are even} \\ 1.53897, & \text{A is odd} \\ 0.80822, & \text{Z and N are odd} \end{cases} \tag{2}$$

Here $B_f = B_f^{LDM} + \delta U_{g.s.}$ is the fission barrier, which was calculated as a sum of the liquid-drop barrier and the ground-state shell correction. The coefficients of the systematics (2) were determined by the fitting to the experimental data and to the rather realistic theoretical predictions[17,18] for the region of $100 \leq Z \leq 120$ and $140 \leq N \leq 190$.

The results of such calculations are shown in the left part of Fig. 4, where the theoretical values of $T^{\alpha}_{1/2}$, $T^{\beta}_{1/2}$ and/or $T^{fis}_{1/2}$ were replaced by experimental ones if known. In accordance with our calculations, the most stable superheavy nuclei located at the island of stability (which could be searched for in nature) are the isotopes of Darmstadtium ($Z = 110$) and/or Copernicium ($Z = 112$) with the neutron number $N \sim 180$. To produce these nuclei in the neutron capture process one needs to bypass the two area of short-living fissile nuclei, namely, the Fermium gap ($Z = 100$) and the region of Z=106÷108 and A∼270.

To test our model we first described available data on the fast neutron capture process realized in nuclear explosions. In this case the time of neutron capture, $\tau_n = (n_0\sigma_{n\gamma})^{-1} \sim 1$ μs $<< T_{1/2}(Z, A)$, is much shorter than half-lives of the produced nuclei (up to the neutron drip line). Keeping only the first two terms of the r.h.s. of Eq.(1), we get the following analytical solution (with initial conditions $N_{Z,A}(t = 0) = 1$ and $N_{Z,A+k}(t = 0) = 0$ at $k > 0$ where k is the number of captured neutrons)

$$N_{Z,A+k} = \frac{x^k}{k!} e^{-x}. \tag{3}$$

This relation can be used for clear understanding of the fast neutron capture process and for a preliminary estimation of relative yields of heavy nuclei synthesized in such a process. Here $x = n\sigma_{n\gamma}$, $n = n_0\tau$ is the total neutron fluence (neutrons per square centimeter) and τ is the duration of explosive neutron irradiation. Thus, the dimensionless quantity $x = n\sigma_{n\gamma}$ is a key factor characterizing the neutron capture process. In nuclear explosions the neutron fluence reaches the values of 10^{25} cm^{-2}, so that $x \sim 1$ and more than 10 neutrons could be captured during one exposure with time duration of about 1 μs.[11]

In Fig. 5 the experimental data on the yield of transuranium nuclei in the test thermonuclear explosion “Mike”[19] are compared with those calculated by Eqs.(1) assuming 1 μs neutron exposure of 1.3×10^{24} neutrons/cm^2 with subsequent one-month decay time. Note that elements 99 and 100 (Einsteinium and Fermium) were first discovered just in debris of the “Mike” explosion. As can be seen, in this case the Fermium gap does not influence the yields of nuclei with $Z > 100$ which follow roughly the relation (3).

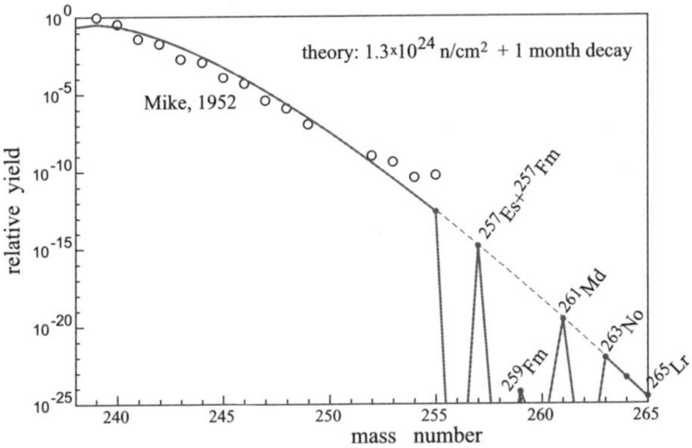

Fig. 5. Experimental (open dotes) and calculated relative yields of heavy nuclei in the test nuclear explosion "Mike".[19]

The idea to use "sequential neutron flux exposure" by generating two or several nuclear explosions in close proximity of each other to increase the resulting charge number of the synthesized nuclei was proposed many years ago.[20] Those time this idea was discussed by the experts (such as Edward Teller) who concluded that technically it could be realized. Here we study quantitatively such a possibility illustrated in the upper part of Fig. 6. In the bottom part of this figure the probabilities of heavy element formation are shown for one, three and ten subsequent short-time (1 μs) neutron exposures of 10^{24} n/cm^2 each following one after another with time interval of 10 seconds with final one month waiting (needed to perform some experimental measurements).

Our results demonstrate for the first time that multiple rather "soft" nuclear explosions could be really used for the production of noticeable (macroscopic) amount of neutron rich long-lived superheavy nuclei. Leaving aside any discussions on possibilities of such processes and associated technical problems, we want to emphasize a sharp increase of the probability for formation of heavy elements with $Z \geq 110$ in the multiple neutron irradiations (enhancement by several tens of orders of magnitude, see Fig. 6). This probability is high enough for some superheavy elements (see the region above the dotted line in Fig. 6) to perform their experimental identification.

It would be interesting also to study the same process of multiple neutron exposures realized in pulsed nuclear reactors. The pulse duration here could be much longer than in nuclear explosions (up to few milliseconds).

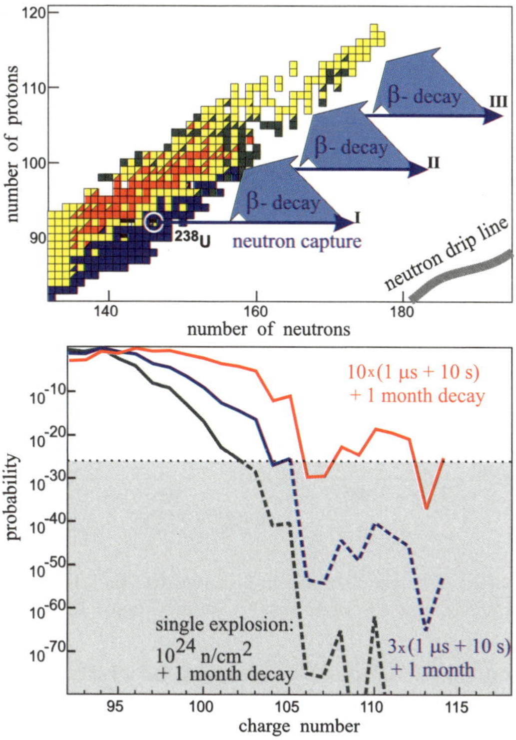

Fig. 6. Schematic picture for multiple neutron irradiation of initial ^{238}U material (up) and probability for formation of heavy nuclei (bottom) in such process (one, three and ten subsequent explosions). Dotted line denotes the level of few atoms.

In spite of that, the neutron fluence usually does not exceed 10^{16} n/cm^2 in existing nuclear reactors ($n_0 \sim 10^{19}$ n/cm^{2}s during one millisecond pulse). Thus, the quantity $x = n\sigma_{n\gamma}$ is about 10^{-8}, the time of neutron capture $\tau_n = (n_0\sigma_{n\gamma})^{-1} \sim 10^5$ s, and only the nearest long lived isotopes (A+1 or A+2) of irradiated elements may be formed during the pulse, see formula (3). Multi–pulse irradiation here corresponds, in fact, to the "slow" neutron capture process, in which new elements with larger charge numbers are situated close to the line of stability and finally reach the Fermium gap where the process stops (see Fig. 4). The result of numerical solution of Eq.(1) for neutron capture process in an ordinary pulsed reactor is shown in Fig. 7 by the dashed line. The probability for formation of heavy elements with $Z > 100$ is negligibly small here independent of a number of pulses and total time of irradiation.

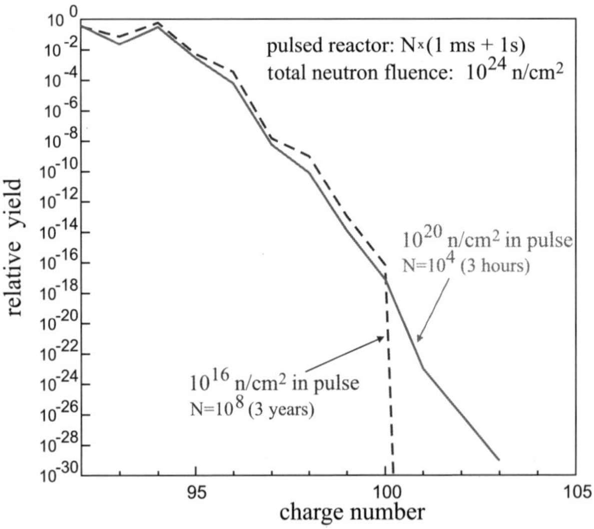

Fig. 7. Relative yields of heavy elements in ordinary (dashed line) and in high–intensity pulsed reactors (solid line) at the same total neutron fluence 10^{24} n/cm^2.

The situation may change if one would be able to increase somehow the intensity of the pulsed reactor. The neutron fluence in one pulse and frequency of pulses should be high enough to bypass the both gaps of short-lived nuclei on the way to the island of stability (see left part of Fig. 4). Thus, the specification of the high–intensity pulsed reactors of next generation depends strongly on properties of heavy neutron rich nuclei located to the right of these gaps. These nuclei are not discovered yet (see Fig. 3), and undoubtedly certain experimental efforts should be made to resolve this problem.

Using our theoretical estimations for the decay properties of these nuclei (see above) we have found that increase of the neutron fluence in the individual pulse by about three orders of magnitude as compared with existing pulsed reactors, i.e. up to 10^{20} netrons/cm^2, could be quite sufficient to bypass the both gaps (see the solid curve in Fig. 7). The details of the calculations and estimations of the yields of long-lived superheavy nuclei produced in neutron capture processes will be published elsewhere.

3. Summary

We have shown for the first time that a macroscopic amount of the long-living superheavy nuclei located at the island of stability may be really produced in multiple (rather "soft") nuclear explosions, if such processes would be realized technically. This goal could be also reached by using the pulsed nuclear reactors of next generation, if their neutron fluence per pulse will be increased by about three orders of magnitude. The experimental study of the decay properties of heavy nuclei located along the beta-stability line (and to the right of it, see Fig. 3) is extremely important for a more accurate analysis of the neutron capture processes (including the astrophysical ones[21]) in the upper part of the nuclear map.

We are indebted to the DFG – RFBR collaboration and to GSI for support of our studies.

References

1. Yu.Ts. Oganessian et al., *Phys. Rev. Lett.* **104**, 142502 (2010).
2. Yu.Ts. Oganessian, V.K. Utyonkov et al., *Phys. Rev.* C**74**, 044602 (2006).
3. V.I. Zagrebaev, M.G. Itkis, and Yu.Ts. Oganessian, *Phys. At. Nucl.* **66**, 1033 (2003).
4. V.I. Zagrebaev, *Nucl. Phys.* A**734**, 164 (2004).
5. P. Möller, J. R. Nix, W. D. Myers, W. J. Swiatecki, *At. Data Nucl. Data Tables* **59**, 185 (1995).
6. V.I. Zagrebaev and W. Greiner, *Phys. Rev.* C **78**, 034610 (2008).
7. U. Mosel and W. Greiner, *Z. Phys.* **222**, 261 (1969); see also *Memorandum zur Errichtung eines gemeinsamen Ausbildungszentrums für Kernphysik Hessischer Hochschulen* (1966).
8. T. Bürvenich, M. Bender, J.A. Maruhn, and P.-G. Reinhard, *Phys. Rev.* C **69**, 014307 (2004).
9. V.I. Zagrebaev, Yu.Ts. Oganessian, M.G. Itkis and W. Greiner, *Phys. Rev.* C **73**, 031602 (2006).
10. D.W. Dorn, *Phys. Rev.* **126**, 693 (1962).
11. G.T. Seaborg, *Ann. Rev. Nucl. Sci.*, **18**, 119 (1968).
12. A.S. Krivochatsky, Yu.F. Romanov, *Production of transuranium and actinide elements by neutron irradiation (in Russian)*, Atomizdat, Moscow, 1969.
13. A. Botvina, I. Mishustin, V. Zagrebaev and Walter Greiner, *Int. J. Mod. Phys.* E**19**, 2063 (2010).
14. A. Sobiczewski, Z. Patyk, and S. Čwoik, *Phys. Lett.* **B224**, 1 (1989).
15. C.S. Wu and S.A. Moszkowski, *Beta decay* (John Wiley & Sons, New York, 1966), p. 183.
16. W.J. Swiatecki, *Phys. Rev.* **100**, 937 (1955).
17. R. Smolańczuk, *Phys. Rev.* **C56**, 812 (1997).
18. A. Staszczak, Z. Łojewski, A. Baran, B. Nerlo-Pomorska, K. Pomorski, in *Dynamical Aspects of Nuclear Fission*, Casta-Papernicka, (1996), 22.
19. H. Diamond et al., *Phys. Rev.* **119**, 2000 (1960).
20. H.W. Meldner, *Phys. Rev. Lett.* **28**, 975 (1972).
21. I.V. Panov, I.Yu. Korneev and F.-K. Thielemann, *Phys. At. Nucl.* **72**, 1026 (2009).

SOFT X-RAY REFLECTION AND STRONG AND WEAK FIELD LIMIT DETERMINATION IN NARROW-LINE SEYFERT 1 GALAXIES

Th. BOLLER*

Max-Planck-Institute for Extraterrestrial Physics,
Garching, PSF 1312, 85741 Garching, Germany
**E-mail: bol@mpe.mpg.de*

We have analyzed the XMM-Newton RGS spectral properties of a sample of the most prominent Narrow-Line Seyfert 1 (NLS1) galaxies observed so far with previous and present X-ray missions. We have fitted the merged RGS1/RGS2 first and second order spectra for all available observations from the XMM-Newton Science Archive to test the reflection model interpretation proposed by Ross & Fabian (2005). For objects with more than one observations a combined RGS1/RGS2 first and second order spectrum has been created, resulting in exposure times ranging between about 13 and 780 ks. For the maximum possible RGS statistics we work out the progress and limitations for the spectral fitting of those NLS1 galaxies most observed with XMM-Newton. The objects 1H 0707–495 and Ark 564 can be modelled with the relativistic reflection model by Ross & Fabian (2005). For 1H 0707–495, the reflection spectrum fit suggests an inner disc radius at about 3 R_G. Assuming a standard disc where the inner disc edge is equal to the marginally stable orbit we infer a rapidly rotating black hole with spin parameter of about 0.8. The emissivity index of the power law amounts to 6.3 suggesting a centrally peaked emission. The ionization parameter is about 1000. The soft X-ray lines are therefore strongly broadened by Compton broadening effects and the emission in the strong gravity field limit. Similar values are obtained for Ark 564. The FWHM of the additional soft X-ray lines required in the fit for 1H 0707-495 are much smaller, about 0.1c, which is consistent with an origination in the weaker field limit. We find that for the prototype objects 1H 0707–495 and for Ark 564 only the relativistic reflection model provides a statistically acceptable fit. The obtained parameters from the relativistic reflection model suggest that the observed emission originates from matter very close to the central supermassive black hole.

Keywords: X-rays galaxies; galaxies active; emission lines; galaxies individual.

1. Introduction

XMM-Newton spectral and timing observations of Narrow-Line Seyfert 1 (NLS1) galaxies have been proven as a strong diagnostic tool to probe the accretion flow at the innermost stable orbit around supermassive black holes. In general the X-ray emission of NLS1 galaxies is unusual in showing steep soft and hard X-ray spectra (e.g. Puchnarewicz et al. 1992, Boller et al. 1996, Brandt, Mathur & Elvis 1997) and very rapid variability (e.g. Boller et al. 1997, Gallo et al. 2004). While their steep soft X-ray spectra discovered in previous X-ray missions have been interpreted as a strengthened and shifted accretion disc spectrum (e.g. Leighly 2000, Boller et al. 2002, Boller and Tanaka 2005), XMM-Newton observations have decomposed the X-ray spectra into a strong, relativistically blurred reflection plus power law component (Miniutti & Fabian 2004, Crummy et al. 2006, Fabian et al. 2009, Ponti et al. 2010). The key observational fact revealed by XMM-Newton for further testing the relativistic reflection model was the detection of a sharp spectral drop at around 7 keV in 1H 0707–495 (Boller et al. 2002, Gallo et al. 2004) and at around 8.5 keV in IRAS 13224–3809 (Boller et al. 2003). The relativistic reflection model requires that the inner radius of the accretion disc is very close to a rapidly spinning black hole, which results into strong light bending effects and strong variability of the power law component compared to the reflection component. The relativistic reflection model is further supported by the presence of a 30 s time delay between the soft (less than 1 keV) and hard (greater than 1 keV) light curves detected in 1H 0707-495 (Fabian et al. 2009) and IRAS 13224–3809 (Ponti et al. 2010). A detailed investigation of reflection and absorption components in 1H 0707–495 are presented by Zoghbi et al. (2010). In this paper we work out the presently available power and constrains of the most commonly accepted relativistic reflection model based on the available XMM-Newton RGS statistics for the investigated objects.

2. The RGS NLS1 Sample

The most interesting members of NLS1 galaxies had been proposed for the XMM Guaranteed Time programme (Boller et al. 1998) to extend previous studies of this class based on ROSAT and ASCA observations. The broad band, large effective area and high spectral resolution of XMM-Newton was considered to allow unequalled X-ray studies of NLS1 galaxies. The observations were proposed to examine the shape of the soft X-ray excess and the transition region between the soft X-ray excess and the hard X-ray power law, to study broad and ionized iron K line emission, ionized gas along the line of sight, as well as to obtain uninterrupted studies of rapid and nonlinear X-ray variability, power spectrum analysis, measurements of spectral variability and searches for simultaneous X-ray and optical variability. Except IRAS 13224–3809, PG 1244+026, and Mrk 110, all objects have been observed more than once with XMM-Newton. The cumulative exposure times available in the XMM-Newton Science Archive (XSA) at the end of 2009 range between about 13 and 780 ks. These objects can be considered as the best studied NLS1 galaxies to date. In this paper we investigate the combined RGS1, RGS2 first and second order XMM-Newton spectra of the individual sources.

3. The Spectral Model

The reflection model of Ross & Fabian (2005) provides the most advanced model for fitting AGN spectra over a large energy range from about 0.1 up to 100 keV (e.g. Crummy et al. 2006, Fabian et al. 2009, Miniutti et al. 2009, Ponti et al. 2010). Relativistic blurring is applied by using the *kdblur* code (Laor 1991) available in *XSPEC, version 12.4.0* (Arnaud et al. 1996). In the spectral fitting the emissivity index, the inner radius, the ionization parameter, the normalization of the power law and reflection component, the inclination, the Fe to solar ratio and the neutral absorption value are free parameters. The outer radius is fixed to 100 R_G.

4. 1H 0707–495

4.1. *RGS relativistic reflection model spectral fitting*

The combined the RGS1/RGS2 first and combined second order spectra of the 10 individual observations of 1H 0707–495 available in the XMM-Newton Science Archive (XSA) have been fitted with the relativistic reflection spectrum. The inner radius r_{in} of the emitting disc is (3.02 ± 0.35) R_G, resulting into a black hole spin of about 0.8, indicating emission close to a Kerr black hole. The accretion disc emissivity is high with (5.56 ± 0.67) substantially above the value expected from a Schwarzschild black hole but again consistent with the value obtained from the EPIC pn data by Fabian et al. (2009). The ionization parameter ξ is (1036 ± 51) erg cm s^{-1}. The combination of emission very close to the central black hole and high ionization results into a combined effect of strong relativistic effects and strong Compton broadening. The FWHM values for emission of many unresolved lines centered at around 0.3 and 0.5 keV is of the order of about c/3.

Fig. 1 (left panel) shows the relativistic reflection model fit for the combined RGS1/RGS2 first and second spectra. The reflection component dominates over the the power law component, which is also seen in the EPIC

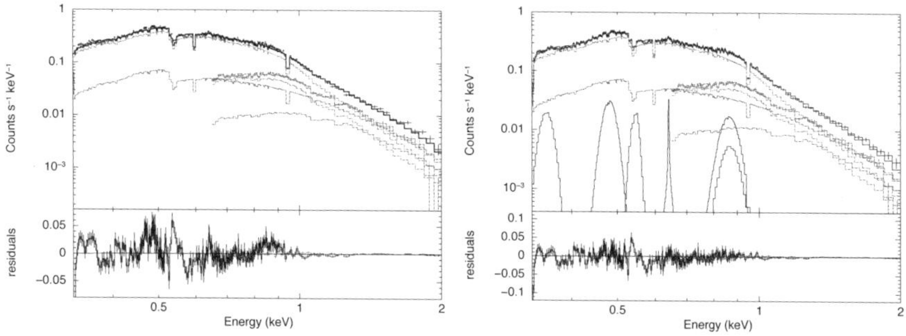

Fig. 1. *Left panel:* Relativistic reflection model fit to the combined RGS1/RGS2 first and second order spectra to 1H 0707–495. The first order combined RGS1/RGS2 data are shown in black, the second order combined RGS1/RGS2 data are shown in green. For both orders the reflection components are marked with cyan, and the power law components with magenta. The inner radius r_{in} is 3.02 ± 0.35 R_G indicating emission very close to the central black hole, accounting for emission in the strong field limit. The emissivity index is high with 5.56 ± 0.67, consistent with the EPIC pn analysis of Fabian et al. (2009). Residua remain in the spectral fit at energies consistent with emission from soft X-ray lines. *Right panel:* Relativistic reflection model fit plus Gaussian line fits to the combined RGS1/RGS2 first and second order spectra. The energies of the Gaussian lines added to the relativistic reflection model are consistent with emission from line blends of C VI, N VII, O VII, O VIII, and Fe L/Ne IX. For the Fe L/ Ne IX Gaussian line fit the first and second orders are shown. The significance of adding these Gaussian lines to the relativistic reflection model is well above the 3 σ limit.

pn data analysis of Fabian et al. (2009). Additional residua remain in the relativistic reflection model fit at line energies consistent with other soft X-ray lines. Gaussian line fits have been added (Fig. 1, right panel) indicating line emission from larger distances to the black hole and emission in the weaker field limit.

4.2. *Comparison with EPIC pn spectral fitting results*

We have merged the 10 individual XMM-Newton EPIC pn observations (the individual event lists, attitude files and response files) for comparison with the RGS spectral fitting results. The model parameters are consistent between the two instruments within the 1 σ errors, except for slight offsets in the photon indices and the normalizations of the power laws, which might be due to calibration uncertainties between the two instruments. In conclusion, reliable spectral fit results can be obtained from the combined RGS1/RGS2 first and second order spectra even in the limited RGS energy band for the case of 1H 0707–495. We note that the inner radius r_{in} of the emitting disc is about 3 R_G indicating emission located very close to the central black hole and consistent with emission lines affected by strong gravity. In the EPIC pn 0.5 to 10 keV band the relative amount of reflection is about 65 per cent. The ionization parameter is high with about 1050 indicating strong Compton broadening of the lines. These combined effects result into a large blurring of many line blends in the Ross & Fabian (2005) relativistic reflection code, consistent with the RGS spectral analysis (c.f. Fig. 1, left panel). In Fig. 2 we show the results from the combined RGS, EPIC pn spectral fit. The higher signal-to-noise in the EPIC pn data reveals the presence of absorption features at 0.74 and 1.27 keV, most probably due to O VII K-shell and Mg K-shell edges. The Fe L/ Ne IX line blend at 0.906 keV is also detected in the EPIC pn data (c.f. Fig. 2 right panel).

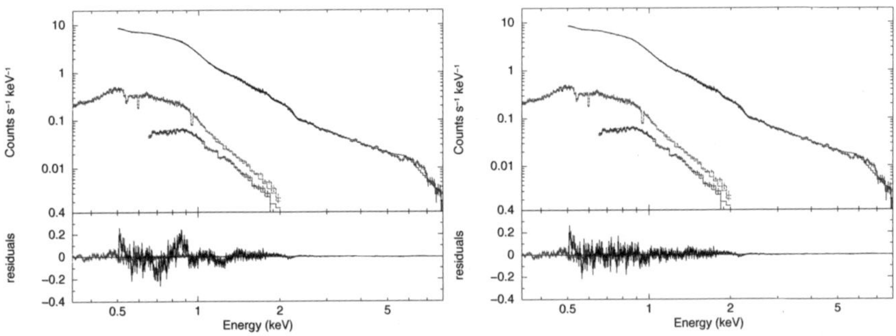

Fig. 2. Left panel: Combined RGS1/RGS2 and EPIC pn relativistic reflection model fit to 1H 0707–495. Right panel: Reflection model fit to the EPIC pn data including Gaussian line fits for an emission line blend of Fe L/ Ne IX at 0.906 keV, and two absorption features at around 0.74 and 1.27 keV, most probably due to a O VII K and a Mg K edge.

4.3. *Additional soft X-ray line emission*

Fig. 1 (left panel) indicates the presence of residua at RGS energies consistent with the emission from additional soft X-ray lines. Positive residua are detected close to the line energies of C VI (0.367 keV), N VII (0.50 keV), O VII (0.563) keV, O VIII (0.653 keV) and the line blend of Fe L/ N IX (0.906 keV). Given the present statistics, only Gaussian line components have been added to the relativistic reflection model. The additional lines originate far away from the inner disc and they are not accounted for in the relativistic reflection model. The high metallicity which is detected in the RGS and EPIC pn analysis, which was already noted in earlier EPIC pn papers on 1H 0707–495 (e.g. Boller et al. 2002, Fabian et al. 2009). This might be due to an enhanced star formation process which is expected in NLS1 galaxies (e.g. Mathur 2000, Boller et al. (2002), Fabian et al. (2009). NLS1 galaxies are expected to be AGN just in the forming, with still a large gas reservoir which leads to an intense star formation process. The significance of the additional soft X-ray lines is well above the 3 σ limit and the line width indicates FWHM values between about 4000 and 25000 km s^{-1}. This suggests emission from larger distances to the black hole of a few 1000 R_G, either from the outer part of the accretion disc or the BLR. These lines probe physical parameters in the weaker field limit. Fig. 1 (right panel) shows the relativistic reflection model plus Gaussian line fits.

4.4. *Model component properties and fitting results*

We have fitted a power law model plus two Gaussian lines to the relativistic reflection component to the model spectrum. At 0.31 keV the FWHM of the line blend is about 75000 km s^{-1}, and at 0.53 keV about 125000 km s^{-1}. We note many individual lines centered around the measured energies contribute to the line broadening due to strong GR plus Compton broadening effects. We have simulated a relativistic reflection model with the fit parameters obtained for 1H 0707–495 and have fitted the line blend at around O VII for the case of Compton broadening (without applying relativistic effects) and for the combined effect of strong GR effects and Compton broadening. The FWHM value for Compton broadening is about 50 eV. For the combined effect of Compton broadening and relativistic effects the FWHM line width is about 100 eV. Ignoring additional blurring by bulk velocity motions, about half of the line width is due to Compton broadening and half due to relativistic effects.

5. Ark 564

5.1. *RGS relativistic reflection model fitting*

We have fitted the combined RGS1/RGS2 first and second order spectra with the relativistic reflection model for 5 individual observations on Ark 564. The individual observations have been combined into a merged spectrum. We note that the innermost radius r_{in} and the disc emissivity index are similar to 1H 0707–495, again pointing to emission of matter in the strong field limit. Also the ionization parameter is high, indicating a substantial contribution of Compton broadening. The Fe to solar ratio is much lower compared to 1H 0707–495. The relativistic reflection model is

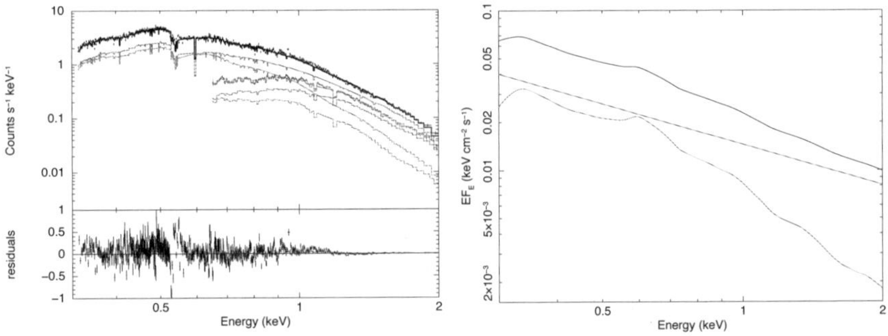

Fig. 3. *Left panel:* Relativistic reflection model fit to the combined RGS1/RGS2 first and second order spectra to Ark 564. The color coding is the same as in the left panel of Fig. 1 (left panel). All parameters are constrained and the reduced χ^2 value is 1.41 for 5363 d.o.f. The inner radius of the emitting disc r_{in} is (2.36 ± 0.15) R_G, indicating emission very close to the central black hole. The emissivity index is high with 6.49 ± 0.84. *Right panel:* Model components obtained from the relativistic reflection model. The parameters obtained for the reflection component indicate emission in the strong field limit. We have fitted a power law and two Gaussian lines to the reflection component and find FWHM values of about 120000 km s^{-1} at the line blend at around 0.33 keV and of about 55000 s^{-1} at around 0.60 keV.

shown in the left panel of Fig. 3. Similar to 1H 0707–495 some residua remain in the fit, however with less significance (which might be due to the lower metallicity compared to 1H 0707–495). The right panel of Fig. 3 shows the reflection and power law model components. We have fitted a power law and two Gaussian lines to the reflection component and find FWHM values of about 120000 km s^{-1} at the line blend at around 0.33 keV and of about 55000 s^{-1} at around 0.60 keV.

6. Remaining Objects

The remaining objects listed in Section 2 were fitted with the relativistic reflection model. Although the relativistic reflection model results into statistically acceptable fits, other models, e.g. simple power law models, can also account for the data. It might be that the spectral features of 1H 0707–495 and Ark 564 are also present in these objects, however with lower statistics and are therefore much more difficult to detect. With the presently available RGS statistics after 10 years of XMM-Newton operation the relativistic reflection model best describes the spectral energy distribution of 1H 0707–495 and Ark 564. For the remaining objects much longer XMM-Newton observation and observation with Astro-H and IXO are required to achieve further progress.

7. Summary

We find that strong soft X-ray reflection is a natural consequence for the existence of a compact corona close to the central black hole disc system, in agreement with the EPIC pn spectral fitting results obtained by Fabian et al. (2009) for 1H 0707–495 and by Ponti et al. (2010) for IRAS 13224–3809. For 1H 707–495 and Ark 564 we find emission of matter close to the central black hole, described by the relativistic reflection model of Ross & Fabian (2005). In addition to emission from relativistically and Compton broadened soft X-ray lines in the strong field limit, for 1H 0707–495 emission lines arising from larger distances from the black hole are detected, extending GR physics to the weaker field limit. Following Müller & Wold (2006) at a distance to the black hole of about 50 R_G the gravitational redshift is only about 0.03. For 1H 0707–495 and Ark 564 simple continuum models and even more complicated models can not explain the data. We conclude that the spectral components predicted by the relativistic reflection model for 1H 0707–495 and Ark 564 might serve as templates for the spectral energy distribution of matter in the innermost centers of AGN. These features might be present in the representative sample of NLS1 galaxies investigated and in most unobscured AGN, however much more difficult to detect, even with the maximum available RGS statistics to date. We argue for the use of RGS data because their energy resolution is better compared to the EPIC pn data allowing the detection of additional soft X-ray lines in the weak field limit. Further, the RGS data are calibrated to energies down to 0.33 keV. In the EPIC pn data calibration uncertainties below 0.5 keV exist due to the redistribution of emission line photons from N VII. Reliable spectral fits are therefore restricted to energies above 0.5 keV. With the current data we cannot uniquely distinguish between models for the remaining objects listed in Section 2. For these objects an e.g. outflowing partially ionized wind model (e.g. Done et al. 2007) might also account for the data. We expect that this ambiguity may vanish by longer XMM-Newton observations and/or observations with the new generation of X-ray telescopes.

Acknowledgments

TB is grateful to Prof. Dr. Dr. H.C. Mult. Walter Greiner for his longstanding support and many fruitful discussions of my scientific work and the teaching activities at the Goethe-University Frankfurt. TB thanks Andreas Müller and Aitor Ibarra Ibaibarriaga for intensive discussions on the content of this paper.

References

1. Arnaud K. A., 1996, in Jacoby G., Barnes J., eds, ASP Conf. Ser. Vol. 101, Astronomical Data Analysis Software and Systems V. Astron. Soc. Pac., San Francisco, p. 17
2. Boller, Th., Brandt, W., Fink, H., A&A, 305, 53, 1996
3. Boller, Th., Brandt, W., Fabian, A.C., Fink, H., MNRAS 289, 393, 1997
4. Boller, Th., Trümper, J., Brandt, W., Fabian, A.C., XMM GT proposal (XMM SSC internal proposal), 1998
5. Boller, Th., Fabian, A. C., Sunyaev, R., Trümper, J., Vaughan, S., Ballantyne, D. R., Brandt, W., MNRAS, 329, 1, 2002
6. Boller, Th., Tanaka, Y., In: Growing black holes: accretion in a cosmological context. Proceedings of the MPA/ESO/MPE/USM Joint Astronomy Conference held at Garching, Germany, 21-25 June 2004. A. Merloni, S. Nayakshin, R. A. Sunyaev (Eds.). ESO astrophysics symposia. Berlin: Springer, ISBN 3-540-25275-4, ISBN 978-3-540-25275-7, 2005, p. 170 - 174
7. Boller, Th., Tanaka, Y., Fabian, A., Brandt, W., Gallo, L., Anabuki, N., Haba, Y., Vaughan, S., MNRAS, 343, 89, 2003
8. Brandt, W.N., Mathur, S., Elvis, M, MNRAS 285, 25, 1997
9. Crummy, J., Fabian, A.C., Gallo, L., Ross, R.R., MNRAS, 365, 1067, 2006
10. Done, C., Gierlinski, M., Sobolewska, M., Schurch, N., ASPC 373, 121, 2007
11. Fabian, A.C. et al., 2009, Nature 459, 540
12. Gallo, L., Tanaka, Y., Boller, Th., Fabian, A.C., Vaughan, S., Brandt, W., MNRAS 353, 1064, 2004
13. Laor, A., ApJ 376, 90, 1991
14. Leighly, K., NewAR 44, 395, 2000
15. Mathur, S., NewAR 44, 469, 2000
16. Miniutti, G., Fabian, A.C., MNRAS 349, 1435, 2004
17. Miniutti, G., Panessa, F., de Rosa, A., Fabian, A. C., Malizia, A., Molina, M., Miller, J. M., Vaughan, S., MNRAS 398, 255, 2009
18. Müller, A., Wold, M., A&A 457, 485, 2006
19. Ponti, G., Gallo, L. C., Fabian et al., MNRAS 913, 2010
20. Puchnarewicz, E.M., Mason, K.O., Cordova, F.A. et al., MNRAS 256, 589, 1992
21. Ross, R.R., Fabian, A.C., MNRAS 358, 211, 2005
22. Zoghbi, A., Fabian, A. C., Uttley, P., Miniutti, G., Gallo, L. C., Reynolds, C. S., Miller, J. M., Ponti, G., MNRAS 401, 2419, 2010

A 100% RENEWABLE POWER SYSTEM FOR EUROPE – LET THE WEATHER AND PHYSICS DECIDE!

MARTIN GREINER*, DOMINIK HEIDE, MORTEN RASMUSSEN, GORM ANDRESEN

Aarhus School of Engineering and Department of Mathematical Sciences, Aarhus University, Ny Munkegade 118, 8000 Aarhus C, Denmark
**E-mail: greiner@imf.au.dk*

The design of sustainable energy systems is no longer only the domain of politics, economics and engineering. Mathematical physics is able to contribute with its generic understanding of everything. A new modelling approach is presented and applied to design a fully renewable European power system. This approach is based on weather data with good spatio-temporal resolution, which is first converted into wind and solar power generation and then used to derive estimates on the optimal mix between the renewable resources and the storage needs.

Keywords: Energy system design; wind power generation; solar power generation; large-scale integration; storage.

1. Introduction

We investigate the design of a future European power system based on a 100% contribution from renewable sources. Some key questions are: How much wind and solar power is good for Europe? Is there an optimal mix between them? What are the required storage and balancing capacities? Where and when do we have to put our investments in order to avoid costly dead-end streets during transitional pathways?

An 8-year-long weather data set with high temporal (1h) and spatial ($47\times48\text{km}^2$) resolution across Europe is the key to a new modelling approach. Its convolution with future-projected wind and solar photovoltaics power locations and capacities reveals how much wind and solar power is generated when and where across the continent. Fig. 1 shows the spatially aggregated total wind and solar power generation time series. The strong fluctuations reveal three characteristic weather time scales: 1 day (intraday), 1 week (synoptic) and 1 year (seasonal). The synoptic time scale is the time it takes for a big weather front to sweep over Europe. Since winds across Europe are stronger in winter than in summer, wind power generation in winter is about two times larger than in summer. The solar power generation has the same seasonal fluctuation strength, but its maximum is shifted to the summer. – Also shown in Fig. 1 is the electricity consumption. The same three time scales show up again. There is more consumption during daylight time than during night hours, on weekdays compared to weekends, and in winter than in summer.

If wind were the only power source in a fictitious 100% wind-only Europe, then the seasonal wind power curve has to be directly compared to the European consumption curve. Although the annual average of wind-power generation and consumption is the same, the seasonal behavior is different. As a consequence an enormous amount of seasonal storage and balancing is required. Over summer the storage and balancing facilities have to feed the deficit. During winter the large wind-power excess is put into the storage. A 100% solar-only scenario will lead to about the same large seasonal storage and balancing needs.

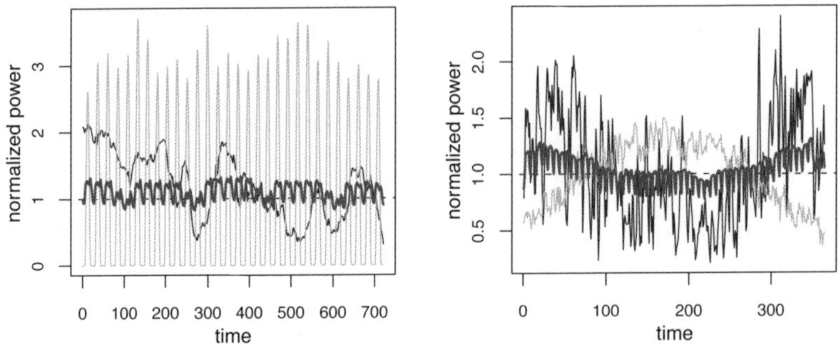

(a) one-hour resolution over one month (b) one-day resolution over one year

Fig. 1. Normalized (blue) wind power generation, (yellow) solar power generation and (red) electricity consumption, with spatial aggregation over Europe.

2. Seasonal Optimal Mix

The seasonal storage needs can be reduced significantly by counterbalancing the seasonal wind power generation with the seasonal solar power generation. Fig. 2 takes 60% from the wind curve and 40% from the solar curve of Fig. 1. The resulting curve is able to nicely follow the seasonal consumption. It is expected that this optimal mix brings seasonal storage needs to a minimum.

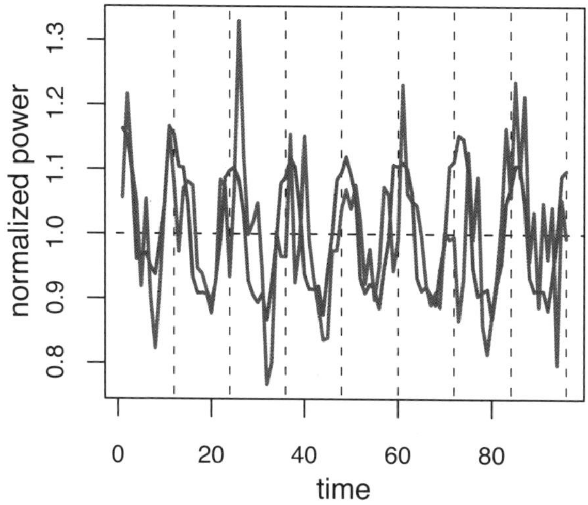

Fig. 2. The seasonal wind- and solar-power generation time series from Fig. 1 are combined with a 60%/40% weighting (green).

It is straightforward to quantify the seasonal optimal mix. Key is the mismatch energy

$$\Delta(t) = \gamma \left(a \frac{W(t)}{\langle W \rangle} + (1-a) \frac{S(t)}{\langle S \rangle} \right) - \frac{L(t)}{\langle L \rangle} . \tag{1}$$

$W(t)$ represents the total European wind power generation at time t, and $\langle W \rangle$ its average. $S(t)$ and $L(t)$ are the respective solar power and consumption time series. The coefficient a represents the relative contribution from wind power generation. For $\gamma = 1$ the annual wind and solar power generation matches the annual consumption. $\gamma > 1$ represents excess generation.

From the mismatch the filling level of the storage with efficiencies η_{in}, η_{out} is constructed:

$$H(t) = H(t-1) + \begin{cases} \eta_{\text{in}} \Delta(t) & \text{if } \Delta(t) \geq 0 , \\ \eta_{\text{out}}^{-1} \Delta(t) & \text{if } \Delta(t) < 0 . \end{cases} \tag{2}$$

The storage energy capacity becomes

$$E_H = max_t \left(H(t) - min_{t'>t} H(t') \right) . \tag{3}$$

This quantity is shown in Fig. 3 as a function of the wind fraction a and $\gamma = 1$. It comes with a minimum at $a \approx 0.6$. This nicely confirms the intuition expressed in Fig. 2.

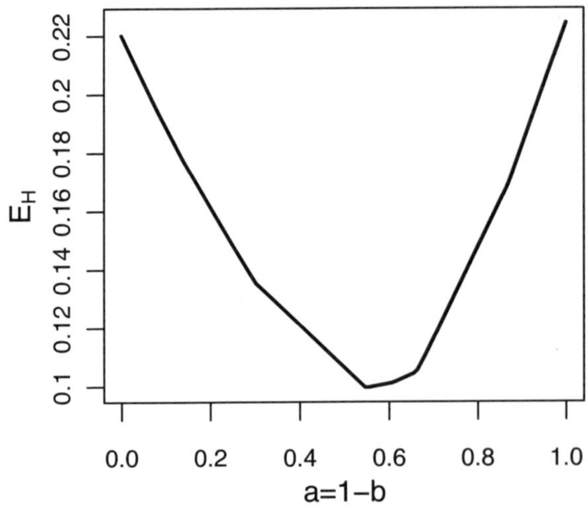

Fig. 3. Energy capacity for an idealized storage ($\eta_{\text{in}} = \eta_{\text{out}} = 1$) and $\gamma = 1$. The optimal mix $a \approx 0.6$ is defined by the minimum.

The current annual electricity consumption of Europe amounts to 3200 TWh. If 60% of this were to come from wind power, then it will require about 750 GW installed wind power capacity. Today Europe has 70-80 GW. In terms of area fully filled with wind turbines, 750 GW correspond to 75000 km^2, which is about one-fifth of Germany. If 40% of the annual European consumption were to come from solar photovoltaic power, it will require an area of about 5000 km^2 fully covered with solar panels. Germany alone has a potential of 1200 km^2 on roof-tops.

3. Seasonal Storage Needs

We also learn from Fig. 3, that for the minimizing optimal mix the storage energy capacity amounts to 10% of the annual consumption. This corresponds to five weeks with no wind and sunshine across Europe. In absolute numbers this is 320 TWh. For comparison, a typical one-family house has an annual consumption of 4000 kWh. Thus, 320 TWh/4 MWh equals to 80 million houses.

320 TWh of storage energy capacity is completely out of reach for Europe. Ways have to be found to reduce this. One possibility is to build more wind turbines and solar panels. Fig. 4 shows the dependence of the required energy storage capacity as a function of the excess generation $\gamma \geq 1$. As

can be seen, it decreases rapidly. For an ideal storage the energy capacity reduces from 320 TWh at $\gamma = 1$ to 16 TWh at $\gamma = 1.5$.

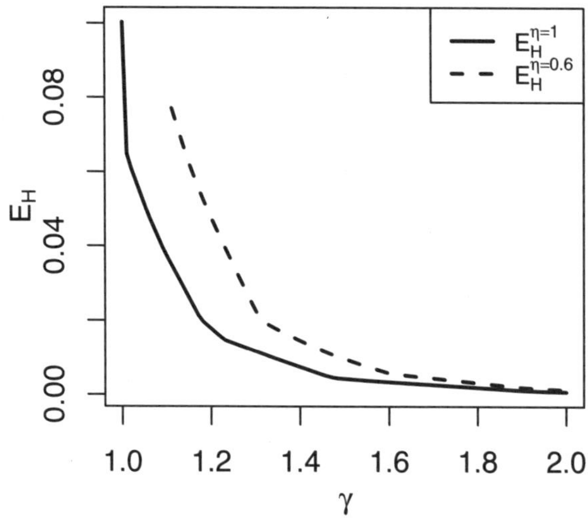

Fig. 4. With $\gamma > 1$ the required storage energy capacity decreases. The solid /dashed curves represent ideal / hydrogen storage with efficiencies $\eta_{\mathrm{in}} = \eta_{\mathrm{out}} = 1$ and 0.6, respectively.

As of now Europe has no significant seasonal storage. The energy capacities of pumped hydro ($\eta_{\mathrm{in}} = \eta_{\mathrm{out}} = 0.9$) and compressed air ($\eta_{\mathrm{in}} = \eta_{\mathrm{out}} = 0.8$) are several orders of magnitude below 20-50 TWh, and no room for more. Hydrogen storage could become an option. With $\eta_{\mathrm{in}} = \eta_{\mathrm{out}} = 0.6$ it has a rather low efficiency, but its energy density is very large. Once the excess generation is around $\gamma \approx 1.5$ the required energy capacity amounts to 35 TWh; see again Fig. 4. This translates into a volume of $0.2\mathrm{km}^3$, which is the size of 100 large soccer stadiums. Hydrogen has to be stored underground in salt caverns. In Europe those are mainly in North Germany and the North Sea. Besides their current usage for the gas and oil reserve, there is still enough capacity to include these 100 soccer stadiums.

4. Outlook

The hydro lakes in Scandinavia and the Alps represent another form of storage, that can be used to balance the European mismatch between renewable electricity generation and consumption. If those together with the possible hydrogen storage in North Germany would serve as the European battery, then a lot of power transmission is required across the continent. Other important topics focus around the questions, what to do with the excess wind and solar power generation and how to design pathways from our current energy system towards a fully sustainable one. Some results can already be found in the three references listed below.[1–3]

References

1. D. Heide, L. von Bremen, M. Greiner, C. Hoffmann, M. Speckmann and S. Bofinger, *Renewable Energy* **35**, 2483 (2010).
2. D. Heide, L. von Bremen, M. Greiner and C. Hoffmann, *Renewable Energy* **36**, 2515 (2011).
3. D. Heide, Statistical physics of power flows on networks with a high share of fluctuating renewable generation (2010), PhD thesis, Johann-Wolfgang-Goethe Universität, Frankfurt, Germany.

HAPPY ISLAND*

LARRY McLERRAN

Brookhaven National Laboratory and Riken Brookhaven Center,
Upton, NY, 11973-5000, USA

I discuss the phase diagram for QCD in the baryon chemical potential and temperature plane. I argue that there is a new phase of matter different from the deconfined Quark Gluon Plasma: Quarkyonic Matter. Quarkyonic Matter is confined and exists at densities parametrically large compared to the QCD scale, when the number of quark colors, N_c is large. I motivate the possibility that Quarkyonic Matter is in an inhomogeneous phase, and is surrounded by lines of phase transitions, making a Happy Island in the μ_B-T plane. I conjecture about the geography of Happy Island.

1. Quarkyonic Matter

In a typical phase diagram of QCD, one will see two phases: Confined Matter, and the Deconfined Quark Gluon Plasma.[1] In some more detailed cartoons, one will also see Color Superconductivity and the Liquid Gas Nuclear matter phase transition.[2,3] The latter two phases make small changes to the bulk properties of nuclear matter, and I shall not discuss them in detail below. The transition between confined matter and deconfined matter is however a big deal, because the number of degrees of freedom changes rapidly. At low temperature and density, the important degrees of freedom are pions, $N_{dof} = 3$. In deconfined matter, there are 16 gluon and around $20-30$ quark degrees of freedom. The energy density scaled by the temperature to the fourth power, ϵ/T^4 is proportional to the number of degrees of freedom and therefore makes a large change. This change takes place in a relatively narrow range of temperature, $\Delta T \sim 30 - 40\ MeV$.

The arguments for such a a confinement-deconfinement transition are strong at finite temperature and zero baryon chemical potential. There is abundant lattice Monte Carlo data that support such a picture.[4] On the

*Invited talk presented at Symposium on Nuclear Physics in GOA, India, Nov. 28-Dec. 2, 2010

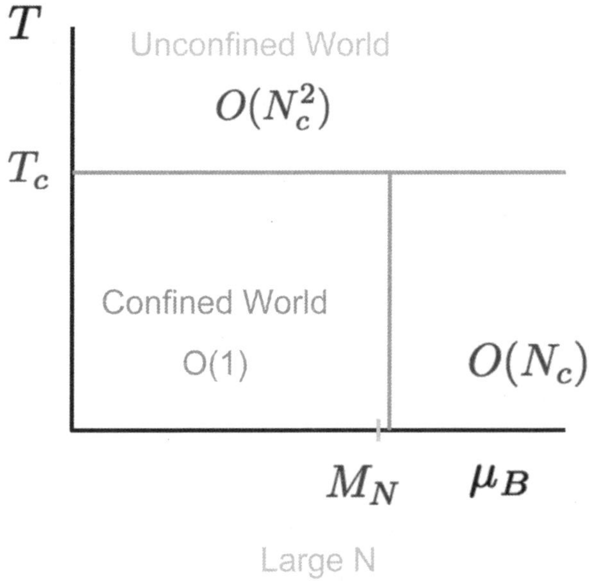

Fig. 1. The phase diagram of QCD in the large N_c limit.

other hand at finite baryon number chemical potential, little is known. It is assumed by many that in the phase diagram there remains a a rapid transition between a confined and a deconfined world.

Rob Pisarski and I have recently provided arguments that the phase structure at finite density is quite different from this simple picture.[5,6] To understand the physics issues consider a world where the number of quark colors is large. In such a world, the baryon mass is $M_{baryon} \sim N_c\Lambda_{QCD}$, where $\Lambda_{QCD} \sim 200\ MeV$ is the dimensional scale of QCD. At low temperature and density, there are no baryons in the system since $e^{(\mu_B - M_N)/T} \sim e^{-N_c}$, for $\mu_B < M_N$. Also, the low temperature world is always confined. This is because in the limit of a large number of colors, quark loops are not important and do not contribute to the potential measured as a function of separation for two heavy quarks used as probes. Since quark loops do not affect the confining potential, the deconfinement temperature is independent of μ_B.

The resulting phase diagram is shown in Fig. 1 . Note that the threshold for high density matter is shifted from the nucleon mass due to interactions of nucleons. The number of degrees of freedom in the unconfined world is of order N_c^2 since gluons have this number of degrees of freedom. The confined

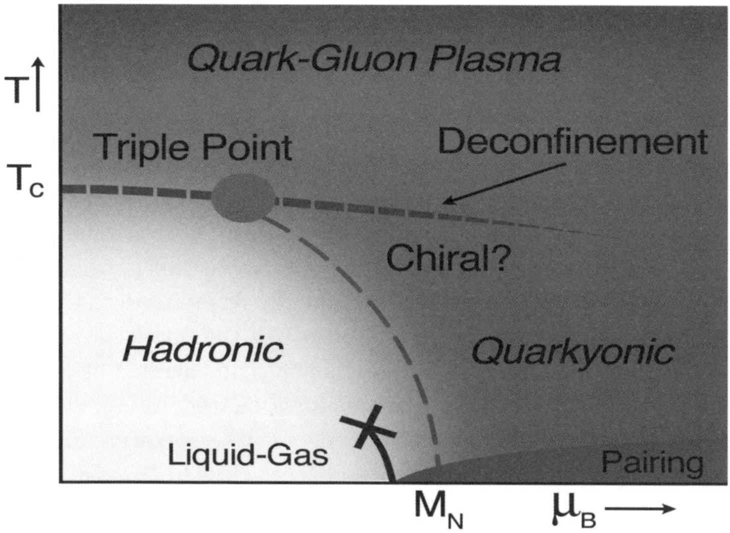

Fig. 2. A somewhat realistic phase diagram of QCD.

world has $O(1)$ degrees of freedom. Quarkyonic matter has only N_c degrees of freedom, because gluons are permanently confined into glueballs.

Quark-holes pairs and gluons are bound into hadrons in Quarkyonic matter. Also, baryonic excitations near the Fermi surface should be bound into baryons. However deep in the Fermi sea, the interactions of quarks are energetic and the confining part of the interaction should not play a significant role. We may think of the system approximately as a Fermi gas of free quarks, with all thermal and Fermi surface excitations permanently confined. The name Quarkyonic is given because this matter has both the properties of free quarks, and confined baryons.

Generation of mass is associated with the breaking of chiral symmetry. In the next section we shall argue how chiral symmetry breaking and restoration appears in the presence of Quarkyonic matter.

A somewhat realistic plot of what we expect for $N_c = 3$ is shown in Fig. 2.

Because of the three phases, there should be an approximate triple point in the QCD phase diagram. This is the analog of the liquid-gas-solid triple point of water. The triple point may or may not be a critical end point, depending upon dynamical details. The triple point should correspond to rapid changes in the properties of the system as approached in any direction in the $\mu_B - T$ plane.

2. Chiral Symmetry Breaking and the Emergence of Happy Island

Early mean field studies of chiral symmetry breaking suggested that chiral symmetry should be restored in Quarkyonic Matter.[7,8] These studies assumed a spatially homogenous chiral condensate. At finite density, one should in fact expect chiral condensates that break translational invariance.

Chiral condensation is generated by pairing between quark and hole pairs. At zero density and temperature, the hole is in the negative energy Dirac sea and corresponds to an anti-quark. At finite density, the same pairing can occur, with little cost in energy, if the quark has an energy a little above the Fermi surface, and the hole has an energy a little below the Fermi surface. The energy loss, for massless quarks, in extracting such a pair is the same as the vacuum situation at low temperature and density.

Such a quark-quark hole pair has an energy close to $2\mu_B$ since they both come from near the Fermi surface. Pair formation into a meson state can take place if the pair has small relative momentum but total momentum near $2\mu_B$. The condensate takes place into modes of high net momentum. The pair has a finite DeBroglie wavelength, and therefore its condensation will involve the breaking of translational invariance. This is the mechanism of charge density waves familiar in condensed matter physics.

It can be argued that the dynamics of pair formation is essentially 1+1 dimensional, and that the dynamics should reduce to the 1+1 dimensional 't Hooft model.[9] There is a $2N_F$ Goldstone symmetry, where the extra factor of 2 arises from a degeneracy under spin. The chiral condensation occurs through a rotation in space of $\overline{\psi}\psi$ with $\overline{\psi}\sigma^{0i}\psi$ where i is in the direction of condensation. Such a condensation macroscopically breaks parity. This condensation is referred to as the chiral spiral.

The formation of the condensation in the end is probably not 1 dimensional.[10] The Fermi surface breaks into patches with differing spatial orientations of the Fermi surface. The number of patches increases with density.

It might also be possible that there are other types of translationally non-invariant condensates that form. this is suggested from a comprehensive analysis of mean field actions.[11]

The breaking of translation invariance signals a real phase transition, since an order parameter acquires an expectation value. This means that there is probably some isolated island in the $\mu_B - T$ plane.

This island is called Happy Island. On the left hand side of Happy island, there is a sharp change in the number of degrees of freedom. We would

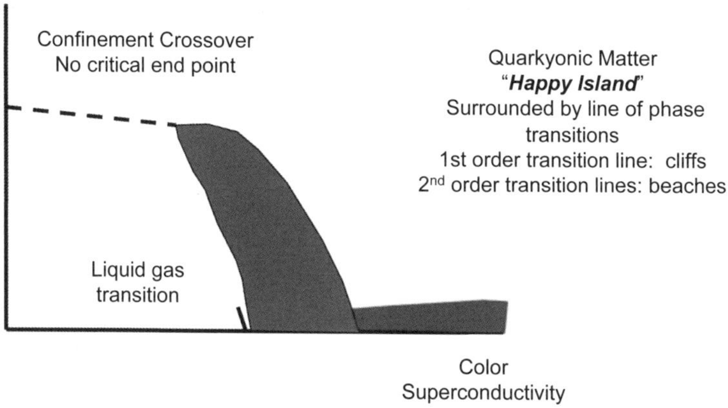

Fig. 3. Happy Island in the $\mu_B - T$ plane.

expect this part of Happy Island to have cliffs. On the right hand side of Happy Island, there should be a relatively weak confinement-deconfinement phase transition, corresponding to beaches. As one walks across Happy Island climbing from the beaches to the cliffs, one might have stair cases corresponding to the phase transitions associated with the patch structure on the Fermi surface.

Of course, history has shown that the discovery of new land masses is often done for reasons that turn out later to be not quite correct. In this spirit one should expect surprises as we explore more.

3. Experimental Hints for Happy Island

The existence of Quarkyonic Matter leads to the prediction of a triple point in the $\mu_B - T$ plane. Also freeze out curves should parallel the boundary of Quarkyonic matter at low T and high μ_B, and parallel the deconfinement transition at low μ_B and high T. This is because along these transition lines, the energy density jumps by roughly an order of magnitude in a narrow band of either T or μ_B. Thus in expansion, the matter may freeze out at μ_B and T values characteristic of such a transition.

The deconfinement transition should occur at some temperature fixed that is independent of μ_B. The Quarkyonic transition should occur roughly where $(\mu_B - M_N)/T \sim cons.$ In Fig. 4 the parameter for freezeout as determined by fits to particle abundances is compared to such simple model considerations.[12] Ratios of particle abundance should show singular behaviour as one approaches the triple point. This indeed occurs, as shown

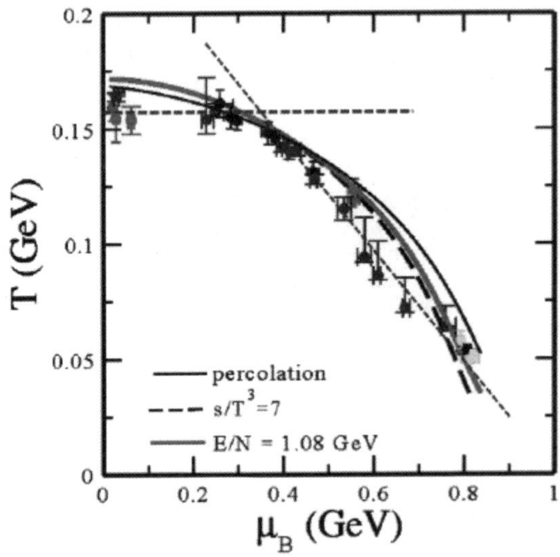

Fig. 4. The freeze out curve for μ_B and T compared to a simple model.

in Fig. 5,[13] The typical center of mass energy where this occurs is near 10 GeV. This suggest heavy ion experiments at center of mass energies of this order and smaller may be able to access the physics of Happy Island.

4. Summary

The physics of Happy Island is rich and unexplored. Little is really known from theory, as speculative conclusions rely heavily on model considerations. Reliable lattice gauge computation is difficult if not impossible to do at high density and low temperature. On the experimental side we have hints of phase boundaries and a triple point not accounted for simply by the existence of a confinement-deconfinement phase transition. What precisely to look for is not known, and a phenomenology of Happy Island is yet non-existent. Nevertheless the arguments for new physics in this region of the μ_B-T plane are compelling. The possible existence of inhomogeneous phases might ultimately have interesting consequences for the physics of neutron stars.

To infer the existence of Happy Island and Quarkyonic Matter, one has used a large number of colors approximation. For mesons, it is widely believed that such an approximation is reasonable. For baryons there are

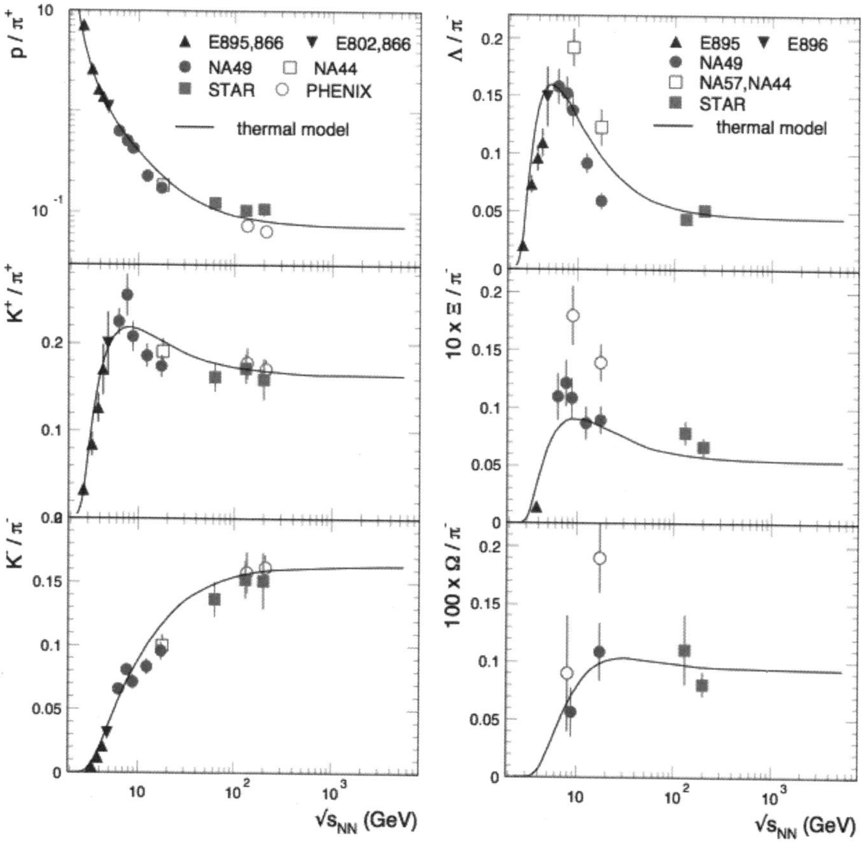

Fig. 5. Ratios of particle abundances.

serious doubts. The issue for us, whether or not quark loops affect the confinement potential is crucial lattice Monte Carlo data and charmonium potential models suggest that in vacuum quark pairs provide little modification to a linear potential out to a distance of the order of a Fermi. This suggests that at least for some range of baryon density, quark pairs provide little modification of the confining potential on distance scales corresponding to confinement.

Acknowledgments

One of the pleasures of doing physics is the combination of imaginative thought and tangible realization embodied in both the ideas we study and the people who study them. This meeting is in honor of a a wonderful friend and colleague, Walter Greiner. Walter did much of the seminal work on the superheavy nuclei and the Island of Stability, concepts that must have sounded as wild and speculative as the idea of a Happy Island for high density strongly interacting matter. In this spirit of the imagination, my wife Alice and I have written the following poem for Walter:

An Ode to Dragons
For Walter

A dragon is a fearsome beast
With massive jaws and fiery breath;
And those who cannot match his strength
May hide themselves and wish his death

A dragon is adventuresome
Unfazed by ergodicity.
He does not keep to precincts ruled
By order and simplicity.

We see few dragons in these days
Yet might remember as a dream
A glimpse of huge wings in the sky
And scales with a metallic gleam...

A dragon lives for endless years,
And many are the tales that tell
Of maidens rescued just in time–
Yet did they wish to end the spell?

A dragon leads a lonely life
And could be kinder than most know,
For it is hard to be a friend
To one who see you as a foe.

A child with no fear in his heart
May feel a lifting toward the sky
As far above, a dragon shares
Just how it feels to soar and fly.

by Alice and Larry McLerran
2010

The research of L. McLerran is supported under DOE Contract No. DE-AC02-98CH10886. I also thank Debades Bandyopadhyay for the wonderful job he did in organizing this meeting.

References

1. N. Cabibbo, G. Parisi, Phys. Lett. **B59**, 67-69 (1975).
2. M. G. Alford, K. Rajagopal, F. Wilczek, Phys. Lett. **B422**, 247-256 (1998). [hep-ph/9711395].
3. R. Rapp, T. Schafer, E. V. Shuryak, M. Velkovsky, Phys. Rev. Lett. **81**, 53-56 (1998). [hep-ph/9711396].
4. G. Boyd, J. Engels, F. Karsch, E. Laermann, C. Legeland, M. Lutgemeier, B. Petersson, Nucl. Phys. **B469**, 419-444 (1996). [hep-lat/9602007].
5. L. McLerran, R. D. Pisarski, Nucl. Phys. **A796**, 83-100 (2007). [arXiv:0706.2191 [hep-ph]]
6. Y. Hidaka, L. D. McLerran, R. D. Pisarski, Nucl. Phys. **A808**, 117-123 (2008). [arXiv:0803.0279 [hep-ph]]
7. L. Y. .Glozman, R. F. Wagenbrunn, Mod. Phys. Lett. **A23**, 2385-2388 (2008). [arXiv:0802.0276 [hep-ph]].
8. L. McLerran, K. Redlich, C. Sasaki, Nucl. Phys. **A824**, 86-100 (2009). [arXiv:0812.3585 [hep-ph]].
9. T. Kojo, Y. Hidaka, L. McLerran, R. D. Pisarski, Nucl. Phys. **A843**, 37-58 (2010). [arXiv:0912.3800 [hep-ph]].
10. T. Kojo, R. D. Pisarski, A. M. Tsvelik, Phys. Rev. **D82**, 074015 (2010). [arXiv:1007.0248 [hep-ph]].
11. M. Buballa, D. Nickel, Acta Phys. Polon. Supp. **3**, 523-532 (2010). [arXiv:0911.2333 [hep-ph]].
12. A. Andronic, D. Blaschke, P. Braun-Munzinger, J. Cleymans, K. Fukushima, L. D. McLerran, H. Oeschler, R. D. Pisarski *et al.*, Nucl. Phys. **A837**, 65-86 (2010). [arXiv:0911.4806 [hep-ph]].
13. A. Andronic, P. Braun-Munzinger, J. Stachel, Nucl. Phys. **A772**, 167-199 (2006). [nucl-th/0511071].

SIGNATURES OF THE STRONGLY INTERACTING QGP IN RELATIVISTIC HEAVY-ION COLLISIONS

E.L. BRATKOVSKAYA* and O. LINNYK

Institut für Theoretische Physik,
Goethe-Universität Frankfurt am Main,
Frankfurt, Germany
** E-mail: Elena.Bratkovskaya@th.physik.uni-frankfurt.de*

W. CASSING and V.P. KONCHAKOVSKI

Institut für Theoretische Physik,
Justus-Liebig-Universität Giessen,
Giessen, Germany

The dynamics of partons, hadrons and strings in relativistic nucleus-nucleus collisions is analyzed within the novel Parton-Hadron-String Dynamics (PHSD) transport approach, which is based on a dynamical quasiparticle model for partons (DQPM) matched to reproduce recent lattice-QCD results - including the partonic equation of state - in thermodynamic equilibrium. The transition from partonic to hadronic degrees of freedom is described by covariant transition rates for the fusion of quark-antiquark pairs or three quarks (antiquarks), respectively, obeying flavor current-conservation, color neutrality as well as energy-momentum conservation. In order to explore the space-time regions of 'partonic matter' the PHSD approach is applied to nucleus-nucleus collisions from 20 A GeV up to RHIC energies of 21 A TeV.

1. Introduction

The 'Big Bang' scenario implies that in the first micro-seconds of the universe the entire state has emerged from a partonic system of quarks, antiquarks and gluons – a quark-gluon plasma (QGP) – to color neutral hadronic matter consisting of interacting hadronic states (and resonances) in which the partonic degrees of freedom are confined. The nature of confinement and the dynamics of this phase transition has motivated a large community for several decades and is still an outstanding question of todays physics. Early concepts of the QGP were guided by the idea of a weakly interacting system of partons which might be described by perturbative QCD (pQCD). However, experimental observations at the Relativistic Heavy Ion Collider (RHIC) indicated that the new medium created in ultrarelativistic Au+Au collisions is interacting more strongly than hadronic matter and consequently this concept had to be severely questioned. Moreover, in line with theoretical studies in Refs. 1–3 the medium showed phenomena of an almost perfect liquid of partons[4,5] as extracted from the strong radial expansion and the scaling of elliptic flow $v_2(p_T)$ of mesons and baryons with the number of constituent quarks and antiquarks.[4]

The question about the properties of this (nonperturbative) QGP liquid is discussed controversially in the literature and dynamical concepts describing the formation of color neutral hadrons from colored partons are scarce. A fundamental issue for hadronization models is the conservation of 4-momentum as well as the entropy problem because by fusion/coalescence of massless (or low constituent mass) partons to color neutral bound states of low invariant mass (e.g. pions) the number of degrees of freedom and thus the total entropy is reduced in the hadronization process.[6–8] This problem - a violation of the second law of thermodynamics as well as the conservation of four-momentum and flavor currents - has been addressed in Ref. 9 on the basis of the DQPM employing covariant transition rates for the fusion of 'massive' quarks and antiquarks to color neutral hadronic resonances or strings. In fact, the dynamical studies for an expanding partonic fireball in Ref. 9 suggest that the latter problems have come to a practical solution.

A consistent dynamical approach - valid also for strongly interacting systems - can be formulated on the basis of Kadanoff-Baym (KB) equations[10] or off-shell transport equations in phase-space representation, respectively.[10–12] In the KB theory the field quanta are described in terms of dressed propagators with complex selfenergies. Whereas the real part of the selfenergies can be related to mean-field potentials (of Lorentz scalar, vector or tensor type), the imaginary parts provide information about the lifetime and/or reaction rates of time-like 'particles'.[13] Once the proper (complex) selfenergies of the degrees of freedom are known the time evolution of the system is fully governed by off-shell transport equations (as described in Refs. 10,13). The determination/extraction of complex selfenergies for the partonic degrees of freedom has been performed before in Refs. 14,15 by fitting lattice QCD (lQCD) 'data' within the Dynamical QuasiParticle Model (DQPM). In fact, the DQPM allows for a simple and transparent interpretation of lattice QCD results for thermodynamic quantities as well as correlators and leads to effective strongly interacting partonic quasiparticles with broad spectral functions. For a review on off-shell transport theory and results from the DQPM in comparison to lQCD we refer the reader to Ref. 13.

Since the actual implementations in the PHSD transport approach have been presented in detail in Ref. 16 we here report again on the actual description of hadronization and present results for Pb +Pb collisions at SPS energies in comparison to experimental data from the NA49 Collaboration.

2. Hadronization in PHSD

The hadronisation, i.e. the transition from partonic to hadronic degrees of freedom, is described in PHSD by local covariant transition rates as introduced in Ref. 9 e.g. for $q + \bar{q}$ fusion to a meson m of four-momentum $p = (\omega, \mathbf{p})$ at space-time point $x = (t, \mathbf{x})$:

$$\frac{dN_m(x,p)}{d^4x d^4p} = Tr_q Tr_{\bar{q}}\ \delta^4(p - p_q - p_{\bar{q}})\ \delta^4\left(\frac{x_q + x_{\bar{q}}}{2} - x\right)$$
$$\times \omega_q\ \rho_q(p_q)\ \omega_{\bar{q}}\ \rho_{\bar{q}}(p_{\bar{q}})\ |v_{q\bar{q}}|^2\ W_m(x_q - x_{\bar{q}}, (p_q - p_{\bar{q}})/2)$$
$$\times N_q(x_q, p_q)\ N_{\bar{q}}(x_{\bar{q}}, p_{\bar{q}})\ \delta(\text{flavor, color}). \qquad (1)$$

In Eq. (1) we have introduced the shorthand notation,

$$Tr_j = \sum_j \int d^4x_j \int \frac{d^4p_j}{(2\pi)^4}\ , \qquad (2)$$

where $\sum_j$ denotes a summation over discrete quantum numbers (spin, flavor, color); $N_j(x,p)$ is the phase-space density of parton j at space-time position x and four-momentum p. In Eq. (1) δ(flavor, color) stands symbolically for the conservation of flavor quantum numbers as well as color neutrality of the formed hadron m which can be viewed as a color-dipole or 'pre-hadron'. Furthermore, $v_{q\bar{q}}(\rho_p)$ is the effective quark-antiquark interaction from the DQPM (displayed in Fig. 10 of Ref. 15) as a function of the local parton $(q + \bar{q} + g)$ density ρ_p (or energy density). Furthermore, $W_m(x,p)$ is the dimensionless phase-space distribution of the formed 'pre-hadron', i.e.

$$W_m(\xi, p_\xi) = \exp\left(\frac{\xi^2}{2b^2}\right)\ \exp\left(2b^2(p_\xi^2 - (M_q - M_{\bar{q}})^2/4)\right) \qquad (3)$$

with $\xi = x_1 - x_2 = x_q - x_{\bar{q}}$ and $p_\xi = (p_1 - p_2)/2 = (p_q - p_{\bar{q}})/2$. The width parameter b is fixed by $\sqrt{\langle r^2 \rangle} = b = 0.66$ fm (in the rest frame) which corresponds to an average rms radius of mesons. We note that the expression (3) corresponds to the limit of independent harmonic oscillator states and that the final hadron-formation rates are approximately independent of the parameter b within reasonable variations. By construction the quantity (3) is Lorentz invariant; in the limit of instantaneous 'hadron formation', i.e. $\xi^0 = 0$, it provides a Gaussian dropping in the relative distance squared $(\mathbf{r}_1 - \mathbf{r}_2)^2$. The four-momentum dependence reads explicitly (except for a factor 1/2)

$$(E_1 - E_2)^2 - (\mathbf{p}_1 - \mathbf{p}_2)^2 - (M_1 - M_2)^2 \leq 0 \tag{4}$$

and leads to a negative argument of the second exponential in (3) favoring the fusion of partons with low relative momenta $p_q - p_{\bar{q}} = p_1 - p_2$.

Related transition rates (to Eq. (1)) are defined for the fusion of three off-shell quarks ($q_1 + q_2 + q_3 \leftrightarrow B$) to color neutral baryonic (B or $\bar{B}$) resonances of finite width (or strings) fulfilling energy and momentum conservation as well as flavor current conservation using Jacobi coordinates (cf. Ref. 16).

On the hadronic side the PHSD transport approach includes explicitly the baryon octet and decouplet, the 0^-- and 1^--meson nonets as well as selected higher resonances as in HSD.[17] Hadrons of higher masses (> 1.5 GeV in case of baryons and > 1.3 GeV in case of mesons) are treated as 'strings' (color-dipoles) that decay to the known (low-mass) hadrons according to the JETSET algorithm.[18] We discard an explicit recapitulation of the string decay and refer the reader to the original work[18] or Ref.[19]

3. Application to Nucleus-Nucleus Collisions

In this Section we employ the PHSD approach to nucleus-nucleus collisions at moderate relativistic energies. It is of interest, how the PHSD approach compares to the HSD[17] model (without explicit partonic degrees-of-freedom) as well as to experimental data. In Fig. 1 we show the transverse mass spectra of π^-, K^+ and K^- mesons for 7% central Pb+Pb collisions at 40 and 80 A·GeV and 5% central collisions at 158 A·GeV in comparison to the data of the NA49 Collaboration.[20] Here the slope of the π^- spectra is only slightly enhanced in PHSD relative to HSD which demonstrates that the pion transverse motion shows no sizeable sensitivity to the partonic phase. However, the $K^\pm$ transverse mass spectra are substantially hardened with respect to the HSD calculations at all bombarding energies - i.e. PHSD is more in line with the data - and thus suggest that partonic effects are better visible in the strangeness-degrees of freedom. The hardening of the kaon spectra can be traced back to parton-parton scattering as well as a larger collective acceleration of the partons in the transverse direction due to the presence of repulsive vector fields for the partons. The enhancement

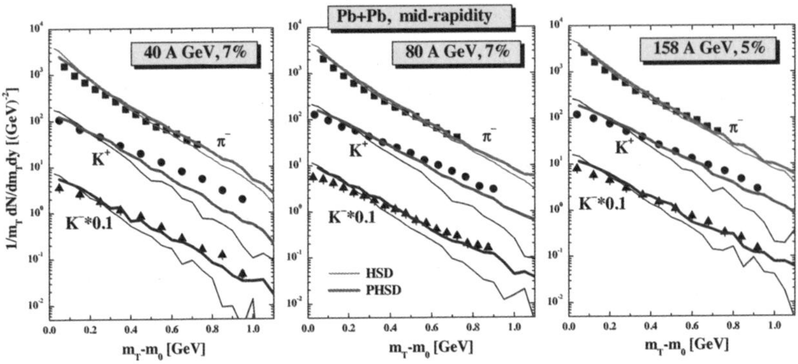

Fig. 1. The π^-, K^+ and K^- transverse mass spectra for central Pb+Pb collisions at 40, 80 and 158 A·GeV from PHSD (thick solid lines) in comparison to the distributions from HSD (thin solid lines) and the experimental data from the NA49 Collaboration.[20]

of the spectral slope for kaons and antikaons in PHSD due to collective partonic flow shows up much clearer for the kaons due to their significantly larger mass (relative to pions). We recall that in Refs. 21 the underestimation of the $K^\pm$ slope by HSD (and also UrQMD) had been suggested to be a signature for missing partonic degrees of freedom; the present PHSD calculations support this early suggestion. Moreover, the PHSD calculations for RHIC energies show a very similar trend - the inverse slope increases by including the partonic phase.

The strange antibaryon sector is of further interest since here the HSD calculations have always underestimated the yield.[22] Our detailed studies in Ref.[16] show that the HSD and PHSD calculations both give a reasonable description of the $\Lambda + \Sigma^0$ yield of the NA49 Collaboration;[23] both models underestimate the NA57 data[24] by about 30%. An even larger discrepancy in the data from the NA49 and NA57 Collaborations is seen for $(\bar{\Lambda} + \bar{\Sigma}^0)/N_{wound}$; here the PHSD calculations give results which are in between the NA49 data and the NA57 data whereas HSD underestimates the $(\bar{\Lambda} + \bar{\Sigma}^0)$ midrapidity yield at all centralities.

The latter result suggests that the partonic phase does not show up explicitly in an enhanced production of strangeness (or in particular strange mesons and baryons) but leads to a different redistribution of antistrange quarks between mesons and antibaryons. In fact, as demonstrated in Ref. 16, we find no sizeable differences in the double strange baryons from HSD and PHSD – in a good agreement with the NA49 data – but observe a large enhancement in the double strange antibaryons for PHSD relative to HSD.

4. Summary

In this contribution we have addressed relativistic collisions of Pb+Pb at SPS energies in the PHSD approach which includes explicit partonic degrees of freedom as well as dynamical local transition rates from partons to hadrons (1). The hadronization process conserves four-momentum and all flavor currents and slightly increases the total entropy since the 'fusion' of rather massive partons dominantly leads to the formation of color neutral strings or resonances that decay microcanonically to lower mass hadrons. Since this dynamical hadronization process increases the total entropy the second law of thermodynamics is not violated (as is the case for simple coalescence models incorporating massless partons).

The PHSD approach has been also applied to nucleus-nucleus collisions from 40 to 160 A·GeV as well as for RHIC energies in order to explore the space-time regions of 'partonic matter'.[16] We have found that even central collisions at the top SPS energy of ~158 A· GeV show a large fraction of non-partonic, i.e. hadronic or string-like matter, which can be viewed as a 'hadronic corona'.[25] This finding implies that neither purely hadronic nor purely partonic 'models' can be employed to extract physical conclusions in comparing model results with data. On the other hand - studying in detail Pb+Pb reactions at SPS energies in comparison to the data[16] - it is found that the partonic phase has only a very low impact on the longitudinal rapidity distributions of hadrons but a sizeable influence on the transverse-mass distribution of final kaons due to the partonic interactions. The most pronounced effect is seen on the production of multi-strange antibaryons due to a slightly enhanced $s\bar{s}$ pair production in the partonic phase from massive time-like gluon decay and a more abundant formation of strange antibaryons in the hadronization process. This enhanced formation of strange antibaryons in central Pb+Pb collisions at SPS energies by hadronization supports the early suggestion by Braun-Munzinger and Stachel[26,27] in the statistical hadronization model - which describes well particle ratios from AGS to RHIC energies.

At RHIC energies our PHSD calculations show also a good reproduction of the hadron transverse mass and rapidity spectra. Furthermore, the elliptic flow v_2 is well described for $Au+Au$ reactions at $\sqrt{s} = 200$ GeV as a function of centrality as well as of transverse momenta up to $p_T \simeq 1.5$ GeV/c.

Work supported in part by the "HIC for FAIR" framework of the "LOEWE" program and by DFG.

References

1. E. Shuryak, Prog. Part. Nucl. Phys. 53 (2004) 273.
2. M. H. Thoma, J. Phys. G 31 (2005) L7; Nucl. Phys. A 774 (2006) 307.
3. A. Peshier and W. Cassing, Phys. Rev. Lett. 94 (2005) 172301.
4. I. Arsene *et al.*, Nucl. Phys. A 757 (2005) 1; B. B. Back *et al.*, Nucl. Phys. A 757 (2005) 28; J. Adams *et al.*, Nucl. Phys. A 757 (2005) 102; K. Adcox *et al.*, Nucl. Phys. A 757 (2005) 184.
5. T. Hirano and M. Gyulassy, Nucl. Phys. A 769 (2006) 71.
6. R. C. Hwa and C. B. Yang, Phys. Rev. C 67 (2003) 034902; V. Greco, C. M. Ko and P. Levai, Phys. Rev. Lett. 90 (2003) 202302.
7. R. J. Fries, B. Müller, C. Nonaka and S. A. Bass, Phys. Rev. Lett. 90 (2003) 202303.
8. Z.-W. Lin *et al.*, Phys. Rev. C 72 (2005) 064901.
9. W. Cassing and E. L. Bratkovskaya, Phys. Rev. C 78 (2008) 034919.
10. S. Juchem, W. Cassing and C. Greiner, Phys. Rev. D 69 (2004) 025006; Nucl. Phys. A 743 (2004) 92.
11. W. Cassing and S. Juchem, Nucl. Phys. A 665 (2000) 377; *ibid* A 672 (2000) 417.
12. Y. B. Ivanov, J. Knoll and D. N. Voskresensky, Nucl. Phys. A 672 (2000) 313.
13. W. Cassing, E. Phys. J. ST 168 (2009) 3.
14. W. Cassing, Nucl. Phys. A 791 (2007) 365.
15. W. Cassing, Nucl. Phys. A 795 (2007) 70.
16. W. Cassing and E. L. Bratkovskaya, Nucl. Phys. A 831 (2009) 215.
17. W. Cassing and E. L. Bratkovskaya, Phys. Rep. 308 (1999) 65.
18. H.-U. Bengtsson and T. Sjöstrand, Comp. Phys. Commun. 46 (1987) 43.
19. T. Falter, W. Cassing, K. Gallmeister, and U. Mosel, Phys. Rev. C 70 (2004) 054609.
20. C. Alt et al., NA49 Collaboration, Phys. Rev. C 66 (2002) 054902; Phys. Rev. C 77 (2008) 024903.
21. E. L. Bratkovskaya, S. Soff, H. Stöcker, M. van Leeuwen, and W. Cassing, Phys. Rev. Lett. 92 (2004) 032302.
22. J. Geiss, W. Cassing and C. Greiner, Nucl. Phys. A 644 (1998) 107.
23. T. Anticic *et al.*, NA49 Collaboration, Phys. Rev. C 80 (2009) 034906.
24. F. Antinori *et al.*, Phys. Lett. B 595 (2004) 68; J. Phys. G: Nucl. Phys. 32 (2006) 427.
25. J. Aichelin and K. Werner, Phys. Rev. C79 (2009) 064907.
26. P. Braun-Munzinger *et al.*, Phys. Lett. B 365 (1996) 1; *ibid.* B 465 (1999) 15; *ibid.* B 518 (2001) 41.
27. A. Andronic, P. Braun-Munzinger and J. Stachel, Nucl. Phys. A 772 (2006) 167.

MICROSCOPIC ORIGIN OF THE SHEAR RELAXATION TIME IN CAUSAL DISSIPATIVE FLUID DYNAMICS

GABRIEL S. DENICOL

Institut für Theoretische Physik, Goethe University, 60438 Frankfurt am Main, Germany

HARRI NIEMI

Frankfurt Institute for Advanced Studies (FIAS), 60438 Frankfurt am Main, Germany

JORGE NORONHA

Instituto de Física, Universidade de São Paulo, C.P. 66318, 05315-970 São Paulo, SP, Brazil

DIRK H. RISCHKE

Institut für Theoretische Physik, Goethe University, and
Frankfurt Institute for Advanced Studies (FIAS), 60438 Frankfurt am Main, Germany

In this paper we show how to compute the shear relaxation time from an underlying microscopic theory. We prove that the shear relaxation time in Israel-Stewart-type theories is given by the inverse of the pole of the corresponding retarded Green's function, which is nearest to the origin in the complex energy plane. Consequently, the relaxation time in such theories is a microscopic, and not a macroscopic, i.e., fluid-dynamical time scale.

Keywords: Dissipative relativistic fluids; transient dynamics; relaxation time.

1. Introduction

Relativistic fluid dynamics has been applied with success to describe the dynamics of hot and dense matter created in relativistic heavy-ion collisions at RHIC, and recently, at LHC.[1] Despite this success, the derivation of relativistic fluid dynamics from an underlying microscopic theory is still an open problem and is under intense theoretical investigation. This is mainly due to nontrivial consequences imposed by causality and stability in relativistic fluids.[3,4]

In this paper we discuss a general method to compute all linear transport coefficients (such as the shear viscosity and its corresponding relaxation time) associated with the dissipative behavior of fluids. In particular, we prove that this prescription gives a value for the shear relaxation time that, under the 14-moment approximation, coincides with the one derived from kinetic theory. This shows that transient dynamics is determined by the slowest microscopic and not by the fastest fluid-dynamical time scale. Our approach[5] depends only on generic analytical properties of retarded Green's functions.

2. Definitions and Power-Counting Scheme

Let us consider the following linear relation between a dissipative current $J(X)$ and its thermodynamical force $F(X)$ in Fourier space,

$$\tilde{J}(Q) = \tilde{G}_R(Q)\,\tilde{F}(Q)\,, \tag{1}$$

where we define the Fourier transformation in the following way,

$$\tilde{A}(Q) = \int d^4X \exp(iQ\cdot X)\,A(X)\;, \tag{2}$$

$$A(X) = \int \frac{d^4Q}{(2\pi)^4} \exp(-iQ\cdot X)\,\tilde{A}(Q)\;. \tag{3}$$

Here, we adopt the following notation for the four-momentum and space-time coordinate, $Q = (\omega, \mathbf{q})$ and $X = (t, \mathbf{x})$, respectively. In this article, we discuss particle diffusion, in which J is the diffusion current of a density n and $F \sim \partial n$, and shear flow, in which J is the shear stress tensor $\pi^{\mu\nu}$ and F the shear tensor $\sigma^{\mu\nu} \equiv \nabla^{\langle\mu}u^{\nu\rangle}$, with the fluid four-velocity u^μ, $\nabla^\mu \equiv \Delta^{\mu\nu}\partial_\nu$, $\Delta^{\mu\nu} = g^{\mu\nu} - u^\mu u^\nu$, and $A^{\langle\mu\nu\rangle} \equiv \frac{1}{2}(\Delta^{\mu\alpha}\Delta^{\nu\beta} + \Delta^{\mu\beta}\Delta^{\nu\alpha} - \frac{2}{3}\Delta^{\mu\nu}\Delta^{\alpha\beta})A_{\alpha\beta}$.

Equation (1) contains all the information about the linearized underlying microscopic theory in the form of the retarded Green's function $\tilde{G}_R(Q)$. When the system exhibits a clear separation between the typical microscopic and macroscopic scales, λ and ℓ, respectively, it is generally assumed that the dynamics of the system can be described in terms of a finite number of macroscopic variables. The macroscopic scale is associated with the variations of the thermodynamic force $F \sim \ell^{-1}$ and its derivatives $\partial^n F \sim \ell^{-n-1}$. The microscopic scale λ is contained in the poles and derivatives of $\tilde{G}_R(Q)$ and would be, for example, the mean-free path in dilute gases. The challenge is to determine the equation of motion satisfied by the current J in terms of the microscopic information contained in the retarded Green's function $\tilde{G}_R(Q)$ when such a separation of scales exists.

3. Gradient Expansion

The separation between microscopic and macroscopic scales is normally used to determine the macroscopic dynamics of the dissipative current J in terms of an expansion in powers of Knudsen number, $\mathrm{Kn} \equiv \lambda/\ell \ll 1$. The simplest realization of this idea is known as the gradient expansion.[6] In the gradient expansion the dissipative currents are expressed solely in terms of gradients of fluid-dynamical variables. In practice, the gradient expansion is implemented by expanding the retarded Green's function, $\tilde{G}_R(Q)$, in a Taylor series around the origin, $Q = 0$,

$$\tilde{G}_R(Q) \sim \tilde{G}_R(0, \mathbf{0}) + \partial_\omega \tilde{G}_R(0, \mathbf{0})\, \omega + \dots \,, \tag{4}$$

where the terms resulting from the expansion in $\mathbf{q}$ are not explicitly denoted. This expansion leads to the following equation of motion for J

$$J\,(X) = CF\,(X) + C_1 \partial_t F\,(X) + \dots \,, \tag{5}$$

where $C = \tilde{G}_R\,(0, \mathbf{0})$, $C_1 = \partial_\omega\, \tilde{G}_R\,(Q)\Big|_{\omega, \mathbf{q}=0}$, ... We remark that the gradient expansion usually does not involve time derivatives of the thermodynamic force. This is not a problem since one can in principle use the conservation laws to replace time derivatives by spatial gradients.

Apart from overall factors which ensure the correct dimensionality, $C \sim \lambda$ and $C_1 \sim \lambda^2$, respectively. Thus, the term CF is of first order in Kn while $C_1 \partial_t F\,(t, x)$ is of second order in the Knudsen number, Kn.[2] If $\mathrm{Kn} \ll 1$, higher-order gradients can in principle be omitted and one obtains a closed macroscopic theory for J. Navier-Stokes theory[2] corresponds to truncating the series at first order while keeping higher-order corrections leads to the Burnett equations.[7]

This method, however, leads to unstable fluid-dynamical theories. In the non-relativistic case, the second-order truncation, i.e., the Burnett equation, is unstable.[8] In the relativistic case, Navier-Stokes theory itself is unstable.[3,4] Recently, it has been shown that this instability in relativistic theories is intimately related to the violation of causality.[4] Such problems prohibit the application of these theories to describe any realistic system.

4. Computing the Relaxation Time from the Poles of $\tilde{G}_R$

In the derivation of Eq. (5), two main assumptions were employed: the separation between the microscopic and macroscopic scales and the gradient expansion itself. The first assumption is essential for the coarse-graining procedure that defines fluid dynamics and cannot be disregarded. On the other hand, the second assumption is solely based on the idea that the Taylor expansion is able to describe the retarded Green's function in the whole macroscopic/mesoscopic frequency domain. This need not be true, since it is known that Green's functions are not necessarily analytic in the whole complex plane. In fact, these functions can have singularities in the complex ω plane, which necessarily restricts the domain of validity of the Taylor expansion.

In order to illustrate the physical meaning of such singularities, it is convenient to consider the generalized diffusion equation for J, also known as the Maxwell-Cattaneo equation.[9] In this case, as in the Israel-Stewart theories of fluid dynamics,[10] the dissipative current satisfies a relaxation equation. In the linear domain, this can be written as,

$$\tau_R \frac{\partial}{\partial t} J + J = DF. \tag{6}$$

In these theories there are two transport coefficients: the relaxation time τ_R and the diffusion coefficient D. In Fourier space, Eq. (6) can be written in the form $\tilde{J}(\omega) = \tilde{G}_R(\omega)\tilde{F}(\omega)$, where

$$\tilde{G}_R(\omega) = \frac{D}{\tau_R} \frac{i}{\omega + i/\tau_R}. \tag{7}$$

It is important to notice the existence of a pole of $\tilde{G}_R(\omega)$ on the imaginary axis. The existence of this pole is directly related to the existence of a non-zero relaxation time and, consequently, to the causal structure of the theory.[4] It is also clear from Eq. (6) that Fick's law and, consequently, the diffusion equation are recovered in the limit of vanishing relaxation time, $\tau_R \to 0$. Mathematically, this limit pushes the pole to infinity, rendering $\tilde{G}_R(\omega)$ an analytic function in the whole complex ω plane. Only in this case a Taylor expansion near the origin has an infinite radius of convergence and is able to provide a good description of $\tilde{G}_R(\omega)$.

However, in transient theories of fluid dynamics, where the relaxation time is necessarily nonzero, can a Taylor expansion still be used? Strictly speaking, the range of the Taylor expansion is limited by the existence of the pole. As was previously shown, the pole is directly related to the relaxation time and, consequently, the Taylor expansion is unable to probe time scales of the order of the relaxation time.

5. The Role of the Analytical Structure of $\tilde{G}_R(\omega)$

Next, we consider a general analytical structure for $\tilde{G}_R(Q)$. We assume that $\tilde{G}_R(Q)$ has N poles in the complex ω plane, $\omega_1(\mathbf{q}), \ldots, \omega_N(\mathbf{q})$, while the structure in the $\mathbf{q}$ plane is analytic. Note that, in principle, the number of singularities does not need to be finite. Also, in many cases $\tilde{G}_R(Q)$ contains branch cuts, especially in the UV region, $\omega/\omega_1 \gg 1$, which is sensitive to temperature-independent properties of the system. We only consider the region in the complex ω plane without branch cuts. Then, $\tilde{G}_R(Q)$ can be expressed in the following way

$$\tilde{G}_R(Q) = \sum_{i=1}^{N} \frac{f_i(Q)}{\omega - \omega_i(\mathbf{q})} = \frac{\Xi(Q)}{[\omega - \omega_1(\mathbf{q})] \cdots [\omega - \omega_N(\mathbf{q})]}, \tag{8}$$

where $f_i(Q)$ and $\Xi(Q)$ are analytic functions in the complex ω plane. We write the right-hand side of Eq. (8) in a polynomial form,

$$1+\Phi_1(\mathbf{q})\,(-i\omega)+\ldots+\Phi_N(\mathbf{q})\,(-i\omega)^N = (-1)^N \frac{[\omega - \omega_1(\mathbf{q})] \cdots [\omega - \omega_N(\mathbf{q})]}{\omega_1(\mathbf{q}) \cdots \omega_N(\mathbf{q})}, \tag{9}$$

where the $\omega_i(\mathbf{q})$ are the zeros of the polynomial on the left-hand side and we introduced the functions $\Phi_m(\mathbf{q})$

$$\Phi_m(\mathbf{q}) = (-i)^m \sum_{1 \le i_1 \ldots < i_m \le N} \frac{1}{\omega_{i_1}(\mathbf{q}) \cdots \omega_{i_m}(\mathbf{q})}. \tag{10}$$

Combining Eqs. (8) and (9), we obtain

$$\left[1 + \Phi_1(\mathbf{q})\,(-i\omega) + \ldots + \Phi_N(\mathbf{q})\,(-i\omega)^N\right] \tilde{G}_R(Q) = \frac{(-1)^N\, \Xi(Q)}{\omega_1(\mathbf{q}) \cdots \omega_N(\mathbf{q})}. \tag{11}$$

After taking the inverse Fourier transform and expanding the functions $\Phi_m(\mathbf{q})$ in a Taylor series around $\mathbf{q} = 0$, we obtain a differential equation satisfied by $G_R(X - Y)$. It should be clear that this will be a linear differential equation of order N in time. The equation of motion for J can be obtained by multiplying Eq. (11) by $\tilde{J}$ and taking the inverse Fourier transform. See Ref. 5 for details. The result is

$$\chi_N \partial_t^N J + \ldots + \chi_1 \partial_t J + J = D_0 F + \ldots + D_N \partial_t^N F + \mathcal{O}\left(\partial_t^{N+1} F, \partial_{\mathbf{x}}\right). \tag{12}$$

Here, we omitted all terms involving spatial derivatives. We introduced the transport coefficients χ_m and D_m,

$$\chi_m = \Phi_m(\mathbf{0}) = (-i)^m \sum_{1 \le i_1 \ldots < i_m \le N} \frac{1}{\omega_{i_1}(\mathbf{0}) \cdots \omega_{i_m}(\mathbf{0})}, \tag{13}$$

$$D_m = i^m \frac{(-1)^N}{m!} \frac{\partial_\omega^m\, \Xi(\omega, \mathbf{0})|_{\omega=0}}{\omega_1(\mathbf{0}) \cdots \omega_N(\mathbf{0})}. \tag{14}$$

If there is a clear separation of scales, i.e., $\mathrm{Kn} \ll 1$, it becomes possible to simplify the right-hand side of Eq. (12). In this case, $D_0 F \gg D_1 \partial_t F \gg D_2 \partial_t^2 F$, and higher derivatives of F can be neglected. However, in order to obtain a relaxation-type equation of motion for J, additional approximations are required. For this purpose, one must impose the limiting procedure, $\chi_i \to 0$, $i \geq 2$. In such a limiting procedure, all the poles of the retarded Green's function except the pole nearest to the origin (in the following referred to as the "first pole") are pushed to infinity.[5] Assuming that this can be done, we obtain the following equation of motion for J

$$\tau_R \partial_t J + J = D_0 F + D_1 \partial_t F + D_2 \partial_t^2 F + \mathcal{O}\left(D_3 \partial_t^3 F, \partial_{\mathbf{x}}\right) , \qquad (15)$$

where one can show that

$$\begin{aligned} \tau_R &= \frac{1}{i\omega_1(\mathbf{0})}, \\ D_0 &= \tilde{G}_R(0, \mathbf{0}) , \\ D_1 &= i\partial_\omega \, \tilde{G}_R(\omega, \mathbf{0})\Big|_{\omega=0} + D_0 \tau_R . \end{aligned} \qquad (16)$$

Note that the time reversal symmetry of the retarded Green's function requires the first pole to lie on the imaginary axis, such that τ_R is real, as it should be. It is also important to remark that Eq. (15) describes the long- (but not asymptotically long-) time evolution of the system. This is the physical origin of the limiting procedure previously implemented and it is the only limit in which a relaxation equation can be obtained and, consequently, a relaxation time can be defined. If one wants to describe even shorter time scales, it becomes necessary to include more terms $\sim \chi_i$ in the discussion. Then, the equations of motion will become more complicated. Naturally, the gradient expansion is recovered when all poles are pushed to infinity and, in this sense, the gradient expansion can be interpreted as an asymptotic solution of the more general equation of motion.[5]

Note that this limiting procedure is only allowed when the first pole lies on the imaginary axis. On the other hand, if the first pole and, for reasons of symmetry, its counterpart on the other side of the imaginary axis, have nonzero real parts one cannot disregard one of the poles while keeping the other. Then, the long-time evolution of the dissipative current will be characterized by a second-order differential equation and will display oscillatory motion. Therefore, in this case the fluid-dynamical motion cannot be reduced to a simple relaxation equation, even in the small-frequency domain.

6. Application: The Linearized Boltzmann Equation

The discussion presented above is general and can, in principle, be applied to any type of theory. In this section, we apply this formalism to kinetic theory. In particular, we calculate the shear viscosity and relaxation time coefficients for a weakly coupled classical gas of hard spheres via the Boltzmann equation.[5]

We perform all our calculations in the local rest frame of the fluid element, where $u^\mu = (1,0,0,0)$. Since we restrict our discussion to shear viscosity, we consider the simplified scenario in which the fluid does not accelerate, $u^\mu \partial_\mu u^\nu \equiv 0$, and does not expand, $\partial_\mu u^\mu \equiv 0$, and where temperature, $T = \beta_0^{-1}$, and chemical potential, $\mu = \alpha_0/\beta_0$, are constant. With all these simplifications, the linearized Boltzmann equation can be written as[5]

$$\partial_t \delta f_{\mathbf{k}} + \mathbf{v} \cdot \nabla f_{\mathbf{k}} - \hat{C} \delta f_{\mathbf{k}} = \mathcal{S}\left(X, K\right), \tag{17}$$

where $\delta f_{\mathbf{k}} = f_{\mathbf{k}} - f_{0\mathbf{k}}$, with $f_{0\mathbf{k}} = \exp\left(\alpha_0 - \beta_0 E_{\mathbf{k}}\right)$, $\mathbf{v} = \mathbf{k}/E_{\mathbf{k}}$, and $\mathcal{S} = \beta_0 f_{0\mathbf{k}} E_{\mathbf{k}}^{-1} k^{\langle\mu} k^{\nu\rangle} \sigma_{\mu\nu}$. Here, we introduced the collision operator, $\hat{C}$, defined as

$$\begin{aligned} \hat{C} \delta f_{\mathbf{k}} = & \frac{1}{\nu E_{\mathbf{k}}} \int dK' dP dP' W_{\mathbf{k}\mathbf{k}' \to \mathbf{p}\mathbf{p}'} f_{0\mathbf{k}} f_{0\mathbf{k}'} \\ & \times \left(\frac{1}{f_{0\mathbf{p}'}} \delta f_{\mathbf{p}'} + \frac{1}{f_{0\mathbf{p}}} \delta f_{\mathbf{p}} - \frac{1}{f_{0\mathbf{k}'}} \delta f_{\mathbf{k}'} - \frac{1}{f_{0\mathbf{k}}} \delta f_{\mathbf{k}} \right), \end{aligned} \tag{18}$$

where $W_{\mathbf{k}\mathbf{k}' \to \mathbf{p}\mathbf{p}'}$ is the transition rate, ν is the symmetry factor ($\nu = 2$ for identical particles) and $dK = d^3\mathbf{k}\, E_{\mathbf{k}}^{-1}/(2\pi)^3$.

Using Eqs. (17) and (18), it is possible to express the Fourier transform of the shear stress tensor, $\tilde{\pi}^{\mu\nu}(Q)$, in terms of the Fourier transform of the shear tensor, $\tilde{\sigma}_{\alpha\beta}(Q)$, in the form $\tilde{\pi}^{\mu\nu} = G_R^{\mu\nu\alpha\beta} \tilde{\sigma}_{\alpha\beta}$,[5] where

$$\tilde{G}_R^{\mu\nu\alpha\beta}\left(\omega, \mathbf{q}\right) = \int dK\, k^{\langle\mu} k^{\nu\rangle} \frac{1}{i\omega - i\mathbf{v}\cdot\mathbf{q} - \hat{C}} \beta_0\, f_{0\mathbf{k}} E_{\mathbf{k}}^{-1} k^{\langle\alpha} k^{\beta\rangle}. \tag{19}$$

In order to compute this Green's function, we define a function $B^{\alpha\beta}(Q,K)$ which satisfies

$$\left(i\omega - i\mathbf{v}\cdot\mathbf{q} - \hat{C}\right) B^{\alpha\beta}(Q,K) = \mathcal{S}(Q,K) = \beta_0 E_{\mathbf{k}}^{-1} k^{\langle\alpha} k^{\beta\rangle} f_{0\mathbf{k}}. \tag{20}$$

As shown in the previous section, in order to extract the shear viscosity and relaxation time coefficients, it is sufficient to consider the case $\mathbf{q} = 0$. Then, the dependence of $B^{\alpha\beta}$ on K can be expressed via the following expansion,

$$B^{\alpha\beta}(\omega, K) = f_{0\mathbf{k}}\, k^{\langle\alpha} k^{\beta\rangle} \sum_{n=0}^{\infty} a_n(\omega) E_{\mathbf{k}}^n. \tag{21}$$

Substituting Eq. (21) into Eq. (19), we obtain the following expression for $\tilde{G}_R^{\mu\nu\alpha\beta}(\omega, \mathbf{0})$,

$$\tilde{G}_R^{\mu\nu\alpha\beta}(\omega, \mathbf{0}) = \sum_{n=0}^{\infty} a_n(\omega) \int dK\, k^{\langle\mu} k^{\nu\rangle} k^{\langle\alpha} k^{\beta\rangle} E_{\mathbf{k}}^n f_{0\mathbf{k}}$$
$$= 2\Delta^{\mu\nu\alpha\beta} \sum_{n=0}^{\infty} I_{n+4,2}\, a_n(\omega) \,, \tag{22}$$

where we introduced the thermodynamic integral

$$I_{nq} = \frac{1}{(2q+1)!!} \int dK\, f_{0\mathbf{k}} E_{\mathbf{k}}^{n-2q} \left(m^2 - E_{\mathbf{k}}^2\right)^q . \tag{23}$$

Thus, the relation between $\tilde{\pi}^{\mu\nu}$ and $\tilde{\sigma}^{\mu\nu}$ can be cast into a more convenient form $\tilde{\pi}^{\mu\nu}(\omega, \mathbf{0}) = 2\tilde{G}_R(\omega, \mathbf{0})\, \tilde{\sigma}^{\mu\nu}(\omega, \mathbf{0})$, with $\tilde{G}_R(\omega, \mathbf{0})$ being given by

$$\tilde{G}_R(\omega, \mathbf{0}) = \sum_{n=0}^{\infty} I_{n+4,2}\, a_n(\omega). \tag{24}$$

The whole frequency dependence of $\tilde{G}_R(\omega, \mathbf{0})$ is contained in the functions $a_n(\omega)$. The problem is then reduced to determining the analytic properties of $a_n(\omega)$. The way to solve this problem is to substitute the expansion (21) into Eq. (20), multiply by $E_{\mathbf{k}}^m k^{\langle\mu} k^{\nu\rangle}$, and integrate over dK. Then one obtains

$$\sum_{n=0}^{\infty} \left(-i\omega\mathcal{D}^{mn} + \mathcal{A}^{mn}\right) a_n(\omega) = \beta_0 I_{m+3,2} \,, \tag{25}$$

where we defined the matrices

$$\mathcal{A}^{mn} \Delta^{\mu\nu\alpha\beta} = -\frac{1}{2} \int dK E_{\mathbf{k}}^m k^{\langle\mu} k^{\nu\rangle} \hat{C}(K)\, f_{0\mathbf{k}} E_{\mathbf{k}}^n k^{\langle\alpha} k^{\beta\rangle}, \tag{26}$$
$$\mathcal{D}^{mn} = \frac{1}{5!!} \int dK\, f_{0\mathbf{k}} E_{\mathbf{k}}^{m+n} \left(m^2 - E_{\mathbf{k}}^2\right)^2 . \tag{27}$$

The solution of this equation can be formally expressed as,

$$a_m(\omega) = \beta_0 \sum_{n=0}^{\infty} \left[\left(-i\omega\mathcal{D} + \mathcal{A}\right)^{-1}\right]^{mn} I_{n+3,2}. \tag{28}$$

Thus, we obtain

$$\tilde{G}_R(\omega, \mathbf{0}) = \beta_0 \sum_{m=0}^{\infty} \sum_{n=0}^{\infty} I_{m+4,2} \left[\left(-i\omega\mathcal{D} + \mathcal{A}\right)^{-1}\right]^{mn} I_{n+3,2}. \tag{29}$$

The poles of this function can be obtained from the roots of the determinant $\det(-i\omega\mathcal{D} + \mathcal{A}) = 0$. Thus, the relaxation times and viscosity coefficients are determined by inverting and finding eigenvalues of infinite-dimensional matrices. In practice, however, the expansion (21) has to be truncated and one never actually deals with infinite matrices. This is a common approach and happens quite often when dealing with the collision operator. Let us try the simplest case and consider only one term in the expansion (this corresponds to using the Israel-Stewart 14-moment approximation in the moments method). Then all matrices become numbers and everything simplifies considerably. In this case, the retarded Green's function has only one pole, $\omega_0 = -i\mathcal{A}^{00}/I_{42}$.[5]

Since in this case the thermodynamic force is given by $F = 2\sigma^{\mu\nu}$, the shear viscosity coefficient is given by D_0. Thus, the shear viscosity and relaxation time are given by

$$\eta = \tilde{G}_R(0,\mathbf{0}) = \beta_0 \frac{I_{42} I_{32}}{\mathcal{A}^{00}}, \tag{30}$$

$$\tau_\pi = \frac{1}{i\omega_0} = \frac{I_{42}}{\mathcal{A}^{00}}. \tag{31}$$

This implies that $\eta/\tau_\pi = \beta_0 I_{32}$. In the massless limit, for a gas of hard spheres, $\mathcal{A}^{00} = (3/5)\, I_{42}\, n_0\, \sigma$, with σ being the total cross-section.

Exactly the same results were obtained in Ref.[12] within a completely different approach. This clearly demonstrates that the relaxation time in Israel-Stewart theories, which determines the time scale related to transient dynamics, is indeed a microscopic scale. In other words, the relaxation time is determined by the inter-particle scattering rate and not by an arbitrary fluid-dynamical time scale.

7. Conclusion

In this work we proved that the existence of singularities in the retarded Green's functions considerably restricts the applicability of the gradient expansion as a method to extract macroscopic dynamics. As a matter of fact, the causal structure of relativistic fluid dynamics hinges on the presence of poles in the Green's functions.

We computed the first pole of the retarded Green's function associated with shear stress in a weakly coupled classical gas of hard spheres. We proved that the inverse of this pole is directly related to the relaxation time that appears in Israel-Stewart theory. Consequently, the relaxation time in Israel-Stewart theories corresponds to a microscopic time scale and such theories are indeed describing the real transient dynamics of the fluid.

In summary, the true relaxation time is always given by the first pole of the retarded Green's function. In general, the location of this pole cannot be found using a truncated Taylor expansion around the origin or, in other words, via the gradient expansion. This result should be expected since the transient behavior of a given system cannot be unambiguously extracted from its asymptotic solution given by the gradient expansion.

The authors thank T. Kodama, T. Koide, A. Ficnar, and G. Torrieri for insightful discussions. H.N. was supported by the ExtreMe Matter Institute EMMI. The authors thank the Helmholtz International Center for FAIR within the framework of the LOEWE program for support.

References

1. See, for instance, P. Huovinen, P. V. Ruuskanen, Ann. Rev. Nucl. Part. Sci. **56**, 163-206 (2006) and refs. therein.
2. L. D. Landau and E. M. Lifshitz, *Fluid Mechanics*, (Pergamon; Addison-Wesley, London, U.K.; Reading, U.S.A., 1959).
3. W. A. Hiscock and L. Lindblom, Ann. Phys. (N.Y.) **151** 466 (1983), Phys. Rev. **D 31** 725 (1985), Phys. Rev. **D 35** 3723 (1987), Phys. Lett. **A 131** 509 (1988), Phys. Lett. **A 131** 509 (1988); S. Pu, T. Koide, D. Rischke, Phys. Rev. D **81**, 114039 (2010).
4. G. S. Denicol, T. Kodama, T. Koide and Ph. Mota, J. Phys. **G 35**, 115102 (2008).
5. G. S. Denicol, H. Niemi, J. Noronha and D. H. Rischke, arXiv:1102.4780 [hep-th] (2011).
6. S. Chapman and T. G. Cowling, *The mathematical theory of non-uniform gases*, 3rd edition (Cambridge University Press, Cambridge, 1970).
7. D. Burnett, Proc. Lond. Math. Soc. **39**, 385–430 (1935); Proc. Lond. Math. Soc. **40**, 382–435 (1936).
8. A. V. Bobylev, Sov. Phys. Dokl. **27**, 29 (1982).
9. C. Cattaneo, Atti. Semin. Mat. Fis. Univ. Modena **3**, 83 (1948).
10. W. Israel and J. M. Stewart, Phys. Lett. **58**A, 213 (1976); Ann. Phys. (N.Y.) **118**, 341 (1979); Proc. Roy. Soc. London A **365**, 43 (1979).
11. S. R. de Groot, W. A. van Leeuwen and Ch. G. van Weert, *Relativistic kinetic theory - Principles and applications*, (North-Holland, 1980).
12. G. S. Denicol, T. Koide and D. H. Rischke, Phys. Rev. Lett. **105**, 162501 (2010) .

PRODUCTION OF HEAVY AND SUPERHEAVY NUCLEI IN EXPLOSIVE PROCESSES

IGOR N. MISHUSTIN

Frankfurt Institute for Advanced Studies, J.W. Goethe University, D-60438 Frankfurt/Main, Germany
mishustin@fias.uni-frankfurt.de
National Research Center "Kurchatov Institute", 123182 Moscow, Russia

This talk consists of two parts. In the first part I discuss properties of hot stellar matter at sub-nuclear densities which is formed in supernova explosions. I present main steps of the statistical approach to the equation of state and nuclear composition, dealing with an ensemble of nuclear species instead of one "average" nucleus. The emphasis is put on possible formation of heavy and superheavy nuclei. In the second part of the talk I discuss the possibility of synthesizing heavy and superheavy elements by multiple neutron capture reactions in terrestrial experiments. The mass distributions of heavy isotopes are calculated by solving a chain of rate equations for consequtive neutron capture. High neutron fluxes which are needed to reach the "stability islend" of superheavy elements can be provided by multiple nuclear explosions.

Preface

First, my words of respect and appreciation go to Prof. Walter Greiner, an outstanding physicist and a great men. He made fundamental contributions in many areas of modern physics, including quantum electrodynamics, nuclear structure and reactions, heavy-ion physics, cluster physics etc. Generally Walter likes Exotics, i.e. exotic topics, exotic places, exotic people, exotic cars etc. The list of his favorite research topics includes: Super Critical Fields, Super Heavy Elements, Super Dense Matter, Structure of Vacuum, Anti-Matter, Ultra-Relativistic Collisions, Supernova Explosions, Heavy-Ion Therapy etc. This is why I talk today about Exotic nuclei production in Exotic environments.

Happy birthday to you, dear Walter!
Thank you for your enthusiasm and inspiration!

1. Introduction

In the first part of my talk I present some recent results on nuclear composition of hot matter at subnuclear densities and temperatures $T < 10$ MeV, relevant for supernova explosions. As well known, the nuclear composition of matter is extremely important for the physics of supernova explosions. For example, the electron capture on nuclei plays an important role in supernova dynamics.[1] But the electron capture rates are sensitive to the nuclear composition and details of nuclear structure.[2] The neutrino-induced reactions are very sensitive to the nuclear structure effects and properties of weak interactions in nuclei.[3] It is also important that the presence of nuclei favors the explosion via the energy balance,[4] since less energy is used for nuclear dissociation. All these considerations show importance of the nuclear physics input in supernova simulations.

For a realistic description of nucleosyntheis in explosive proesses one can use rich experience accumulated in recent years by studying intermediate-energy nuclear reactions. In particular, multifragmentation reactions provide valuable information about hot nuclei in dense environments. According to present understanding, based on numerous theoretical and experimental studies of multifragmentation reactions, prior to the break-up a transient state of nuclear matter is formed, where hot nuclear fragments exist in equilibrium with free nucleons. This state is characterized by a certain temperature $T \sim 3 - 6$ MeV and a density which is typically 3-5 times smaller than the nuclear saturation density, $\rho_0 \approx 0.15$ fm^{-3}. One can say that a hot intermediate system produced in nuclear multifragmentation reactions represents a small piece of supernova matter. A very good description of such systems is achieved with the Statistical Multifragmentation model (SMM).[5]

The statistical nature of multifragmentation is confirmed by numerous experimental observations, such as evolution of the fragment mass distribution with excitation energy (temperature), fragment correlations revealing the critical behavior, anomaly in the caloric curve, and by many others.[5–8] Recent experiments[9] directly confirm the basic assumption of the SMM, namely, that the primary fragments are hot, their internal excitation energy may reach up to 3 MeV per nucleon. Therefore, properties of these hot nuclei can be extracted from multifragmentation reactions and used for the description of matter under stellar conditions. The steps in this direction were made in our papers.[10–12] A similar model was also used in Ref. 13 where, however, only cold nuclei in long-lived states were considered.

2. Statistical Description of Supernova Matter

In the supernova environment, as compared to the nuclear reactions, several new important ingredients should be taken into consideration. First, the matter at stellar scales must be electrically neutral and therefore electrons should be included to balance positive nuclear charge. Second, energetic photons which are present in hot matter can change nuclear composition via photonuclear reactions. And third, the matter is irradiated by a strong neutrino wind from the protoneutron star.

Below we consider macroscopic volumes of matter consisting of various nuclear species (A, Z), nucleons ($n = (1,0)$ and $p = (1,1)$), electrons (e^-) and positrons (e^+) under condition of electric neutrality. We expect that in this situation an ensemble of various nuclear species will be generated like in a liquid-gas coexistence region, as observed in the multifragmentation reactions. Here the nuclear system is characterized by the temperature T, baryon density ρ_B and electron fraction Y_e (i.e. the ratio of the net electron density to the baryon density). One may expect that new nuclear effects come into force in this environment.

Composition of stellar matter can safely be studied within the Grand Canonical approximation dealing with chemical potentials of the constituents. Generally, the chemical potential of a species i with baryon number B_i, charge Q_i and lepton number L_i, which participates in chemical equilibrium, can be found from the general expression:

$$\mu_i = B_i\mu_B + Q_i\mu_Q + L_i\mu_L \tag{1}$$

where μ_B, μ_Q and μ_L are three independent chemical potentials which are determined from the conservation of total baryon number $B = \sum_i B_i$ electric charge $Q = \sum_i Q_i$ and lepton number $L = \sum_i L_i$ of the system. This gives

$$\begin{aligned} \mu_{AZ} &= A\mu_B + Z\mu_Q \ , \\ \mu_{e^-} &= -\mu_{e^+} = -\mu_Q + \mu_L \ , \\ \mu_\nu &= -\mu_{\tilde{\nu}} = \mu_L \ . \end{aligned} \tag{2}$$

These relations are also valid for nucleons, $\mu_n = \mu_B$ and $\mu_p = \mu_B + \mu_Q$. If ν and $\overline{\nu}$ escape freely from the system, the lepton number conservation is irrelevant and $\mu_L = 0$. In this case two remaining chemical potentials are determined from the conditions of baryon number conservation and electro-neutrality:

$$\rho_B = \frac{B}{V} = \sum_{AZ} A\rho_{AZ} \ , \ \rho_Q = \frac{Q}{V} = \sum_{AZ} Z\rho_{AZ} - \rho_e = 0 \ .$$

Here ρ_{AZ} is the number density of nuclear species (A,Z), $\rho_e = \rho_{e^-} - \rho_{e^+}$ is the net electron density.

For describing an ensemble of nuclear species in thermodynamical equilibrium we use the Grand Canonical version of the SMM,[5,17] properly modified for supernova conditions. After integrating out translational degrees of freedom the density of nuclear species with mass A and charge Z is calculated as

$$\rho_{AZ} = \frac{N_{AZ}}{V} = g_{AZ}\frac{V_f}{V}\frac{A^{3/2}}{\lambda_T^3}\exp\left[-\frac{1}{T}\left(F_{AZ}-\mu_{AZ}\right)\right], \tag{3}$$

were g_{AZ} is the g.-s. degeneracy factor of species (A,Z), $\lambda_T = \left(2\pi\hbar^2/m_N T\right)^{1/2}$ is the nucleon thermal wavelength, $m_N \approx 939$ MeV is the average nucleon mass. V is the actual volume of the system and V_f is so called free volume, which accounts for the finite size of nuclear species. We assume that all nuclei have normal nuclear density ρ_0, so that the proper volume of a nucleus with mass A is A/ρ_0. At low densities the finite-size correction can be taken into account within the excluded volume approximation $V_f/V \approx (1-\rho_B/\rho_0)$.

The internal excitations of nuclear species (A,Z) play an important role in regulating their abundancies. Sometimes they are included through the population of nuclear levels known for nearly cold nuclei.[13] However, in the supernova environment not only the excited states but also the binding energies of nuclei will be strongly affected by the surrounding matter. By this reason, we find it more justified to use another approach which can easily be generalized to include in-medium modifications. Namely, the internal free energy of species (A,Z) with $A>4$ is parameterized in the spirit of the liquid drop model

$$F_{AZ}(T,\rho_e) = F_{AZ}^B + F_{AZ}^S + F_{AZ}^{\rm sym} + F_{AZ}^C \ , \tag{4}$$

where the right hand side contains, respectively, the bulk, the surface, the symmetry and the Coulomb terms. Explicit expressions for first three terms can be found in Ref. 5. In the Coulomb term we include the modification due to the screening effect of electrons. By using the Wigner-Seitz approximation it can be expressed as[14]

$$F_{AZ}^C(\rho_e) = \frac{3}{5}c(\rho_e)\frac{(eZ)^2}{r_0 A^{1/3}} \ , \quad c(\rho_e) = \left[1 - \frac{3}{2}\left(\frac{\rho_e}{\rho_{0p}}\right)^{1/3} + \frac{1}{2}\left(\frac{\rho_e}{\rho_{0p}}\right)\right] ,$$

where $r_0 = 1.17$ fm and $\rho_{0p} = (Z/A)\rho_0$ is the proton density inside the nuclei. The screening function $c(\rho_e)$ is 1 at $\rho_e = 0$ and 0 at $\rho_0 = \rho_{0p}$. Nucleons and light clusters ($A \le 4$) are considered as structureless particles characterized only by mass and proper volume. All further details of the model can be found in our recent paper.[12] The advantage of our approach is that we consider all the fragments present in the matter, contrary to the previous calculations,[15,16] which consider only one "average" nucleus characterizing the liquid phase.

3. Nuclear Composition of Supernova Matter

In numerical calculations we first fix temperature T, baryon density ρ_B and electron fraction Y_e. Then we consider a box containing the baryon number B =1000 and proton number $Z = Y_e \cdot B$. The box volume is fixed by the average baryon density, $V = B/\rho_B$. We use an iterative procedure to find chemical potentuals μ_B and μ_Q. Finally, relative yields of all nuclei with $1 \leq A \leq 1000$ and $0 \leq Z \leq A$ are calculated from Eq. (3). Nuclei with larger masses (A >1000) can be produced only at relatively high densities, $\rho_B > 0.1\rho_0$, which are relevant for the regions deep inside the protoneutron star.

First we consider the case when lepton fraction is fixed as expected inside a neutrinosphere. Figure 2 (left panels) shows the results for lepton fraction Y_L=0.2 and typical temperature T = 3 MeV. Mass distributions are shown in the lower left panel. One can see that the islands of heavy nuclei, $200 < A < 400$, appear at relatively high baryon density, $\rho_B = 0.1\rho_0$, corresponding to conditions in a protoneutron star. These nuclei are very neutron-rich, $Z/A \approx 0.27$. The Z/A ratios are decreasing with A less rapidly than in the nuclear multifragmentation reactions.[18] This can be explained by the screening effect of electrons. The width of the charge distribution at given A is determined by temperature T and γ-coefficient in the symmetry energy: $\sigma_Z \approx \sqrt{AT/8\gamma}$.[17,18] At lower density, $\rho_B = 0.01\rho_0$, the mass distribution is rather flat up to $A \approx 80$ and then decreases rapidly for larger A. For $\rho_B = 10^{-3}\rho_0$ only light clusters are present and the mass distribution drops exponentially.

Let us consider the situation more appropriate for early times of a supernova explosion, when the electron capture reactions did not change significantly the electron fraction of matter. In this case the electron fraction is fixed, and the electron and proton chemical potentials are determined independently, without using the equilibrium relation $\mu_e = -\mu_Q$. Corresponding results for $Y_e = 0.4$ and $Y_e = 0.2$ at $T = 1$ MeV and several baryon densities are presented in Fig. 2 (right top and bottom panels).

One can see that heavy and even superheavy nuclei, $50 < A < 400$, can be formed in this case too. They exist in a very broad range of densities,

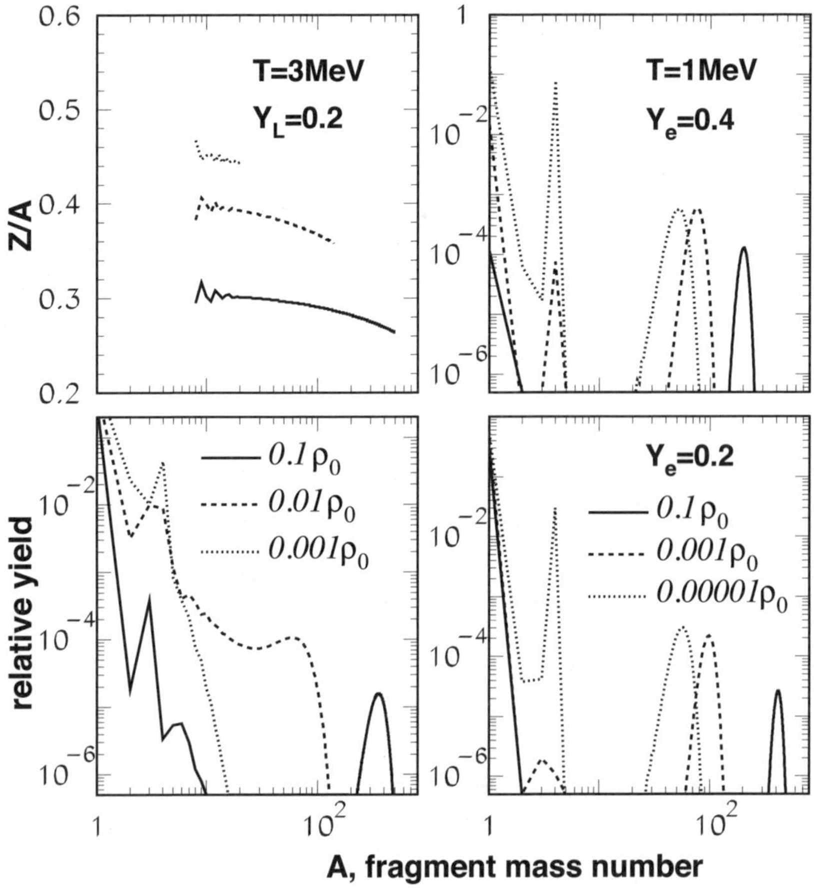

Fig. 1. Mean charge-to-mass ratios (left top panel), and mass distributions of hot primary fragments (other panels) calculated with the SMM generalized for supernova conditions. Left panels are calculations for temperature $T = 3$ MeV and fixed lepton (electrons+neutrinos) fraction Y_L =0.2 per nucleon. Right panels are calculations for temperature $T = 1$ MeV and fixed electron fractions $Y_e = 0.4$ (top) and 0.2 (bottom). Lines show the fragment mass distributions at different baryon densities (in units of the normal nuclear density ρ_0=0.15 fm^{-3}), indicated in the figure.

$0.1\rho_0 > \rho_B > 10^{-5}\rho_0$. At given density the mass distribution has a Gaussian shape. In the $Y_e = 0.4$ case the most probable nuclei, corresponding to the maxima of distributions, have Z/A ratios 0.400, 0.406, and 0.439, for densities $0.1\rho_0$, $10^{-3}\rho_0$, and $10^{-5}\rho_0$, respectively. The Gaussian mass distributions may in some cases justify earlier calculations,[15,16] when only

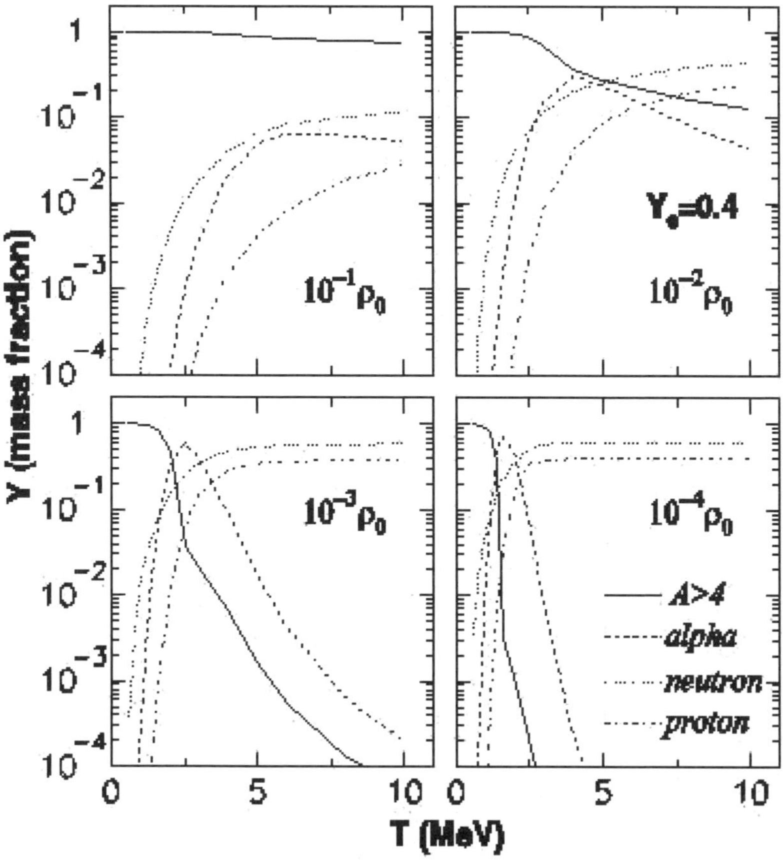

Fig. 2. Mass fractions of different nuclear species as functions of temperature for $Y_e = 0.4$ calculated for different baryon densities (indicated in the panels). Neutrons, protons, α-particles and heavier nuclei (A>4) are shown by dotted, dash-dotted, dashed and solid lines, respewctively.

one heavy nuclear species was assumed at each density. As seen from the bottom panel, changing the electron fraction from 0.4 to 0.2 leads to a significant increase of nuclear masses. Also, the nuclei become more neutron rich: the corresponding Z/A ratios are 0.280, 0.359, and 0.420. Our calculations show that even larger effect can be caused by the reduction of the symmetry energy of hot fragments (see Ref. 19). Changing γ from 25 to 15 MeV would lead to the island of nuclei with masses around 800 instead of 400.

Figure 3 displays the mass fractions of different nuclear species as functions of temperature for several baryon densities and fixed $Y_e = 0.4$. One can see several interesting trends. First, nuclei with $A > 4$ survive at high temperatures only if the baryon density is large enough, $\rho_B > 10^{-2}\rho_0$, At lower densities they are destroyed by hard photons already at $T > 2$ MeV. On the other hand, the fractions of free neutrons and protons increase gradually and dominate at $\rho_B \leq 10^{-2}\rho_0$. A significant change in the trend is observed at $T > 3$ MeV which can be related to the liquid-gas transition in such a matter. It is interesting to note that α-particles may exist abundantly only in a narrow range of temperatures, $2 < T < 4$ MeV (see two lower panels).

From this analysis we conclude that statistical model can explain gross features of element abundances in supernova environments. It provides "seed" nuclei for further nuclear transformations in r- and rp-processes.

Now I jump to completely different physical conditions for synthesizing heavy and superheavy elements which can be created in terrestrial experiments. Namely, in our recent paper[20] we have proposed to use intensive neutron fluxes generated by nuclear explosions. Actually, this method was already partially employed alongside with nuclear reactors in the 60-s.[21–23] However, because of some technical and political constraints it was abandoned later. Nevertheless, this method provides the highest neutron fluxes which are not possible to reach with other terrestrial techniques. In principle this method is suitable for obtaining neutron-rich nuclei up to their drip-lines.

4. Physical Conditions in Nuclear Explosions

For our further calculations we should know the density of neutrons generated in the active zone of an explosion. Let us assume a 100 kt explosion which corresponds to fissioning of $1.45 \cdot 10^{25}$ nuclei of ^{235}U (or ^{239}Pu) with total mass of 5.6 kg. At normal density this amount of U or Pu has the volume of about 300 cm^3. Assuming that about 1.3 new neutrons are produced in average per fission, we can estimate the total number of free neutrons to be $2 \cdot 10^{25}$. This gives the maximum neutron density of about $6 \cdot 10^{22}$ cm^{-3}. However, due to a finite efficiency, the amount of the fissile material and accordingly its volume should be larger. On the other hand, with the implosion assembly method, the fissile material is initially compressed by several times, that will increase accordingly the neutron density. According to Ref. 24, the neutron multiplication process should be over within the time interval of about $1.5 \cdot 10^{-7}$ sec after beginning of the chain reaction. To this time the density of free neutrons increases exponentially and reaches the maximum values estimated above. Many of these fission neutrons may leave the core. However, some of them might be reflected back with an appropriate reflector to keep the neutron density in the core higher. These neutrons have a chance to slow down to thermal energies corresponding to temperatures $T \approx 10$ keV at the early expansion stage. Taking into account all these uncertainties as well as strong time dependence of the neutron yield, in our estimates below we use the neutron density of $4 \cdot 10^{22}$ cm^{-3} for thermal neutrons.

There are several important time scales which influence strongly the efficiency of the neutron capture reaction. Obviously, the goal is to maximize the time τ of the target exposure to the strongest neutron flux. The shortest time scale is the neutron multiplication time τ_m associated with the chain reaction. This time is determined by the properties of the fissile material and is typically in the range of 0.1 μsec, as mentioned above. Such short times are obtained when all available neutrons are used for fission reactions, to maximize the yield of the explosion. However, this time will increase significantly if the active zone will contain more nuclei absorbing neutrons, like ^{238}U (reduced enrichment), bringing the system closer to the critical regime.

The disassembly stage is defined by a condition that the chain reaction is ceased. This happens when most of the fission energy is deposited in the active zone and the matter is transformed into a radiation-dominated hot and dense plasma. Due to the huge internal pressure it would rapidly expand into the vacuum, if nothing is done to confine it. However, by introducing a heavy tamper around the active zone, one can delay the fast expansion of plasma to a few microseconds. This time is needed for the shock wave to propagate through the thick layer of heavy tamper's material, usually ^{238}U. The hydrodynamic time can be estimated as a time required for the rarefaction wave to reach the center of the active zone, $\tau_h \approx R/c_s \lesssim 10\mu$sec, where R is the core radius and c_s is the sound velocity in the core. The tamper can also serve as a reflector and absorber for the neutrons. To explore the whole parameter space, in our calculations we consider the exposure times τ from 0.1 to 3 μsec.

5. Production of Superheavy Elements via Multiple Neutron Capture

Our analysis[20] has shown that at temperatures $T \sim 10$ keV we can simply consider a two-step process, i.e. the capture of a neutron followed by the photon emission, without complications induced by re-emission of neutrons and fission. In this case we can write the set of coupled equations for densities of nuclei with mass numbers $A{=}A_0$, $A_0{+}1$, $A_0{+}2$, ... :

$$\frac{d\rho_{A_0}}{dt} = -\langle v_{rel}\sigma_n\rangle \rho_n \rho_{A_0},$$

$$\frac{d\rho_{A+1}}{dt} = \langle v_{rel}\sigma_n\rangle \rho_n \rho_A - \langle v_{rel}\sigma_n\rangle \rho_n \rho_{A+1}, \qquad (5)$$

which can be solved numerically for any given time dependence of the neutron density ρ_n. Here v_{rel} is the neutron-nucleus relative velocity which is close to the neutron velocity v_n, and σ_n is the neutron capture cross-section, which in the considered energy region is of about 1 bn and independent of the nuclide mass number.* These equations describe the evolution of the ensemble of nuclear species due to the neutron capture reactions with gain $(A-1 \to A)$ and loss $(A \to A+1)$ terms. It is worth noting that final nuclide abundances depend only on the time-integrated neutron flux $\Phi = \int \rho_n v_n dt \approx v_n \rho_n \tau$. According to the analysis of Refs. 22 and 23 typical values of Φ reached in low-yield nuclear explosions are in the range of $3{\cdot}10^{24} \div 10^{25}\ n/\mathrm{cm}^2$.

In Fig. 3 we present the distributions of isotope abundances after exposure of a ^{238}U target to a constant neutron flux during fixed time intervals indicated at the lines. The figure shows that the distribution of neutron-rich isotopes becomes broader with increasing the exposure time, or, equivalently, the integrated neutron flux Φ. According to Refs. 22 and 25 in samples of material extracted from underground explosion sites concentrations of rare nuclei (A=257) on the level of $10^{-11} - 10^{-12}$ were measured by radiochemical methods. Even higher sensitivity can be reached by using mass spectroscopic methods.[26] In previos nuclear explosionns the observed neutron-rich isotopes have not more than 20 absorbed neutrons.[22,25] According to calculations of Fig. 4, where neutron density was fixed at $4{\cdot}10^{22}$ cm^{-3}, to produce such isotopes in observable amounts we would need $\tau \approx 0.3\mu$sec. This corresponds to the integrated neutron flux of about

*Of course, when the neutron number approaches a magic number or the drip-line limit, the capture cross section σ_n will strongly decrease, and the solution of the equations will be more complicated.

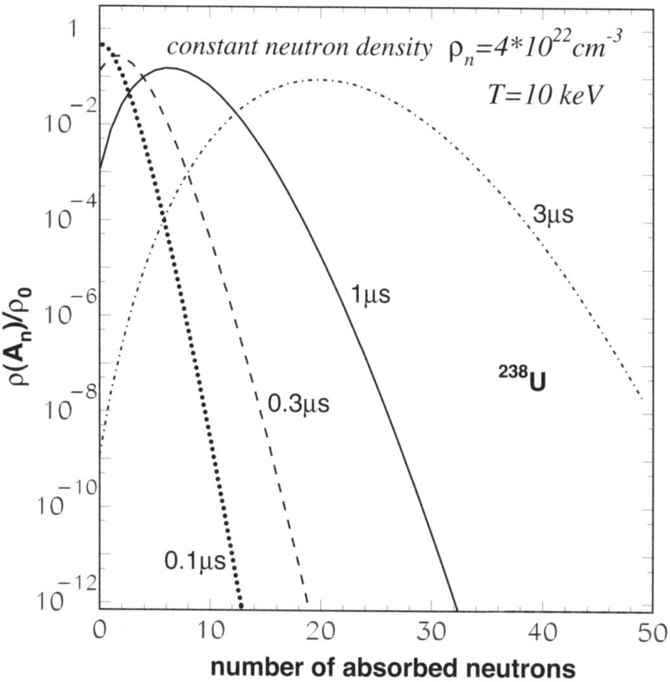

Fig. 3. Densities of nuclei produced after capture of neutrons normalized to the initial density of ^{238}U target. Density and temperature of neutron, as well as exposure times (in microseconds) are shown in the figure.

$2 \cdot 10^{24}\ n/\text{cm}^2$, which within a factor of 2 agrees with the values estimated in Refs. 22 and 25.

From Fig. 4 we conclude that observable concentrations of nuclei with more than 40 absorbed neutrons can be obtained only if the exposure time is longer than 1μsec. Obviously, to increase the number of captured neutrons one should increase the integrated neutron flux Φ, i.e. either by increasing the neutron density ρ_n or by increasing the exposure time τ.[†] To achieve this goal one can try e.g. to increase the neutron multiplication time by introducing neutron absorbers, reflectors, and other construction elements. With the exposure time of 3 μsec the fraction of nuclei captured 50 neutrons would be about 10^{-8}. In this case, even if we take 1 gram of the exotic target

[†]We are taking here about time scales which are still much less than the β-deacy times, i.e. milliseconds.

material like ^{248}Cm, the concentration of nuclides with A≈300 will be 10^{-12} which can be detected by present experimental methods.

6. Conclusions

It is obvious that the nuclear explosions provide the highest fluxes of neutrons under terrestrial conditions which will hardly be feasible with any other experimental technique in the near future. We have demonstrated that irradiation of nuclei with neutrons produced during the first microseconds of a nuclear explosion is a very promising way to synthesize neutron-rich heavy and super-heavy elements. For comparison, modern spallation sources (e.g. ESS) may supply thermal neutrons with densities around 10^{12} cm^{-3} , and nuclear reactors – with 10^{11} cm^{-3}. We believe, there exist technical possibilities to increase the time and intensity of target exposure in the course of a nuclear explosion. The main idea of optimization should be to maximize the neutron density and exposure time and to minimize the energy yield of the explosion.

An interesting possibility which should be considered in more details is generating two or several nuclear explosions within a time delay up to milliseconds in close proximity of each other (see talk by Valery Zagrebaev). This delay may be sufficient for very neutron-rich nuclei produced in the first explosion to undergo multiple β-decay. Then the daughter nuclei will be able to absorb additional neutrons from the second explosion, and in this way increase the resulting mass number, as illustrated in Fig. 4. Instead of nuclear explosions, one can in principle use other technical means, as e.g. pulsed reactors, if the integrated neutron flux during the pulse can be made high enough while the time intervals between the pulses can be reduced to a fraction of second. Even mass production of super-heavy elements can be envisaged in the future.

The methods discussed above can be used not only for synthesizing long-living superheavy elements but also for production of very neutron-rich isotopes in the vicinity of the drip-line. These studies are extremely important for understanding the origin of heavy elements in the Universe. Hot and dense environments created in nuclear explosions are rather similar to late stages of supernova explosions, when heavy nuclei are synthesized by multiple neutron capture in r-process.

If long-lived superheavy elements are indeed produced in the explosions, an important question to be answered is, how to find them among tons of debris. In underground explosions the processed material may be extracted and analysed by radio-chemical and mass-spectroscopic methods. Generally, the methods used for searching for superheavy elements in nature may be applied in this case too.

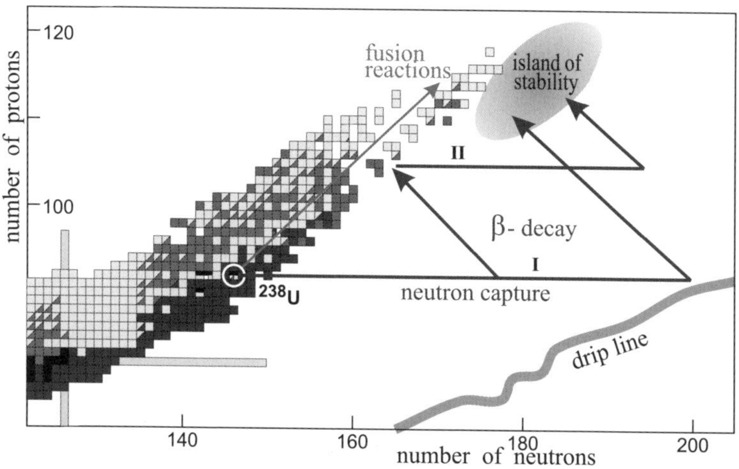

Fig. 4. (Colour online) The upper end of the nuclear chart showing the experimentally investigated isotopes. The colours stand for the decay modes: yellow - α-decay, red - β^+ or electron capture, blue - β^- decay, green - spontaneous fission. Black boxes are stable nuclei. Grey area is the predicted "island of stability". Red arrow represents the heavy-ion fusion reactions used so far. Blue arrows show schematically the proposed explosion-based methods to reach this island: I indicates fast capture of neutrons during first microseconds of the nuclear explosion followed by the β-decay. II indicates second nuclear explosion within few milliseconds after the first one.

This talk is based on joint publications with A. Botvina, W. Greiner and V. Zagrebaev. I thank them cordially and hope for the fruitful collaboration in the future.

References

1. W.R. Hix et al., Phys. Rev. Lett. **91** 201102 (2003).
2. K. Langanke and G. Martinez-Pinedo, Nucl. Phys. A **673**, 481 (2000).
3. C.J. Horowitz, Phys. Rev. D **55**, 4577 (1997).
4. H.A. Bethe, *Rev. Mod. Phys.* **62**, 801 (1990).
5. J.P. Bondorf, A.S. Botvina, A.S. Iljinov, I.N. Mishustin, K. Sneppen, Phys. Rep. **257**, 133 (1995).
6. A.S. Botvina et al., Nucl. Phys. A **584**, 737 (1995).
7. M. D'Agostino et al., Nucl. Phys. A **650**, 329 (1999).
8. R.P. Scharenberg et al., Phys. Rev. C **64**, 054602 (2001).

9. S. Hudan et al., Phys. Rev. C **67**, 064613 (2003)
10. A.S. Botvina and I.N. Mishustin, *Phys. Lett.* **B584**, 233 (2004).
11. A.S. Botvina and I.N. Mishustin, *Phys. Rev.* **C72**, 048801 (2005).
12. A.S. Botvina, I.N. Mishustin, Nucl. Phys. **A843**, 98 (2010).
13. C. Ishizuka, A. Ohnishi, K. Sumiyoshi, *Nucl.Phys.* **A723**, 517 (2003).
14. D.Q. Lamb, J.M. Lattimer, C.J. Pethick and D.G. Ravenhall, Nucl. Phys. A **360**, 459 (1981).
15. J.M. Lattimer, C.J. Pethick, D.G. Ravenhall and D.Q. Lamb, *Nucl. Phys.* **A432**, 646 (1985).
16. J.M. Lattimer and F.D. Swesty, *Nucl. Phys.* **A535**, 331 (1991).
17. A.S. Botvina, A.S. Iljinov, I.N. Mishustin, Sov. J. Nucl. Phys. **42**, 712 (1985).
18. A.S. Botvina, I.N. Mishustin, Phys. Rev. C **63**, 061601 (2001).
19. A.S. Botvina, O.V. Lozhkin and W. Trautmann, Phys. Rev. C **65**, 044610 (2002).
20. Alexander Botvina, , Igor Mishustin, Valeri Zagrebaev and Walter Greiner, Int. J. Mod. Phys. **19**, No. 10, 2063 (2010).
21. G.T. Seaborg, Ann. Rev. Nucl. Sci. **18**, 119 (1968).
22. A.S. Krivochatsky, Yu.F. Romanov, Production of transuranium and actinide elements by neutron irradiation (in Russian), 320 p., Atomizdat, Moscow, 1969.
23. H.W. Meldner, Phys. Rev. Lett. **28**, 975 (1972); H.W. Meldner, A. Cowan, J.R. Nix, and R.W. Stoughton, Phys. Rev. C **13**, 182 (1978).
24. H.A. Sandmeier, S.A. Dupree, G.E. Hansen, Nucl. Sci. Eng. **48**, 343 (1972).
25. George I. Bell, Phys. Rev. **139**, B 1207 (1965).
26. F. Dellinger, O. Forstner, R. Golser, W. Kutschera, A. Priller, P. Steier, A. Wallner, G. Winkler, Nucl. Instr. Meth. B **268**, 1287 (2010).

HOW COLLECTIVE FLOW OBSERVATIONS INDICATE THE QGP FLUID?

L.P. CSERNAI[1,2], Y. CHENG[1], Sz. HORVAT[1], V.K. MAGAS[3], I.N. MISHUSTIN[4], B.R. SCHLEI[5], D.D. STROTTMAN[4,6], and S. ZSCHOCKE[7]

[1] *Department of Physics and Technology, University of Bergen, Bergen, Norway*
** E-mail: csernai@ift.uib.no*

[2] *MTA–KFKI Research Inst. for Particle Physics and Nuclear Physics, 1525 Budapest, Hungary*

[3] *Departament d'Estructura i Constituents de la Matèria, Universitat de Barcelona, Diagonal 647, 08028 Barcelona, Spain*

[4] *Frankfurt Institute for Advanced Studies (FIAS), Ruth-Moufang-Str. 1, D-60438 Frankfurt am Main, Germany*

[5] *GSI Helmholtz Centre for Heavy Ion Research GmbH, Planckstraße 1, 64291 Darmstadt, Germany*

[6] *Dep. de Fisica Teorica and IFIC, Centro Mixto Universidad de Valencia and CSIC, Institutos de Investigación de Paterna, Aptd. 22085, 46071 Valencia, Spain*

[7] *TU Dresden, Institut für Planetare Geodäsie, Lohrmann-Observatorium, D-01062 Dresden, Germany*

The existence of collective dynamics and shock waves in nuclear collisions were predicted in the early '70s by Greiner and colleagues. This was a revolutionary prediction, and it took a decade to prove it beyond doubt. Now at LHC this is the most dominant observable and the first heavy ion result from ALICE.

I would discuss some of the details of the recent advances in the field.

The Constituent Quark Number Scaling of the v_2, for different mesons and baryons is instinctively a proof for collective flow development in quark gluon plasma. Hybrid models with recombination into hadrons can account for the observed scaling, in a limited p_t-range. However, the scaling is not obvious. We follow the rapid and simultaneous hadronization of QGP dynamically, where partons gain weight and the perturbative vacuum field disappears. This dynamical process influences the flow observables and can reproduce the scaling. These studies may provide an insight into the features and dynamics of this intermediate phase between ideal QGP and Hadronic Matter.

PACS number(s): 25.75.Nq, 24.85.+p, 24.10.Nz, 24.10.Pa, 21.65.-f

1. The Beginning of High Energy Flow Studies

In the 1970s the possibility to accelerate heavy nuclei to several several hundred MeV/nucleon energy opened with exciting expectations. One of the the most important of these early ideas is in W. Scheid, H. Müller, and W. Greiner's paper in 1974 in Phys. Rev. Lett.[1] Here the compression of nuclear matter in shock waves was first suggested. The preceding works by Scheid, Ligensa and Greiner discussed nuclear compressibility in such collisions, while in another early subsequent work of H. Baumgardt, *et al.*[2] possible experimental consequences were already discussed.

Still the idea of the strongly coupled collective dynamics was not accepted immediately, and many were opposing these views. It took about a decade until the first real unambiguous result was achieved by Prof. Hans Gutbrod's group: H. Å. Gustafsson, *et al.*[4] This group constructed a large new detector in Berkeley "The Plastic Ball" and this helped them to achieve undebatable results.

Then the collective flow became one of the most widely accepted and used methods to analyse the properties of matter at these high energy densities. New states of matter were discovered by 2000-2001 at the CERN SPS and the BNL RHIC, the Quark-Gluon Plasma (QGP). Just two weeks before this talk the first experimental preprint from the LHC heavy ion programme appeared and the topic was the asymmetry of the collective flow, the v_2 flow parameter, which was found to be bigger than in any other previous experiment.

Although there were expectations for large transparency and weakening flow, this result was not completely unexpected, as with increasing energy density and increasing particle multiplicity the collective processes are also getting stronger, and we get better and better insight into the properties of matter under extreme conditions, similar to those in the early universe.

Recent experiments also indicate that the collective flow is primarily developed in the QGP state. However, the final hadronization and freeze out does not make the direct study of the QGP very easy. We need a precise and detailed theoretical description of all stages of a reaction to be able to extract all possible information from these extremely energetic reactions.

2. The Collective Flow in the QGP Era

The observed Constituent Quark Number Scaling of the flow parameter, v_2, for different mesons and baryons is instinctively a proof for collective flow development in quark gluon plasma. Fluid dynamical models assuming phase and chemical equilibrium during all stages of the reaction cannot account for the observed scaling in general, although special assumptions, like the assumption of constant freeze out temperature, may lead to similar results. Hybrid models with recombination into hadrons are more successful as these models incorporate the description of final out-of-equilibrium processes, and so in a limited p_t-range these models reproduce the data rather well. Here we try to gain more insight into the principles which may govern these processes.

For our description we start out from an initial state based on the initially known data from the reaction and on simple theoretical assumptions. We do not fit our initial state parameters to the measured and observed final state data. The initial state is constructed on a longitudinal flux-tube model, without any transverse motion.[5]

The subsequent explosive fluid dynamical expansion is described by an upgraded Particle in Cell (PIC) numerical model assuming equilibrated QGP, described by the MIT Bag Model equation of state and perfect fluid. See Figure 1. The whole fluid dynamical regime of the process is assumed to take place in the QGP phase. The hadronization happens simultaneously to the final freeze out in a rapid process. Although, the model calculation represents a perfect fluid described by the Relativistic Euler equations, the numerical model still includes a small dissipation, due to the coarse graining by the finite cell size of our calculation. The dissipation caused by the "numerical viscosity" was analysed recently.[6] It shows about 6% entropy increase during the expansion, which approximately equals the results obtained for the minimal viscosity values expected at the phase transition boundary between QGP and Hadronic Matter.[7]

3. Matching Different Stages with Conservation Laws

Different modules are used to describe the stages of the reaction, from an initial state (characterized by flux-tube model), to thermal equilibrated QGP (described by hydrodynamics), and then the final freeze out with simultaneous hadronization.[8] The general transition through a space-time hypersurface (with the normal, $d\sigma_\mu$) must satisfy the conservation laws:

$$[N^\mu d\sigma_\mu] = 0 \quad , \tag{1}$$

$$[T^{\mu\nu} d\sigma_\mu] = 0 \quad , \tag{2}$$

$$[S^\mu d\sigma_\mu] \geq 0 \quad , \tag{3}$$

where the square bracket indicates the change of a quantity across the hypersurface. In most cases we may assume that the energy-momentum

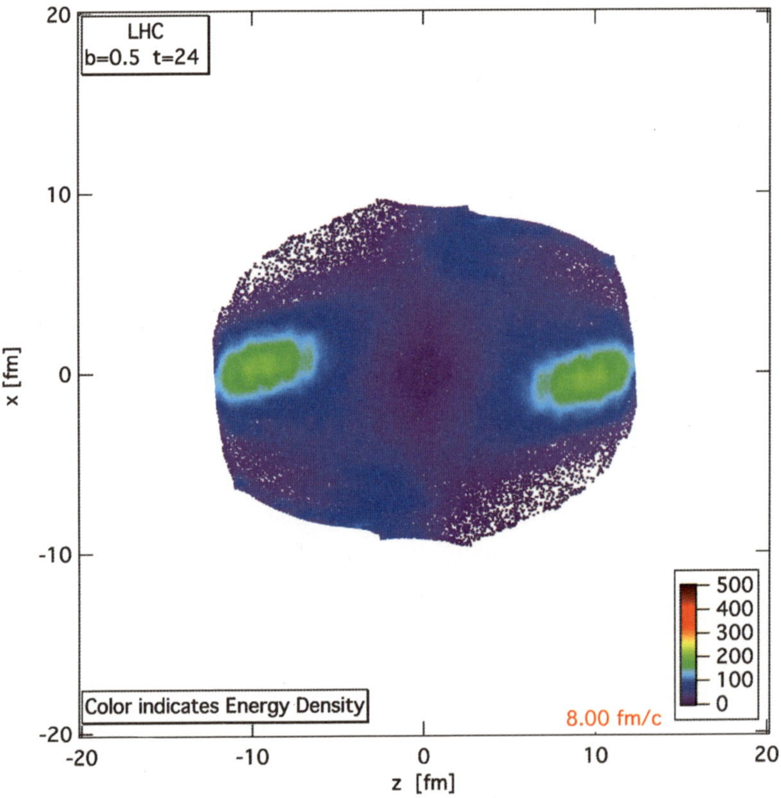

Fig. 1. The result of a Relativistic Fluid Dynamical Calculation with the improved PIC method for the recent LHC Pb + Pb reaction at 1.38 + 1.38 A·TeV energy and impact parameter of $b = 0.5b_{max}$. The size of the grid-cells are $(0.585 \text{ fm})^3$ and at the initial state each grid-cell contained 3^3 - 8^3 Lagrangian fluid cells, called Marker particles, depending on the required fine accuracy. The time-step in the calculation was $dt = 0.043$ fm/c. The calculation is performed in the CM frame. The figure shows the status of the explosive expansion 8.00fm/c after the initial state of longitudinally expanding flux-tubes. The transverse flow in the x-direction is clearly visible.

tensor is of the form: $T^{\mu\nu} = (e + P)u^\mu u^\nu - Pg^{\mu\nu}$ is the representing a perfect fluid[9] in terms of the energy dentity, e, pressure, P, and flow 4-velocity, u^μ. Then, the conserved baryon current j and energy momentum current A^μ crossing the surface element are:

$$j \;= N^\mu d\sigma_\mu \;\; = n\, u^\mu d\sigma_\mu, \tag{4}$$

$$A^\mu \;= T^{\mu\nu} d\sigma_\nu \;\; = w\, u^\mu u^\nu\, d\sigma_\nu - P\, g^{\mu\nu}\, d\sigma_\nu. \tag{5}$$

We can construct two invariant scalar equations by (i) taking the norm, $A^\mu A_\mu$, and (ii) taking the projection to the normal direction, $A^\mu d\sigma_\mu$. These lead to the expressions:

$$\begin{aligned} A^\mu A_\mu &= w^2(u^\mu d\sigma_\mu)^2 + P^2(d\sigma^\mu d\sigma_\mu) - 2Pw(u^\mu d\sigma_\mu)u_\mu g^{\mu\nu} d\sigma_\nu \\ &= w(e-P)(u^\mu d\sigma_\mu)^2 + P^2(d\sigma^\mu d\sigma_\mu)\ , \end{aligned} \tag{6}$$

$$A^\mu d\sigma_\mu = w(u^\mu d\sigma_\mu)^2 - P(d\sigma^\mu d\sigma_\mu)\ . \tag{7}$$

Eliminating the term $w(u^\mu d\sigma_\mu)^2$ from Eqs. (6) and (7), we obtain

$$A^\mu A_\mu = (e-P)A^\mu d\sigma_\mu + e\,P\,(d\sigma^\mu d\sigma_\mu)\ . \tag{8}$$

The above equation can be solved directly if the EoS, $P = P(e)$, is known. In case of finite baryon current equation (4) can be used with Eq. (7), and we obtain:

$$\frac{e+P}{n^2} = \frac{1}{j^2}\ [A^\mu d\sigma_\mu + P(d\sigma^\mu d\sigma_\mu)]\,. \tag{9}$$

Using these results, Eqs. (8) and (9), simple solutions can be found for the transition if the invariant scalars, $A^\mu d\sigma_\mu$ and $d\sigma^\mu d\sigma_\mu$, as well as the baryon current, j, are known in the initial state, together with the final EoS, $P = P(e,n)$. Thus the final quantities can be evaluated by this general solutions, and this method is applicable in matching different stages through a transition surface.

4. The Hybrid Model

In the above described way we can connect the hydrodynamical model with the parton cascade model (PACIAE), transferring the conserved quantities of the fluid, cell by cell to the created partons, satisfying the conservation laws based on the assumption of the thermal equilibrium before and after the transition. The produced partons will be the initialization of the PACIAE model, which is a parton and hadron cascade model based on PYTHIA.[10,11] The subsequent parton evolution, can be simulated by a Monte Carlo method until the parton-parton collisions are ceased, and the partons reach hadronization at the moment of FO. After hadronization, via recombination into the final hadrons, we obtain a configuration of hadrons in spatial and momentum coordinates.

To implement the connection of two models, the hydro code, assuming a QGP fluid, under went a transition at $t = Const.$ Then we evaluated the post transition parton distribution, using the Cooper-Frye formula, generating the partons randomly in local equilibrium in each fluid cell. Their distributions are constrained by the conservation laws in each cell. In the post

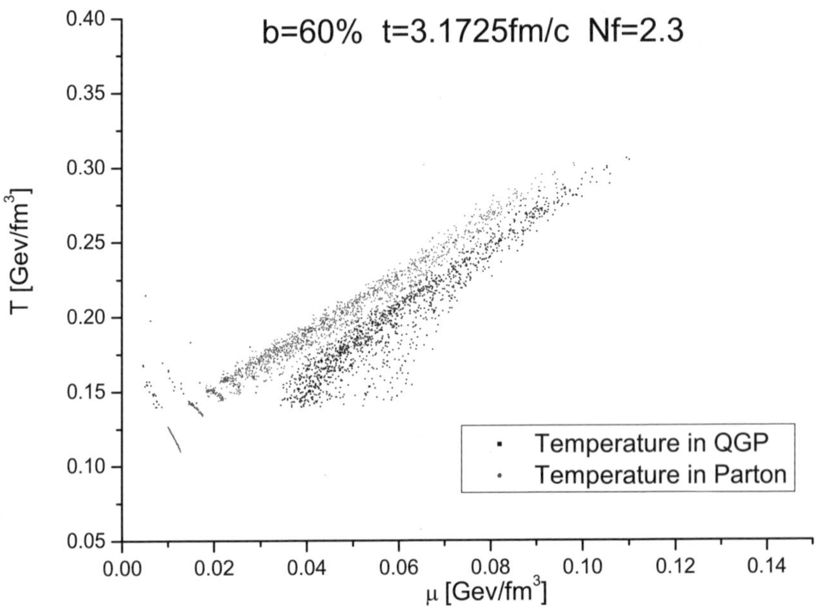

Fig. 2. Transition from QGP fluid to Parton gas phase is shown in a $Au + Au$ reaction at $65 + 65$ A·GeV energy. The final parton distribution (red dots) lies above the QGP distribution (black dots). The (T, μ) distributions are shown at the transition time-step, 3.175 fm/c, for impact parameter $b = 60\%$ of b_{max}, with the effective number of final flavours $N_F = 2.3$.

transition stage, particles in partonic matter include quarks, anti-quarks and massless gluons, so the created partons, u, d, s, can be obtained. The energy and 4-momenta of partons are generated from the fluid cells, just as the baryon number density, satisfying conservation laws. The change of temperature and chemical potential at the transition for each fluid cell is shown in Fig. 2. The parton distribution lies above the QGP distribution, indicating that the average temperature increases in the transition. This is due mainly to the Bag energy transfer from QGP phase to parton phase. At low temperatures and low chemical potentials, the final parton distribution shows a quantized structure. This arises from the PIC method, where the number of the marker particles in a surface fluid cell becomes low, one, two or three, quantizing the distribution of the final parton gas from dilute fluid cells.

This post transition configuration is the first stage of the partonic cascade in the hybrid model (Hydro-PACIAE). Fig. 3 shows the final p_t distribution of charged hadrons, which is almost consistent with the STAR data in central collision. In the intermediate transverse momentum range, the theoretical calculation yields higher values than the experimental data. This may be due to the initial strong flow effects, which lead to final particles flying apart with larger momentum.

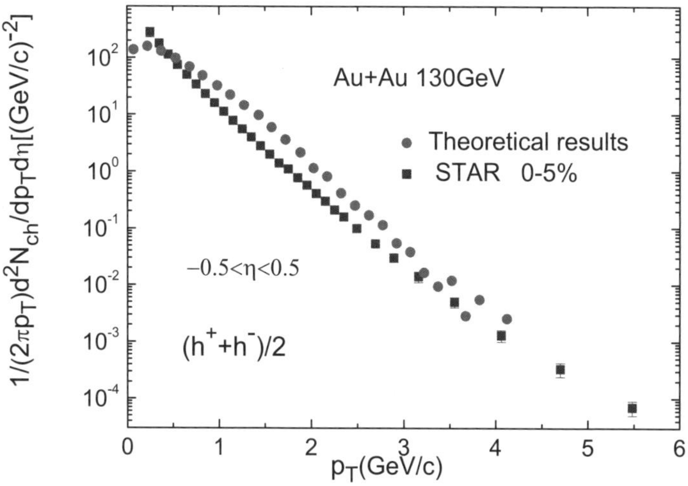

Fig. 3. In the reaction presented in the previous Figure, the p_T distribution from the Hydro-PACIAE model calculation, and the corresponding STAR data in central collision.

5. Final Stages and Measurables

The selection of final freeze out arises self-consistently in a hybrid model, although the choice of the initial state and the transition hypersurface are important.

We may have more explicit insight in simplified freeze out and hadronization models, which may explicitly include non-equilibrium chemical and kinetic freeze out, as well as the study of the hypersurface where this final freeze out and hadronization takes place.

We use VESTA (Volume-Enclosing Surface exTraction Algorithm) to generate temporal projections of freezeout hypersurfaces (FOHS) from the hydrodynamic simulation data.[8,12] For the particular case of 2+1D (i.e., two spatial and one temporal dimensions), VESTA not just only yields visual renderings of FOHS, but it also provides the enabling input to further numerical evaluations for the generation of multi-particle production spectra (*cf.*, Ref.[12] and Refs. therein). In the general four dimensional space-time, VESTA is applied to sequences of 3D spatial simulation data, where each of the 3D data are generated for a fixed time. Then, several isosurfaces are rendered from a set of 3D (spatial) simulation data, and hence allow for the making of animated 3D views, i.e., digital movies. The full 4D FOHS, however, which are neccessary for the proper subsequent numerical evaluations, are generated by us using the STEVE (Space-Time-Enclosing Volume Extraction) algorithm.[13] The choice of this hypersurface has a strong influence on the measurable quantities. E.g. direct choice of fixed temperature freeze out leads to special artefacts, or late t=const. type of surface leads to a central rapidity dip in the $v_2(y)$, arising from central dilute cells, which should have been frozen out earlier.

The constituent quark number scaling of elliptic flow is also studied in a non-equilibrium hadronization and freeze-out model with rapid dynamical transition from ideal, deconfined and chirally symmetric Quark Gluon Plasma, to final noninteracting hadrons.[14] In this transition a Bag model of constituent quarks is considered, where the quarks gain constituent quark mass while the background Bag-field breaks up and vanishes. The constituent quarks then recombine into hadrons, while chemical, thermal and flow equilibrium break down one after the other. The resulting temperatures and flow velocities of baryons and mesons are different. Using a simplified few source model of the elliptic flow, we are able to reproduce the constituent quark number scaling, with assumptions on the details of the non-equilibrium processes.

References

1. W. Scheid, H. Müller and W. Greiner, *Phys. Rev. Lett* **32** (1974) 741; W. Scheid, R. Ligensa and W. Greiner, *Phys. Rev. Lett.* **21** (1968) 1479, and W. Scheid and W. Greiner, *Z. Phys.* **226** (1969) 364.
2. H. Baumgardt, *et al.*, *Z. Phys.* **A273** (1975) 359.
3. G.F. Chapline, M.H. Johnson, E. Teller, and M.S. Weiss, *Phys. Rev.* **D8** (1973) 135.
4. H. Å. Gustafsson, H.H. Gutbrod, et al., *Phys. Rev. Lett.* **52** (1984) 1590.
5. V.K. Magas, L.P. Csernai, D.D. Strottman, *Phys. Rev.* **C64** (2001), 014901, and *Nucl. Phys.* **A 712** (2002) 167-204.
6. Sz. Horvát, V.K. Magas, D.D. Strottman, L.P. Csernai, *Phys. Lett.* **B692** (2010) 277-280.
7. Laszlo P. Csernai, Joseph I. Kapusta, and Larry D. McLerran *Phys. Rev. Lett.* **97** (2006) 152303.

8. Yun Cheng, L. P. Csernai, V. K. Magas, B. R. Schlei, and D. Strottman, "Matching Stages of Heavy-Ion Collision Models," Phys. Rev. **C81** (2010) 064910.
9. L. P. Csernai: Introduction to Relativistic Heavy Ion Collisions (Wiley, New York, 1994).
10. Ben-Hao Sa, Xiao-Mei Li, Shou-Yang Hu, Shou-Ping Li, Jing Feng, and Dai-Mei Zhou. Phys. Rev. **C 75**, (2007) 054912.
11. T. Sjöstrand, Comput. Phys. Commun. **82** (1994) 74.
12. B. R. Schlei, "Volume-Enclosing Surface Extraction," arXiv:1011.1787.
13. B. R. Schlei, "Hyper-Surface Extraction in Four Dimensions," Theoretical Division - Self Assessment, Special Feature, a portion of LA-UR-04-2143, Los Alamos (2004) 168.
14. L.P. Csernai, S. Zschocke, Sz. Horvat, Yun Cheng, and I.N. Mishustin; in Proceedings: "(QCHS09) The IX International Conference on Quark Confinement and Hadron Spectrum - Universidad Complutense, **Madrid**, Aug. 30 - Sep. 3, 2010"; AIP Conference Proceedings, in press.

COLLECTIVE OSCILLATIONS AND DIFFUSE SUPERNOVA NEUTRINO BACKGROUND

KAMALES KAR* and SOVAN CHAKRABORTY**

Saha Institute of Nuclear Physics,
1/AF Bidhanagar, Kolkata 700064, India
**E-mail: kamales.kar@saha.ac.in*
***E-mail: sovan.chakraborty@saha.ac.in*

SANDHYA CHOUBEY

Harish Chandra Research Institute,
Chhatnag Road, Jhushi, Allahabad 211019, India
E-mail: sandhya@hri.res.in

Core-collapse supernova explosions give rise to the emission of a huge flux of neutrinos of all flavors. In this article we describe the phenomenon neutrino-neutrino interaction of these weakly interacting particles at the very high density central region of the stellar core giving rise to non-linear collective oscillations in both the neutrino and antineutrino sectors. The effect of the collective oscillations on the Diffuse Supernova Neutrino Background is elaborated with emphasis on its future detection and the connection of that to neutrino mass hierarchy.

Keywords: Supernova explosions; collective neutrino oscillations; diffuse supernova neutrino background.

1. Introduction

Neutrinos are weakly interacting particles but they play a crucial role in the explosions of core-collapse supernovae (SN). Electron type neutrinos are produced in the presupernova stage or the early stage of stellar collapse of large stars with masses greater than $8M_\odot$. But a much larger number of neutrinos and antineutrinos of all three flavors are pair produced in the central high density region of these stars and they mostly stream out of the star carrying almost all the energy released in the stellar collapse (typically a few times 10^{53} ergs). A small fraction of them, however, get absorbed in the matter near the shock wave and this delayed neutrino heating of the shock region is believed to revive the stalled shock after the shock loses energy due to nuclear dissociation of the matter. This revived shock eventually reaches the edge of the core and causes explosion.

The neutrinos of all flavors cross a huge density gradient when they move out of a core-collapse supernova. Thus the vacuum oscillations[1] of the neutrinos get modified by the matter-induced oscillation or MSW effect.[2,3] But on top of that, at the very high density region in the centre of the stellar core, the neutrino-neutrino interaction gives rise to a non-linear collective oscillation.[4] In this article we shall discuss the effects of the collective oscillation and see how that can modify the diffuse supernova neutrino background (DSNB) i.e. the background of neutrinos formed from all supernova explosions in the universe in the past.[5] We shall in particular look at the problem of the possibility of distinguishing the normal neutrino mass hierarchy from the inverted one from the future observation of DSNB neutrinos.

2. Core Collapse Supernovae and Emission of Neutrinos

Stars with masses larger than $8M_\odot$ go through the different stages of nuclear burning. But once the core of the star produces iron type nuclei after silicon burning, these nuclear reactions stop and the gravitational force helped by deleptonisation of the core by electron capture, makes the star collapse. Upto a density of $\sim 10^{11} g/cc$ neutrinos radiate out of the core and the pressure inside is determined by the relativistic degenerate electron gas. However at higher densities the neutrinos start getting trapped as their mean free path becomes comparable to the radius of the core and eventually the lepton fraction becomes a constant. When the centre of the core attains a density larger than the nuclear matter density, pressure waves start travelling outward.[6,7] These waves initially stop at the sonic point but eventually give rise to a shock wave moving outward. This shock wave fast loses its energy in dissociating nuclei into protons and neutrons, becomes stagnant and then over the timescale of a few seconds gets revived by the mechanism of delayed neutrino heating.

At the late stages of the evolution the central region of the core has a very large number of neutrinos and antineutrinos of all three flavors. A small fraction of these neutrinos gets absorbed by the matter ahead of the shock and the rest comes out of the star in huge numbers. SN1987A was the first and only supernova that had an almost simultaneous detection of these $\nu/\bar{\nu}$ s in terrestrial detectors along with the optical fireworks. However the present day neutrino detectors are well-prepared to detect the SN neutrinos in future nearby explosions and extract the properties of the neutrino not yet known.

3. Collective Neutrino Oscillations for Supernova Neutrinos

In vacuum neutrino oscillations one describes the transformation from the three flavor basis of $|\nu_e >$, $|\nu_\mu >$ and $|\nu_\tau >$ to mass eigenstate basis of $|\nu_1 >$, $|\nu_2 >$ and $|\nu_3 >$ which have mass eigenvalues of m_1, m_2 and m_3 respectively. The oscillation in the 1-2 sector gives rise to the solar neutrino deficit problem and in the 2-3 sector it results in the atmospheric neutrino anomaly. The mass square difference Δ_{12} and Δ_{23} are the important parameters (along with the oscillation angles, θ_{12}, θ_{23} and θ_{13},which define the transformation operator from the flavor basis to the mass basis) and are estimated from the measurements done at the solar and atmospheric neutrino detectors. Here Δ_{ij} is defined as $\Delta_{ij} = m_j^2 - m_i^2$. The sign of Δ_{23} is still not known though it is observed that $\Delta_{12} << |\Delta_{23}|$; the scenario with a positive Δ_{13} is called the Normal Hierarchy (NH) and the one with a negative value is known as the Inverted Hierarchy (IH).

The neutrino Hamiltonian (H_p) at the very high density is given by

$$H_p = H_{vacuum} + H_{MSW} + H_{\nu\nu}. \tag{1}$$

The density matrices in flavor space are written as

$$\rho_p = \sum |\psi_i\rangle\langle\psi_i| \ \ , \ \ \bar{\rho}_p = \sum |\bar{\psi}_i\rangle\langle\bar{\psi}_i| \tag{2}$$

with the evolution equations given by

$$\partial_t \rho_p = -i[H_p, \rho_p] \ \ , \ \ \partial_t \bar{\rho}_p = -i[\bar{H}_p, \bar{\rho}_p]. \tag{3}$$

In equation (1) H_{vacuum} and H_{MSW} are the usual vacuum and matter induced terms in the Hamiltonian. In equation (3) the antineutrino Hamiltonian ($\bar{H}_p$) has exactly the same form as equation (1) with the only change that the vacuum term, H_{vacuum}, picks up a negative sign. The contribution from $\nu\nu$ forward scattering to the leading order of G_F is given by

$$H_{\nu\nu} = \sqrt{2}G_F \int (d^3\vec{q}/(2\pi)^3)(1 - cos\theta_{pq})(n\rho_q - \bar{n}\bar{\rho}_q) \tag{4}$$

where n and $\bar{n}$ are the number densities of neutrinos and antineutrinos respectively and $cos\theta_{pq}$ is the angle between neutrino momenta $\vec{p}$ and $\vec{q}$. This under the 'single Angle Approximation'[8] i.e. assuming that flavor evolution is trajectory independent and neutrinos in any trajectory transform like ν's propagating in radial direction, becomes

$$H_{\nu\nu} = \sqrt{2}G_F D(r) \int q^2 dq(n\rho_q - \bar{n}\bar{\rho}_q) \tag{5}$$

where the geometrical factor $D(r)$ for a launching angle of zero degree, is

$$D(r) = \frac{1}{2}\left(1 - \sqrt{1 - \left(\frac{R_\nu}{r}\right)^2}\right)^2 . \tag{6}$$

In core collapse supernovae the matter density at the centre reaches values $\sim 10^{15}$ g/cc and as a result the neutrinos go through a large density variation while moving out. The collective oscillation normally is over within the first 200 kms whereas the matter induced MSW oscillations take place around radii $10^4 - 10^5$ kms. As they are widely separated spatially we in our calculations assume the two effects independent of each other. One has also seen[9,10] that for shallow neutrino-neutrino interaction potentials the collective oscillation involves the evolution of the flavors (ν_e, ν_x) where the third flavor ν_y does not take part (here ν_x and ν_y are a linear combination of flavors ν_μ, ν_τ and orthogonal to each other). Our calculations of collective oscillations are done in this effective two flavor basis. The six species of neutrinos ν_e, ν_μ, ν_τ and their antiparticles stream out of the respective neutrinospheres. Our calculation use typical energy distribution of neutrino number fluxes of the form

$$F^0_\nu(E) = \left(\frac{L^0_\nu}{\bar{E}}\right) \times \left(\frac{(1+\alpha)^{1+\alpha}}{\Gamma(1+\alpha)\bar{E}}\left(\frac{E}{\bar{E}}\right)^\alpha e^{-(1+\alpha)E/\bar{E}}\right) = \Phi^0_\nu \times \psi(E) \tag{7}$$

where $\psi(E)$ in the above equation stands for the typical energy distributions used in simulations;[11] the total number flux $\Phi^0_\nu = L^0_\nu/\bar{E}$ where L^0_ν and $\bar{E}$ are the luminosity and average energy respectively. The average energies used are $\bar{E}_{\nu_e}= 12$ MeV, $\bar{E}_{\bar{\nu}_e}= 15$ MeV, $\bar{E}_{\nu_x}= \bar{E}_{\nu_y}=18$ MeV with $\alpha_{\nu_x}=\alpha_{\bar{\nu}_x} = \alpha_{\nu_y}=\alpha_{\bar{\nu}_y} = 4$ and $\alpha_{\nu_e}=\alpha_{\bar{\nu}_e} = 3$. One also takes a total of 3×10^{53} ergs of energy coming out in neutrinos and antineutrinos of all flavors. These initial neutrino fluxes in different flavors change first due to collective oscillations and then through the MSW process. The final fluxes that emerge are given by[12,13]

$$F_{\nu_e} = s^2_{12}P_c(\phi^r_{\nu_e}, \phi^r_{\bar{\nu}_e}, E)(2P_{13}-1)(F^0_{\nu_e} - F^0_{\nu_x}) + s^2_{12}(1-P_{13})(F^0_{\nu_e} - F^0_{\nu_x}) + F^0_{\nu_x} \tag{8}$$

$$F_{\bar{\nu}_e} = c^2_{12}\bar{P}_c(\phi^r_{\nu_e}, \phi^r_{\bar{\nu}_e}, E)(F^0_{\bar{\nu}_e} - F^0_{\nu_x}) + F^0_{\nu_x} \tag{9}$$

$$F_{\nu_e} = s^2_{12}P_c(\phi^r_{\nu_e}, \phi^r_{\bar{\nu}_e}, E)(F^0_{\nu_e} - F^0_{\nu_x}) + F^0_{\nu_x} \tag{10}$$

$$F_{\bar{\nu}_e} = c_{12}^2 \bar{P}_c(\phi^r_{\nu_e}, \phi^r_{\bar{\nu}_e}, E)(2P_{13}-1)(F^0_{\bar{\nu}_e} - F^0_{\nu_x}) + c_{12}^2(1-P_{13})(F^0_{\bar{\nu}_e} - F^0_{\nu_x}) + F^0_{\nu_x}. \quad (11)$$

In the above equations c_{12} and s_{12} stand for $cos(\theta_{12})$ and $sin(\theta_{12})$ and P_{13} is the effective jump probability between the neutrino mass eigenstates due to the atmospheric mass squared driven MSW resonance. For θ_{13} large ($sin^2\theta_{13} \geq 0.01$) P_{13}=0, while for smaller values ($sin^2\theta_{13} < 0.01$) P_{13}=1.0. $P_c(\phi^r_{\nu_e}, \phi^r_{\bar{\nu}_e}, E)$ and $\bar{P}_c(\phi^r_{\nu_e}, \phi^r_{\bar{\nu}_e}, E)$ are the neutrino and antineutrino survival probabilities after the collective effects. Recently one has seen that the oscillation probabilities after the collective oscillations depend very sensitively on the neutrino fluxes in different flavors or, in other words, on how the total energy gets divided into the six species.[10,12,14] Normally one assumes an equipartition of energy but there is no deeper reason to believe that this is true in such a turbulent system. This is studied by defining two relative fluxes $a = \Phi^o_{\nu_e}/\Phi^o_{\nu_x}$ and $b = \Phi^o_{\bar{\nu}_e}/\Phi^o_{\nu_x}$. The numerical result for the variation of the conversion probabilities for both neutrinos and antineutrinos are described in detail in Dasgupta et al.[14] and Chakraborty et al.[12] We briefly state here the findings. In the two flavor scenario we study, for the Inverted Hierarchy (IH) one sees that a above a critical energy (split energy E_c), the spectra in both the electron neutrino(ν_e) and antineutrino($\bar{\nu}_e$) sectors end with a complete exchange or swap with ν_x and $\bar{\nu}_x$ respectively. This is known as the spectral swap. On the other hand the Normal Hierarchy (NH) is found to be more stable.[8,15–17] Single spectral split has also been reported for certain combination of luminosities.[12] The implications of this collective oscillation of supernova neutrinos for r-process nucleosynthesis has recently been studied and exclusion plots for the combinations of the relative initial fluxes were obtained for allowed r-process.[12,18] However in the rest of the article we describe the effect of these spectral splits and their dependence on the relative neutrino fluxes for Diffuse Supernova Neutrino Background (DSNB).

4. Diffuse Supernova Neutrino Background

There were efforts to detect the neutrinos forming DSNB and through their non-observation an upper limit on DSNB has been found. In the energy window higher than 19.3 MeV Superkamiokande found this upper limit as 1.2 $\bar{\nu}_e$ cm$^{-2}s^{-1}$ in a water Cerenkov detector[19] and for neutrinos in the window of 25 MeV and 50 MeV the Liquid Scintillation Detector (LSD)[20] found this as 6.8×10^3 ν cm^{-2} s^{-1} both at 90% CL. Let us now try to calculate the flux of neutrinos or antineutrinos in DSNB as a function of energy. This is given as

$$F_\nu'(E_\nu) = \frac{c}{H_0} \int_0^{z_{max}} R_{SN}(z)\, F_\nu(E) \frac{dz}{\sqrt{\Omega_m(1+z)^3 + \Omega_\Lambda}} \,. \tag{12}$$

Here $E_\nu = E/(1+z)$ with E_ν the redshifted energy observed at the detector while E is the energy of the neutrinos at the source. In the above equation H is the Hubble constant, $F_\nu(E)$ gives the flux for each core collapse SN with energy E and $R_{SN}(z)$ is the cosmic SN rate at redshift z. For the standard $\Lambda - CDM$ cosmology, one has $\Omega_m = 0.27$ and $\Omega_\Lambda = 0.73$.[21] The integration is carried out to a maximum redshift of z_{max} that can contribute to DSNB.

Using equation (12) one[22] calculated the number of expected events in **a)** Super-Kamiokande with a fiducial mass of 22.5 kton **b)** Super-Kamiokande with Gadolinium added to water for a more efficient detection **c)** 1 Mton future water detector like Hyper-Kamiokande and **d)** 50 kton scintillation detector LENA for DSNB antineutrinos, for different values of the jump probability P_{13}. This work also predicted the number of events in a 100 kton liquid argon detector for the DSNB neutrino events. The calculations were done for different neutrino/antineutrino spectra as given by the Lawrence Livermore (LL) and Garching (G1, G2) simulations.[11] The finding was that the future generation detectors would be able observe DSNB fluxes and in some cases of small θ_{13} may even be able to distinguish the mass hierarchies.

However all the earlier calculations assumed fixed flux ratios for the emitted neutrinos in each supernova explosion or in other words the total numbers of neutrinos in each flavor was assumed fixed for each case. But it is unlikely that the DSNB flux coming from all past supernovae with different progenitor masses and conditions will have the same flux ratios for each event. A recent work[13] took a more realistic scenario of two distributions for the relative neutrino fluxes 'a' and 'b' defined earlier. These are assumed to be given by two log-normal distributions around their expectation values. It is not expected that in the next few decades one would be able to get observational input on the form of the distribution of the number of supernovae with the relative neutrino fluxes 'a' and 'b'. So to test the dependence on the form of the number distribution, four different distributions were used.[13] The distribution of the parameter 'a' is taken to be $D_\nu(\mu_1, \sigma_1)$ and the parameter 'b' is taken to be $D_{\bar{\nu}}(\mu_2, \sigma_2)$, given by

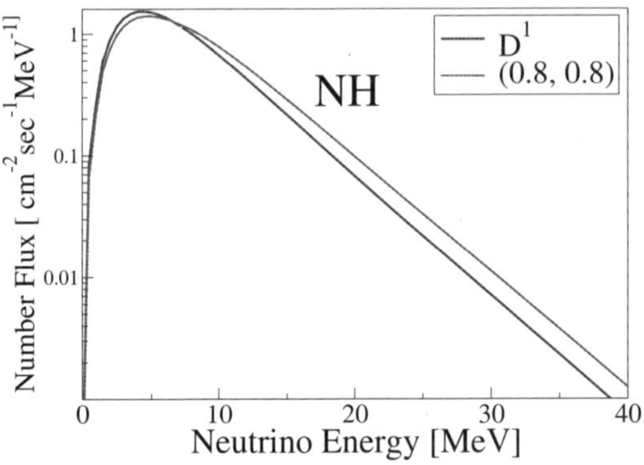

Fig. 1. The $\bar{\nu}_e$ flux for Normal Hierachy for small θ_{13} for the distribution D^1 and with (a,b)=(0.8,0.8). The lower blue curve is for D^1 and the upper green curve is for (0.8,0.8).

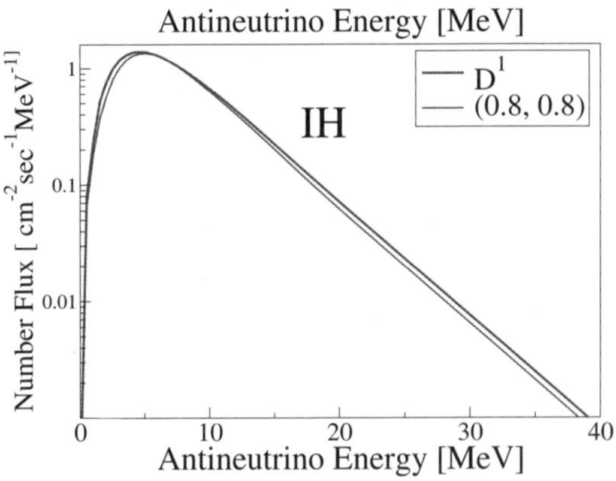

Fig. 2. The $\bar{\nu}_e$ flux for Inverted Hierachy for small θ_{13} for the distribution D^1 and (a,b)=(0.8,0.8). The upper blue curve is for D^1 and the lower violet curve is for (0.8,0.8).

$$D_\nu(\mu_1, \sigma_1) = \frac{e^{-(\frac{log(a)-\mu_1}{2\pi\sigma_1})^2}}{\sqrt{2\pi}\sigma_1 a}; D_{\bar{\nu}}(\mu_2, \sigma_2) = \frac{e^{-(\frac{log(b)-\mu_2}{2\pi\sigma_2})^2}}{\sqrt{2\pi}\sigma_2 b}. \tag{13}$$

i) The distribution for D^1_ν and $D^1_{\bar{\nu}}$ are taken with the equipartition expectation values 1.5 and 1.2 and the variations are within 2-fold uncertainty i.e. within a factor of 1/2 and 2 ii) The distribution for D^2_ν and $D^2_{\bar{\nu}}$ have the same expectation value of 1.0 and the variations are again within the 2-fold uncertainty iii) The distribution for D^3_ν and $D^3_{\bar{\nu}}$ have the same expectation value of 0.8 given by Garching simulations and the variation is again within the 2-fold uncertainty iv) Normalised constant distribution for D^4_ν between 0.45-2.55 and $D^4_{\bar{\nu}}$ between 0.45-2.05.

While varying the the flux ratios one encounters abrupt changes in the collective oscillation probabilities and so the number of events averaged over different distributions are expected to differ from each other and from the case of constant flux ratios. Fig. 1 gives the result of a calculation for the antineutrino flux for NH with the distribution $D^1_{\bar{\nu}}$ compared to fixed (a,b)=(0.8,0.8). Fig. 2 gives the same antineutrino flux for IH again with the distribution $D^1_{\bar{\nu}}$. Finally Fig. 3 gives the number of the positrons re-

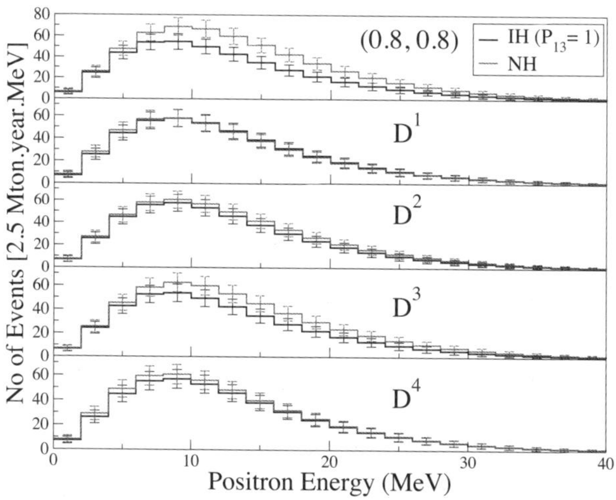

Fig. 3. Number of expected events per year in 2 MeV positron energy bins in SK. The solid black lines show the projected event spectrum for NH while the dashed red lines are IH with $P_{13} = 1$. The errors shown are statistical errors only.

leased by DSNB antineutrinos in SK-like detectors in 2 MeV energy bins for all four distributions compared to the case with (a,b)=(0.8,0.8). The cases with NH and IH are explicitly shown for each case with $P_{13} = 1$.[23] We briefly mention the findings of the study. The ability to distinguish between the hierarchies becomes weaker when one considers the realistic case with different distribution of fluxes. The absolute difference in the number of events for normal (NH) and inverted hierarchy (IH) as a function the parameter 'a', in the case where 'b' is taken equal to 'a', shows a steady decrease for 'a' going from 0.5 to 1.0 and reaching zero for the value of 'a' slightly larger than 1.0. The sensitivity of this difference on the expectation values (μ_1, μ_2) and the widths (σ_1, σ_2) of the distribution is also investigated. One finds that only for low values of the expectation value less than 1 and small widths, upto 0.2 it may be possible to distinguish NH from IH with small systematic errors.

So even with the uncertainty in the distribution of fluxes of neutrinos emitted one should be able to detect DSNB in future large detectors. Then checking with the calculated number of events for some favorable cases one would be able to find the mass hierarchy particularly if one knows the angle θ_{13} more accurately. However better understanding of some of the features of collective oscillation need to be achieved and the accuracy of the 'single angle' approximation need to be ascertained in the near future.

References

1. B. Pontecorvo, *Sov. Phys. JETP* **6** 429 (1956).
2. L. Wolfenstein, *Phys. Rev. D* **17** 2369 (1978).
3. S.P. Mikheyev and A. Yu. Smirnov, *Sov. J. Nucl. Phys.* **42(6)** 913 (1985); *Nuovo Cimento* **9c** 17 (1986).
4. H. Duan, G.M. Fuller and Y.-Z. Qian, *Ann. Rev. Nucl. Part. Sc* **60** 569 (2010) and references therein.
5. J.F. Beacom, *Ann. Rev. Nucl. Part. Sc.* **60** 439 (2010).
6. H.A. Bethe, *Rev. Mod. Phys.* **62** 801 (1990).
7. H.T. Janka, K. Langanke, A. Marek, G. Martinez-Pinedo and B. Mueller, *Phys. Rep.* **442** 38 (2007).
8. H. Duan, G.M. Fuller, J. Carlson and Y.Z. Qian, *Phys. Rev. D* **74** 123004 (2006).
9. B. Dasgupta and A. Dighe, *Phys. Rev. D* **77** 113002 (2008).
10. G. Fogli, E. Lisi, A. Marrone and I. Tamborra, *JCAP* **04** 030 (2009).
11. M.T. Keil, G.G. Raffelt and H.T. Janka, *Astrophys. J.* **590** 971 (2003).
12. S. Chakraborty, S. Choubey, S. Goswami and K. Kar, *JCAP* **1006** 007 (2010).
13. S. Chakraborty, S. Choubey and K.Kar, Phys. Lett. B **702**, 209 (2011).
14. B. Dasgupta, A. Dighe, G.G. Raffelt and A.Yu. Smirnov, *Phys. Rev. Lett.* **103** 051105 (2009).
15. G.G. Raffelt and A. Yu Smirnov, *Phys. Rev. D* **76** 081301 (2007); erratum ibid. **77** 029903 (2008).
16. G.G. Raffelt and A. Yu Smirnov, *Phys. Rev. D* **76** 125008 (2007).
17. G.L. Fogli, E. Lisi, A. Marrone and A. Mirrizi, *JCAP* **0712** 010 (2007).
18. H. Duan, A. Friedland, G.C. McLaughlin and R. Surman, arXiv:1012.0532 [astro-ph]
19. M. Malek et al. [Super-Kamiokande Collaboration], *Phys. Rev. Lett.* **90** 061101 (2003).
20. M. Aglietta et al., *Astropart. Phys.* **1,** 1(1992).

21. C. Amsler et al. [Particle data Group], 'Reviews of Particle Physics', *Phys. Lett.* **667** 1 (2008).
22. S. Chakraborty, S. Choubey, B. Dasgupta and K. Kar, *JCAP* **0809** 013 (2008).
23. S. Chakraborty, Ph. D. thesis, Neutrino Oscillations in Supernovae (Homi Bhabha National Institute, 2010).

TERMINATING ROTATIONAL BANDS IN NUCLEI

INGEMAR RAGNARSSON
Division of Mathematical Physics, LTH, Lund University,
Box 118, SE-221 00 Lund, Sweden
E-mail: ingemar.ragnarsson@matfys.lth.se

Because the angular momentum in nuclear configurations is limited, most rotational bands will terminate. Examples of terminating bands in different mass regions are given and interpreted in the configuration-constrained Cranked Nilsson-Strutinsky (CNS) formalism. Important characteristics of the bands can be read out from the energy vs. angular momentum function but also from the (continuous) variation of the energy with respect to particle number.

Keywords: Terminating bands, maximum angular momentum, Nilsson-Strutinsky formalism, fixed configurations, nuclear mass.

1. Introduction

Rotational bands in nuclei are mainly characterized by their valence particles, i.e. the particles in open N-shells or j-shells, where more valence particles will in general correspond to a larger deformation. With increasing angular momentum, the bands will either stay within such a configuration, or the coupling between the shells will lead to a continuous excitation of particles from the closed shells. If the bands stay within their configuration, it means that their angular momentum is limited and they will finally terminate[1] in a state where the angular momentum of all the valence particles is quantized along the rotation axis. With standard definitions[2] of the (ε_2, γ)-plane and with rotation around the x-axis, the oblate (prolate) shape at termination will correspond to $\gamma = 60°$ $(-120°)$.

In the pure oscillator model with rotation around one of the principal axis, their is a definite division line[3] between configurations which terminate and configurations where the couplings between the N-shells dominate so that that they become more and more deformed with increasing angular momentum, I (with no upper limit on I). The important parameter in this case is the difference of the number of quanta in the two perpendicular directions, or equivalently their deformation at $I = 0$. For example, prolate configurations with a deformation at no rotation of $\varepsilon_2 < 0.499$ will develop towards smaller deformation with increasing I and terminate at $I = I_{max}$.

The nuclear single-particle orbitals are richer than those of the pure oscillator in the sense that they split up into j-shells. Thus, if a configuration is followed with increasing spin, it turns out that if its deformation is smaller than $\varepsilon_2 = 0.3 - 0.4$, it is in general possible to label the orbitals by their N-shell but also to make a further distinction[4] between the high-j intruder shell and the other j-shells. This is illustrated for a ^{108}Sn configuration in Fig. 1. With $Z = 50$, the ground state of this nucleus is spherical but it is

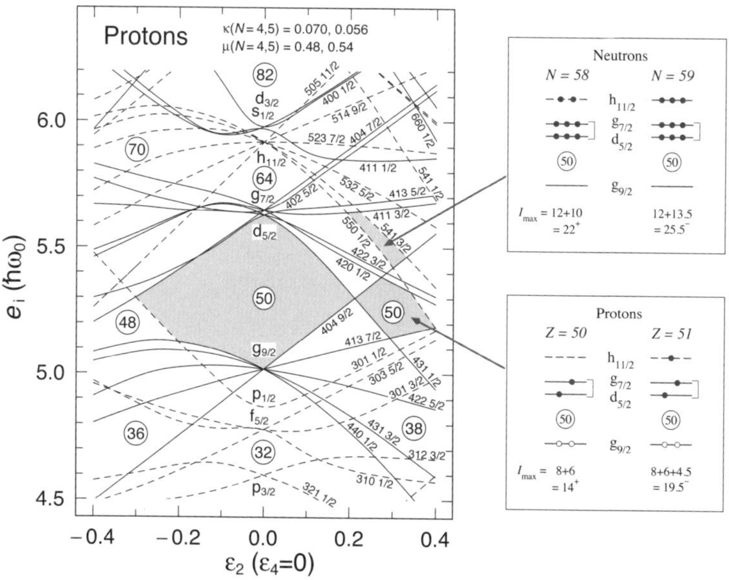

Fig. 1. Nilsson diagram illustrating the gaps for $Z = 50$ and $N = 58$ which lead to the ^{108}Sn configuration shown schematically in the boxes to the right. In those boxes are also shown the resulting configurations if a high-j $h_{11/2}$ proton and neutron is added and the maximum spin value I_{max} for these configurations.

also possible to fix a configuration corresponding to the gaps at $\varepsilon_2 \approx 0.3$ where the important difference compared with the spherical ground state configuration is the two proton holes in the 404 9/2 orbital of $g_{9/2}$ origin and another less important difference is the two neutrons excited to the negative parity $h_{11/2}$ orbital (labelled as 550 1/2). This configuration can then be schematically drawn as in the boxes to the right where the bracket around the $d_{5/2}$ and $g_{7/2}$ subshells indicate that we consider them as one entity. If the particles stay in this configuration with increasing angular momentum (spin), it is evident that the spin is limited where, for example, the maximum contribution of the two holes in $g_{9/2}$ is $9/2 + 7/2 = 8\hbar$ and that of the two particles in the ($d_{5/2}$, $g_{7/2}$) orbitals is $7/2 + 5/2 = 6\hbar$. Similarly, as indicated in Fig. 1, the maximum neutron spin in this ^{108}Sn configuration is $22\hbar$, leading to a total $I_{\max} = 14 + 22 = 36\hbar$ for this band which is observed[5] in the spin range $I = 12 - 32\hbar$. With increasing spin, it becomes favourable to put more particles into high-j orbitals. Such configurations are illustrated for $Z = 51$ and $N = 59$ in Fig. 1. It turns out that when the $Z = 51$ and $N = 58, 59$ configurations of Fig. 1 are combined, bands are formed in ^{109}Sb and ^{110}Sb which are low in energy in an extended spin range and observed[1,6,7] to their $I_{\max}$ values, 41.5^- and 45^+, respectively (see Fig. 2 below).

2. The Configuration-Constrained Cranked Nilsson-Strutinsky (CNS) Formalism

Terminating bands have mainly been interpreted within a cranked Nilsson-Strutinsky formalism[1,8] based on the modified oscillator (MO) potential. In this simple model, it is possible to get an illustrative understanding of the features which are observed and which can be traced back to the number of valence particles and holes in orbitals of high-j and low-j character. The MO potential can be written

$$V_{\text{MO}} = V(\varepsilon_2, \gamma, \varepsilon_4) - \kappa\hbar\,\overset{0}{\omega}_0 \left(2\vec{l}_t \cdot \vec{s} + \mu\left(l_t^2 - \langle l_t^2 \rangle_N\right)\right)$$

where $V(\varepsilon_2, \gamma, \varepsilon_4 = 0)$ corresponds to an anisotropic harmonic oscillator potential, i.e. with ellipsoidal equipotential surfaces. Then with $\varepsilon_4 \neq 0$, it becomes possible to get a realistic description of all nuclear shapes which are relevant for terminating bands. The corresponding single-particle Hamiltonian, h_{MO}, is cranked with the rotational frequency ω around one principal axis, the x-axis, leading to the cranking Hamiltonian, $h^{\omega}_{\text{MO}} = h_{\text{MO}} - \omega j_x$ which is diagonalized

$$h^{\omega}_{\text{MO}} \chi^{\omega}_{\nu} = e^{\omega}_{\nu} \chi^{\omega}_{\nu}.$$

The eigenvalues e^{ω}_{ν} are the single-particle energies in the rotating frame (the routhians) with the energies in the laboratory frame calculated as expectation values $e_{\nu} = \langle \chi^{\omega}_{\nu} | h_{\text{MO}} | \chi^{\omega}_{\nu} \rangle$. The total spin is now calculated from the sum of the expectation values of j_x for the occupied (occ.) orbitals and the shell energy as the difference between the discreet and the smooth energy sums, according to the Strutinsky prescription,[2,9]

$$I = \sum_{occ} \langle j_x \rangle \qquad E_{\rm shell} = \sum_{occ} e_\nu - \left\langle \sum e_\nu \right\rangle .$$

A very important feature of the formalism is the unique possibilities to label the orbitals and thus to *fix configurations*. First the diagonalization is carried out in a rotating harmonic oscillator basis, where the principal quantum number $N_{\rm rot}$ is preserved. Then, with the reasonable approximation to neglect the coupling between the $N_{\rm rot}$ shells from the hexadecapole, the $\vec{l}_t \cdot \vec{s}$ and l_t^2 terms, $N_{\rm rot}$ will be pure also in the modified oscillator. Furthermore, within the $N_{\rm rot}$ shells, it turns out to be relatively straightforward[4] to split up the orbitals into those of high-j character (the intruder orbitals corresponding to the highest j for each N-shell) and the remaining orbitals referred to as being of low-j character. In order to make such a distinction, it is important to remove the so called virtual interactions between the single-particle states and form diabatic orbitals whose wave-functions vary smoothly with rotational frequency, see Fig. 4 of Ref. 8 or Fig. 4 of Ref. 1. A configuration can now be labelled by the number of particles of signature $\alpha = 1/2$ and $\alpha = -1/2$[a] of high-j and low-j character, respectively, in the different $N_{\rm rot}$ shells. For such a configuration, the total energy is obtained[11] as the sum of the rotating liquid drop energy and the shell energy. If this energy is minimized with respect to the hexadecapole deformation, it becomes possible to draw total energy surfaces (often referred to as potential energy surfaces) in the (ε_2, γ)-plane. The energy which should be compared with experiment is obtained from a minimization with respect to all deformation degrees of freedom, ε_i,

$$E_{\rm tot}(Z,N,I) = \min_{\varepsilon_i} \left[E_{\rm rld}(Z,N,I,\varepsilon_i) + E_{\rm shell}(Z,N,I,\varepsilon_i) \right] .$$

The rotating liquid drop energy $E_{\rm rld}$ is obtained from the sum of the static liquid drop energy $E_{\rm rld}$, calculated according to the Lublin-Strasbourg[12] drop (LSD) parameterization (with the average pairing energy $\langle E_{\rm pair} \rangle$ subtracted), and a rigid rotation energy,

$$E_{\rm rld}(Z,N,I,\varepsilon_i) = E_{\rm ld}(Z,N,\varepsilon_i) + \frac{\hbar^2 I(I+1)}{2\mathcal{J}_{\rm rig.}(Z,N,\varepsilon_i)} .$$

[a]The signature (exponent) quantum number α is preserved for systems which are symmetric with respect to a 180° rotation around the x-axis.[10]

The rigid body moment of inertia $\mathcal{J}_{\mathrm{rig.}}$ is calculated with a radius parameter $r_0 = 1.16$ fm and a diffuseness parameter $a = 0.6$ fm. These values have been obtained from a fit[11] to experimental $\langle r^2 \rangle$-values. In the calculations presented below, so called standard single-particle parameters are used in the $A \sim 60$ region, with some modifications introduced long ago of these parameters in the $A \sim 110$ and $A \sim 155 - 160$ regions, see Ref. 1.

It is important to note that with the new features which were introduced in Ref.11, it is now possible to calculate the total energy for any value of the nuclear spin and compare with experimental energies. This, is analogous to the calculation of nuclear masses.[13] Thus, in a similar way as it is illustrative to show experimental and calculated ground state masses relative to a static liquid drop energy (microscopic energies), it is illustrative to show observed and calculated energies at high spin relative a rotating liquid drop energy, see Fig. 2. The difference between experiment and calculations is then shown in the lower panel of this figure, suggesting that in a similar way as for ground state masses, it is in general possible to reproduce the energies within an accuracy of approximately ± 1 MeV. Note that this only refers to the high-spin region because pairing is neglected in the CNS formalism.

3. Examples of Terminating Bands in Different Mass Regions

It is instructive to exemplify the evolution with spin for the nucleus ^{158}Er which shows one of the best examples of terminating bands and is the first heavy nucleus where such bands were discovered.[15,16] Fig. 3 taken from Ref. 17 shows observed energies in the yrast region relative to a rigid rotation reference. The low spin states are formed from collective rotation with a band-crossing at $I \approx 12$ when the spin vectors of two high-j neutrons are aligned and another more smooth band-crossing around $I = 30$ when two high-j proton spin vectors are aligned.[b] Then, around $I = 40$, the collective yrast band is crossed by bands which come down in energy at high spin and terminates in aligned non-collective states at $I = 46^+, 48^-$ and 49^-. It turns out that these spin values correspond to the highest spin states which can be built from the valence particles outside a $^{146}_{64}$Gd$_{82}$ core; 4 $h_{11/2}$ protons, 2 or 3 $i_{13/2}$ neutrons and the remaining 6 or 5 valence neutrons in the ($h_{9/2}$,$f_{7/2}$) orbitals (cf. Fig. 6 below). The different modes of rotation are also shown schematically, first with the 'aligned' high-j neutrons and protons and then with all 12 valence particles rotating around the x-axis

[b]In the CNS formalism without pairing presented here, the alignment of the high-j particles will be more gradual.

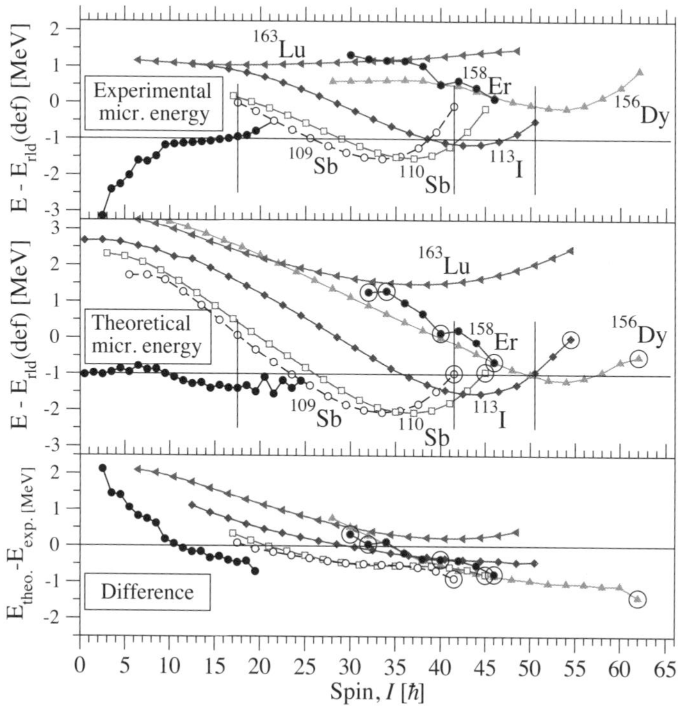

Fig. 2. Typical examples of high-spin bands shown relative to a rotating liquid drop reference for experimental bands in the upper panel, the calculated configurations assigned to these bands in the middle panel, with the difference between experiment and calculations in the lower panel. The low spin states of ^{109}Sb are included mainly to show the increasing difference between experiment and calculations with decreasing spin because of the neglect of pairing correlations. The smooth terminating bands for ^{109,110}Sb correspond to the configurations shown in Fig. 1 while the ^{156}Dy band is shown also in Fig. 7. The favoured termination band of ^{158}Er is seen also in Fig. 3 while the ^{163}Lu band[14] corresponds to the triaxial superdeformed (TSD) band. The general agreement between calculations and experiment is seen from the fact that in the high spin range, the differences in the lower panel are within ±1 MeV and they are relatively constant for the different bands.

(or more accurately, quantized along this axis) in the terminating state. Higher spin states can then be formed from core-breaking excitations as is also illustrated with one proton hole and thus 13 valence particles outside the core. Such states, which have been identified in more detail in Ref. 18, can be built in several different ways but they are all relatively much higher in energy than the terminating states. Then it is also illustrated how the valence space states are by-passed by a collective rotational band which is

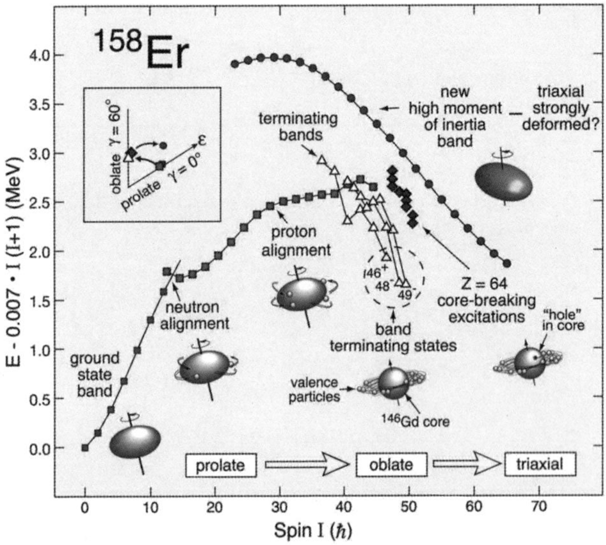

Fig. 3. Pictorial illustration of the yrast energies and corresponding coupling schemes in different spin regions of ^{158}Er.

thus considerably more deformed with several particles excited from the ^{146}Gd core. Furthermore, the shape change is illustrated in an inset, from prolate collective rotation ($\gamma = 0°$) at low spin over triaxial shapes to non-collective oblate ($\gamma = 60°$) at termination.

Terminating bands are most easily understood as formed from configurations with several but not too many valence particles outside closed shells or in such configurations but with few valence holes in the core. This is illustrated in Fig. 4 where it is indicated that well-developed terminating bands are found in regions with a several valence particles outside $^{56}_{28}$Cr$_{28}$ ($A \approx 60$), $^{100}_{50}$Sn$_{50}$ ($A \approx 110$) and $^{146}_{64}$Gd$_{82}$ ($A \approx 155 - 160$). In the $A \approx 110$ region, it is clearly the bands with two holes in the $Z = 50$ proton core which dominates (with valence space bands observed close to termination only in the heavier nuclei ^{113}I (Ref. 19) and ^{114}Xe (Ref. 20)) while there is a mixture of both type of bands in the $A \approx 60$ and $A \approx 155 - 160$ regions.

If a terminating band is built in the valence space with most particles in high-j or intermediate-j orbitals, it is in general high above yrast at low spin but it comes down in the yrast region a few spin units before termination with the terminating state very low in energy. Thus, when drawn relative to a rotating reference, it will come down in energy with increasing spin

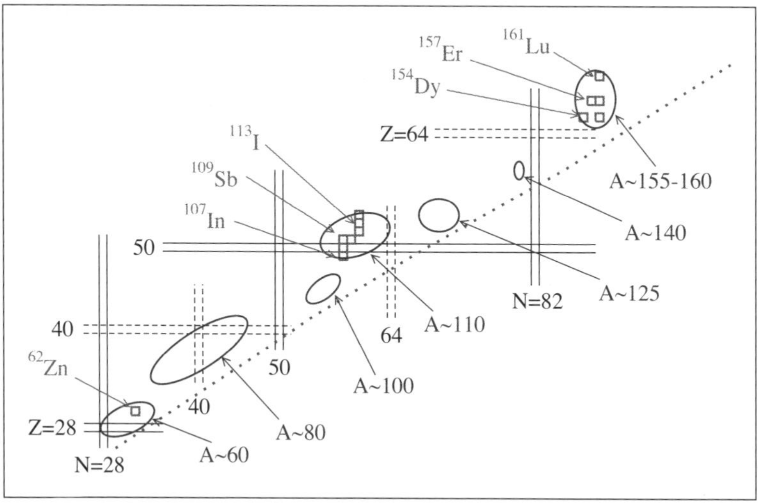

Fig. 4. Regions where terminating bands have been studied. It is especially the $A \sim 60$, $A \sim 110$ and $A \sim 155 - 160$ regions with a few particles outside the closed ^{56}Ni, ^{100}Sn and ^{146}Gd cores which are discussed here. Well-developed terminating bands are also found in the $A \sim 80$ region (see e.g Refs. 23–26) where the maximum spin in the $(N, Z) = 28 - 50$ valence space is below $40\hbar$ and thus well within experimental reach. Another interesting region is $A \sim 100$, see e.g. Refs. 27,28. Terminating states have also been identified in the $A \sim 125$[29] and $A \sim 140$[30,31] regions with particles outside the ^{114}Sn core and holes outside the ^{146}Gd core, respectively, but the corresponding bands are less well developed.

(favoured termination) as seen for the ^{158}Er configuration in Fig. 2. A few holes in the core on the other hand will make the terminating state less favoured but such configurations are often low in energy a few spin units before termination as seen for the ^{109,110}Sb and ^{113}I bands in Fig. 2. Such bands are referred to as unfavoured terminations where it should be noted the terms 'favoured' and 'unfavoured' refer to the relative energy cost to build the last spin units before termination but not to the excitation energy of the band.

The $A \approx 110$ region indicated in Fig. 4 is a very nice example of so called smooth terminating bands, i.e. bands which can be followed in an extensive spin range where they gradually become less collective and terminate in a non-collective fully aligned state. One can for example follow how these bands, with two $g_{9/2}$ proton holes (see Fig. 1), develop smoothly with mass number along the chain indicated in Fig. 4. Starting from the nucleus ^{107}In (Ref. 21) where the most favoured configuration has two $h_{11/2}$ high-j particles, the chain continues over ^{108}Sn (Ref. 5), ^{109,110}Sb (Refs. 6,7), ^{112}Te (Ref. 22), ^{113}I (Ref. 19) and finally ^{114}Xe (Ref. 20) with six $h_{11/2}$ particles. In both experiment and calculations, it is the bands in ^{109,110}Sb (shown in Fig. 2) which are lowest in energy and these are the only cases where the terminating state has been observed, i.e. the bands in the other nuclei of this chain have been observed up to one or two states before termination as exemplified for ^{113}I in Fig. 2. Furthermore, life-times have been measured[32] for the bands in ^{108}Sn and ^{109}Sb, confirming the decreasing collectivity with increasing spin.

In a similar way as terminating bands are formed with a few valence particles outside the ^{100}Sn core, such bands are formed with a few valence particles outside ^{56}Ni. In ^{62}Zn, as illustrated in Fig. 5, it is possible to follow how the terminating structures develop, starting from the ground state configuration, with successive excitations of particles to the $g_{9/2}$ orbitals and from the $f_{7/2}$ orbitals (across the $Z = 28$ gap). With one, two and three particles excited to the $g_{9/2}$ orbitals, the bands become more and more down-sloping (favoured terminations). Then with holes in the core, the energy curves turn over at the highest spin values (unfavoured terminations). The general agreement between calculations and experiment shows convincingly that the assignments are correct. On the other hand, one notes that the difference between calculations and experiment for the structures which terminate at $(20, 21^+)$ and $(23, 24^-)$, are around -0.6 MeV and $+0.2$ MeV, respectively. As there is one more $g_{9/2}$ neutron in the latter structure, this fact indicates that the $g_{9/2}$ neutron subshell is placed too high in energy with present parameters. Note also that the highest spin band shown with $I_{\mathrm{max}} = 30^+$ remains somewhat collective at its I_{max} value, indicating that this band does not terminate as has been discussed in more detail for the bands in ^{74}Kr.[26] This indicates a strong coupling between all the $N = 3$ j-shells in a similar way as the coupling between the N-shells of the rotating oscillator prevents termination for strongly deformed configurations as discussed above.

It is also instructive to consider the excitations across the $Z = 64$ gap for terminating bands in ^{154,156}Dy. This is illustrated schematically in Fig. 6 where it is shown how favoured terminations are expected if the $Z = 64$ and $N = 82$ cores of ^{154}Dy are closed but how the terminations are expected to become more and more unfavoured with more particles excited from the $Z = 64$ core. Comparing with experiment,[35] it turns out that with more protons excited it is more favourable to combine with the $N = 90$ configurations observed[36] in ^{156}Dy. Thus, as seen in Fig. 7, all the pro-

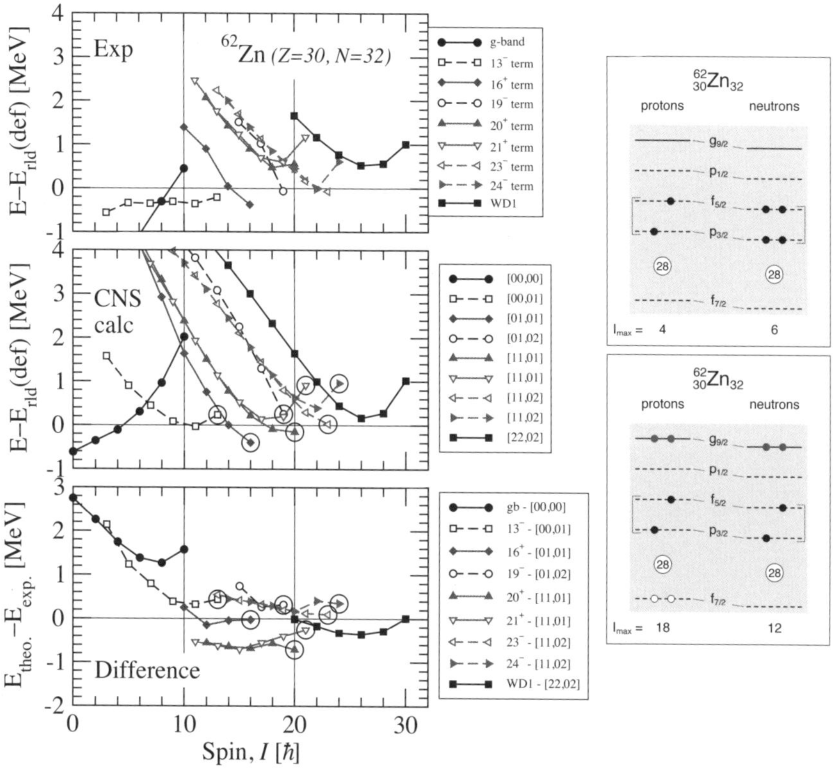

Fig. 5. Same as Fig. 2 but for ^{62}Zn configurations in the $I = 0-30$ spin range, with the configurations for the ground band and $I_{\rm max} = 30$ configurations illustrated to the right. The observed bands labelled by their terminating spin values are taken from Ref. 33 while the band observed to $I = 30$ labelled as WD1 (well-deformed band 1) is taken from Ref. 34. The calculated configurations are labelled as $[p_1p_2, n_1n_2]$ where $p_1(n_1)$ is the number of $f_{7/2}$ proton (neutron) holes and $p_2(n_2)$ is the number of $g_{9/2}$ protons (neutrons).

ton configurations of Fig. 6 are observed[35,36] experimentally, those with no, one or two holes in ^{154}Dy and those with three or four holes in ^{156}Dy. The fact that the differences in the lower panel come close together gives strong support to our interpretation even though one should also note that the highest observed spin states in ^{156}Dy are somewhat tentative. The suggested termination at $I = 62^+$ is quite special because it is the highest spin observed in normal-deformed bands. Furthermore, the proton spin is $29\hbar$ which comes quite close to the highest spin which can be built from the $Z = 50 - 82$ orbitals with all five j-shells half-filled, $I_{\rm max} = 33$. It would

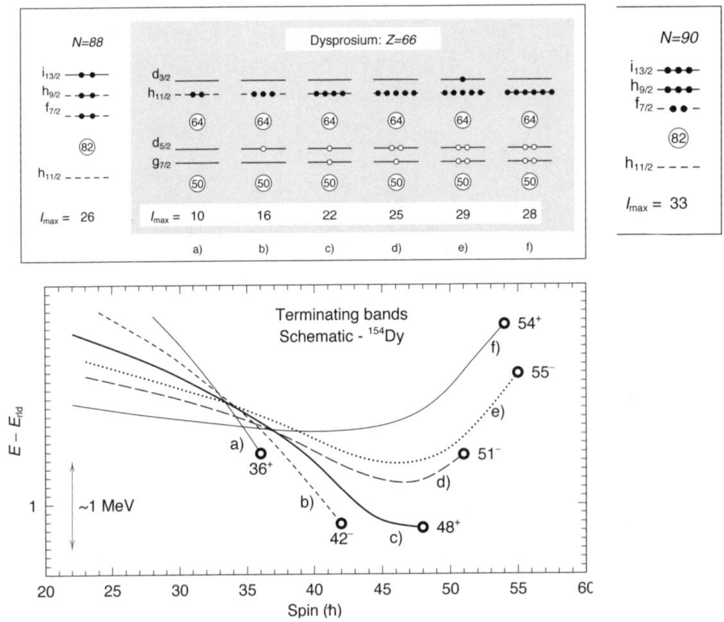

Fig. 6. Schematic illustration of the bands which are expected in ^{154}Dy from the configurations in the upper panel. Also a possible $N = 90$ configuration is shown.

thus be very interesting to get a more definite understanding of these bands in the $I \approx 60$ region of ^{156}Dy. Another interesting feature is the more collective band which is observed in ^{154}Dy. This band has a natural explanation as formed with two neutron holes in the $N = 82$ core.

4. Outlook

In this short review, we have pointed out that a very important feature of valence space configurations in essentially all nuclei with mass numbers up to $A = 160 - 165$ is that their angular momentum (their spin) is limited and this spin value, I_{max} is within experimental reach. This fact puts very important constraints on possible theoretical assignments, i.e. from comparisons with the maximum spin value in different configurations and from the behaviour of the energy vs. spin curve close to I_{max}, it is in most cases possible to get a unique theoretical interpretation. The interpretation becomes easier if the spin value is reasonably high because pairing interactions are then of minor importance and the configurations are understood from the filling of different j-shells or groups of j-shells. In the CNS ap-

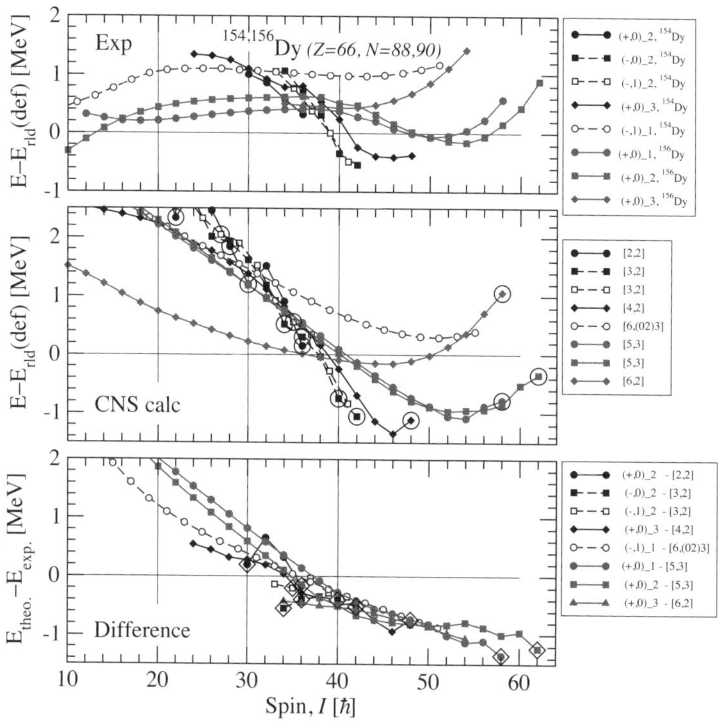

Fig. 7. Same as Fig. 2 but for bands in ^{154,156}Dy. The calculated bands are labelled by the number of $h_{11/2}$ protons and $i_{13/2}$ neutrons with the number of $N = 4$ and $h_{11/2}$ neutron holes in parentheses when different from zero. Thus, for example the [5,3] bands have the proton configurations d) and e) combined with the $N = 90$ configuration shown in Fig. 6.

proach,[1,8,11] special methods have been introduced to assign the orbitals to one j-shell or a group of j-shells according to their main amplitudes and this approach has been very successful in the interpretation of terminating bands. Here we have exemplified with the continuous evolution of specific configurations with spin or particle number in the $A \approx 60$, $A \approx 110$ and $A \approx 150 - 155$ regions of nuclei.

With the general success of band termination calculations as a background, there are several open questions. For example, in different regions, there are terminating structures which are predicted low in energy but have not been observed, as exemplified by the discussion about the $A = 80$ region in Ref. 1. At the highest spin values, an important question is how the terminating states are fed and what comes next as discussed above for ^{157,158}Er. Comparisons with nuclei in the $A = 60$ region are instructive because, for example in ^{62}Zn, superdeformed bands with known spin and excitation energies and with some similarities to terminating structures have been observed[34] beyond the bands shown in Fig. 5. It would also be very interesting to learn more about the properties of the ^{156}Dy bands in the $I = 60$ spin range discussed above and the bands in ^{131,132}Ce approaching $I = 70$. As discussed in Ref. 37, these bands which also show some features of termination are observed to very high rotational frequencies where the j-shells become more mixed and particle instability might become important, see e.g. Ref. 38.

Terminating bands have been investigated mainly in the CNS approach based on the modified oscillator potential. Exceptions are some early CNS studies[39,40] based on the Woods-Saxon potential and studies based on the cranked relativistic mean field (CRMF), see e.g. Ref. 25, where, however, it has been difficult to follow the bands all the way to the non-collective terminating state. Furthermore, shell model calculations[41] have been carried out but essentially only for nuclei with $A < 60$ where also comparisons between CNS calculations and SM calculations, see e.g. Ref. 42, show a general consistency. In any case, considering the importance of band terminations in heavier nuclei, more systematic studies in other models than CNS appear well motivated.

It has turned out to be difficult to follow bands to termination in self-consistent pairing calculations, see e.g. Ref. 43. Recently, however, a formalism based on a mesh in the pairing gap Δ and the Fermi level λ has been developed,[44] where it has been possible to get smooth energy surfaces in the full (ε_2, γ)-plane, including the non-collective axis. One goal is to calculate the total binding energy in the full spin range and thus to develop a realistic 'mass model' in the full (Z, N, I)-space, adding a new degree of freedom when determining the parameters of such models.

This project is supported by the Swedish Natural Science Research Council. The author is grateful for support from the Royal Physiographic Society in Lund.

References

1. A.V. Afanasjev, D.B. Fossan, G.J. Lane and I. Ragnarsson, Phys. Rep. **322**,1 (1999).
2. G. Andersson *et al.*, Nucl. Phys. A **268**, 205 (1976).
3. T. Troudet and R. Arvieu, Z. Phys. A **291**, 183 (1979).

4. I. Ragnarsson, V. P. Janzen, D. B. Fossan, N. C. Schmeing and R. Wadsworth, Phys. Rev. Lett. **74**, 3935 (1995).
5. R. Wadsworth *et al.*, Phys. Rev. C **53**, 2763 (1996).
6. H. Schnare *et al.*, Phys. Rev. C **54**, 1598 (1996).
7. G. J. Lane *et al.*, Phys. Rev. C **55**, R2127 (1997).
8. T. Bengtsson and I. Ragnarsson, Nucl. Phys. A **436**, 14 (1985).
9. V.M. Strutinsky, Nucl. Phys. A **95**, 420 (1967).
10. A.L. Goodman, Nucl. Phys. A **230**, 466 (1974).
11. B.G. Carlsson and I. Ragnarsson, Phys. Rev. C **74**, 011302(R) (2006).
12. K. Pomorski and J. Dudek, Phys. Rev. C **67**, 044316 (2003).
13. P.Möller, J.R. Nix, W.D. Myers and W.J. Swiatecki, Atomic Data Nucl. Data Tables **59**, 185 (1995).
14. S.W. Ødegård *et al.*, Phys. Rev. Lett. **86**, 5866 (2001).
15. P.O. Tjøm *et al.*, Phys. Rev. Lett. **55**, 2405 (1985).
16. J. Simpson *et al.*, Phys. Lett. B **327**, 187 (1994).
17. E.S.. Paul *et al.*, Phys. Rev. Lett. **98**, 012501 (2007).
18. A.O. Evans *et al.*, Phys. Rev. Lett. **92**, 252502 (2004).
19. K. Starosta *et al.*, Phys. Rev. C **64**, 014304 (2001).
20. E.S. Paul *et al.*, Phys. Rev. C **65**, 051308(R) (2002).
21. E. Ideguchi *et al.*, Phys. Rev. C **81**, 034303 (2010).
22. E.S. Paul *et al.*, Phys. Rev. C **55**, 014308(R) (2007).
23. I. Ragnarsson and A.V. Afanasjev, Proc. Conf. Nucl. Structure on the limits, ANL, Argonne, July 22-26, 1996, ANL/PHY-97/1, p. 184.
24. B.G. Carlsson and I. Ragnarsson, Phys. Rev. C **70**, 024303 (2004).
25. A.V. Afanasjev and S. Frauendorf, Phys. Rev. C **71**, 064318 (2005).
26. J.J. Valiente-Dobón *et al.*, Phys. Rev. Lett. **95**, 232501 (2005).
27. J. Gizon *et al.*, Phys. Lett. B **410**, 95 (1997).
28. J. Timár *et al.*, Phys. Rev. C **62**, 044317 (2000).
29. Purnima Singh *et al.*, Phys. Rev. C **82**, 034301 (2010) and references therein.
30. C.M. Petrache *et al.*, Phys. Rev. C **74**, 034304 (2006).
31. Somen Chanda *et al.*, Phys. Rev. C **79**, 054332 (2009).
32. R. Wadsworth *et al.*, Phys. Rev. Lett. **80**, 1174 (1998).
33. C.E. Svensson *et al.*, Phys. Rev. Lett. **80**, 2558 (1998).
34. J. Gellanki *et al.*, Phys. Rev. C **80**, 051304(R) (2009).
35. W.C. Ma *et al.*, Phys. Rev. C **65**, 034312 (2002).
36. F.G. Kondev *et al.*, Phys. Lett. B **437**, 35 (1998).
37. E.S. Paul *et al.*, Phys. Rev. C **71**, 054309 (2005).
38. M. Cerkaski *et al.*, Phys. Lett. **70B**, 9 (1977); **72B**, 149 (1977).
39. W. Nazarewicz, J. Dudek, R. Bengtsson, T. Bengtsson and I. Ragnarsson, Nucl. Phys. A **435**, 397 (1985).
40. J. Dudek and W. Nazarewicz, Phys. Rev. C **31**, 298 (1985).
41. E. Caurier, G. Martínez-Pinedo, F. Nowacki, A. Poves, and A.P. Zuker, Rev. Mod. Phys. **77**, 427 (2005).
42. A. Juodagalvis, I. Ragnarsson and S. Åberg, Phys. Rev. C **73**, 044327 (2006).
43. T. Bengtsson, Nucl. Phys. A **512** (1990) 124.
44. B.G. Carlsson *et al.*, Phys. Rev. C **78**, 034316 (2008)

LESSONS FROM 25 YEARS ON THE DRIPLINES: STILL A CATWALK?

J.S. VAAGEN[1], S.N. ERSHOV[2], and M.V. ZHUKOV[3]

[1] *Department of Physics and Technology, University of Bergen, Bergen, Norway*
E-mail: JanS.Vaagen@ift.uib.no

[2] *Joint Institute for Nuclear Research, 141980 Dubna, Russia*
E-mail: ershov@theor.jinr.ru

[3] *Fundamental Physics, Chalmers University of Technology, S-41296 Göteborg, Sweden*
E-mail: f2bmz@chalmers.se

A quarter of a century's concerted international work in halo physics has resulted in an extended nuclear paradigm encompassing the limits of existence of cold nuclei and also structures beyond - continuum structures of open (nuclear) quantum systems. Realistic working models, based on cluster constituents, have sprung out of the very nature of halo phenomena, in particular from the three-body Borromean property of two-neutron halos; the lack of low-lying binary breakup channels. This has provided transparency and possibility for insight into new quantum behaviour, also in continua at driplines. Breakup spectra and progressively exclusive correlation cross sections can be computed and show, where relevant data exist, that general agreement is encouraging. Progress in studies of two-proton emitters has provided another pathway beyond driplines, where again few-body theory appears relevant.

PACS number(s): 21.60.Gx, 24.10.Eq, 24.50.+g, 25.60.Gc

1. Getting on the Catwalk – A Historical Remark

When Rutherford's nuclear atom model was introduced in the spring of 1911, a century ago, it was met with indifference and scarcely considered to be a theory of the constitution of the atom. The new conception of the atom was not mentioned in the proceedings of the 1911 Solvay Congress (where Rutherford participated), nor was it widely discussed in the physics journals. Thus the nuclear atom did not make it straight to the catwalk of physics of those days.

This we learn from Helge Kragh's book "Quantum Generations - A History of Physics in the Twentieth Century"[1] , and he gives reasons for this.

In its initial version the model, based on scattering experiments, made a separation between heavy and light matter, but it had little to say about the structure of the electron veil, thus provided little guidance for ongoing atomic experiments. Although tiny (but heavy), Rutherford foresaw a complex nature for the new "heart of matter", but he could say less about its inner workings. Revealing the secrets of the nucleus - (a name apparently first used by Nicholson), called for new acceleration principles and penetrating probes to be developed.

By 1913 Rutherford's model was underpinned by improved data on alpha scattering, but still it had little new to add on the atomic architecture. The fate of the model changed with the arrival of a young Dane, Niels Bohr, in Manchester. He turned Rutherford's picture of the nuclear atom into a proper, albeit still ad hoc, "planetary" theory. Ten years later he obtained the 1922 Nobel Prize in Physics "for his services in the investigation of the structure of atoms and of the radiation emanating from them". No mention of "the Heart of Matter"!

Twenty five years after Rutherford's nuclear atom was introduced, finally a model of the nucleus itself - the liquid drop model - had consolidated its position. This model came to play an important role in the scientific life of Walter Greiner, who happened to be born just at that time. Thus, adding another 25 years takes us to the early 1960s when young Greiner's contributions to the unified nuclear model helped elucidate the interplay between collective - liquid drop inspired - nuclear motion and nuclear single particle motion, originated by Bohr and Mottelson and Rainwater. For this theory the latter three got the 1975 Nobel Prize in Physics, the last one in nuclear physics in spite of the last half century's ventures into new nuclear territories using accelerators to study the nuclear stuff under extreme conditions.

Another 25 years takes us to the mid 1980s. By then hydrodynamic concepts for heavy-ion collisions had become legitimate - other speakers in this Goa conference will have more to say about Greiner's contributions and present status. Our focus will be on the 1985 article[2] by Isao Tanihata and his team at Berkeley, reporting pioneering studies of reactions with secondary beams, which showed the viability of dripline studies and became of paradigmatic significance for our understanding of the heart of matter. Thus, they found the neutron rich ^{11}Li to have a vastly larger interaction cross section than neighbouring nuclei. The challenge of understanding this and similar consecutive observations, became amplified in the wider nuclear theory community after the Scandinavian experimentalists P. Gregers Hansen and Bj`orn Jonson in 1987 put forward an ad hoc binary halo the-

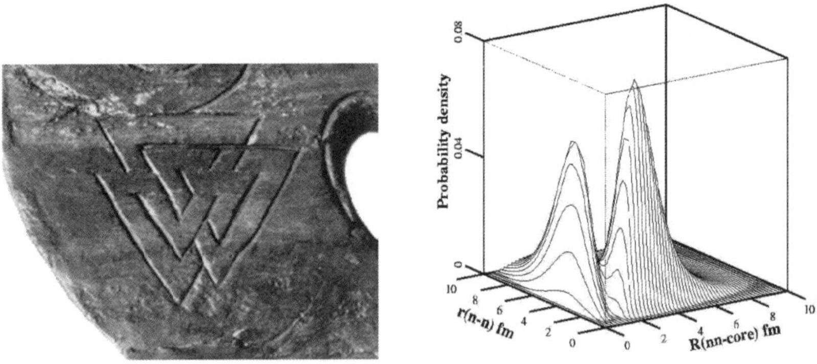

Fig. 1. Left: The Borromean logo in the viking (Odin's triangle walknot) version. Right: The ^{6}He ground state density.

ory,[3] conceived according to the legend in a lunchroom conversation at the Niels Bohr Institute with the late Danish theorist Jens Bang. Only Björn Jonson can tell what really happened. It took a few more years before the halo assumption was properly underpinned experimentally, and the concept of three-body Borromean halo nuclei[4] coined and explored for the most exotic, two-neutron halos, bringing the field to the catwalk, not only of nuclear physics, but in physics more generally, extending the nuclear paradigm. Figure 1 shows the now famous ring logo of the field, in the version of the vikings (go to Google for more), and the density distribution of the ground state of ^{6}He, our theoretical test bench Borromean nucleus. This density distribution was directly explored in a Dubna two-neutron transfer study,[5] which made headlines in physics ranking a decade ago.

The asymptotic behaviour of a Borromean three-body halo system is in the hyperradius similar to that of a two-body system in its binary separation. This is a conceptually simplifying feature, and expansion in hyperspherical harmonics has proven very useful for genuine insight in Borromean nuclei.

2. On the 2010 Catwalk

Twenty five years of history is illustrated in Fig. 2, which also displays two newcomers on the dripline catwalk, ^{22}C and ^{24}O, both with 16 neutrons and not the old magic 20, and which appear to be the last hadronically stable carbon and oxygen isotopes. They have made headlines in a variety of modern media and certainly put nuclear physics on the catwalk.

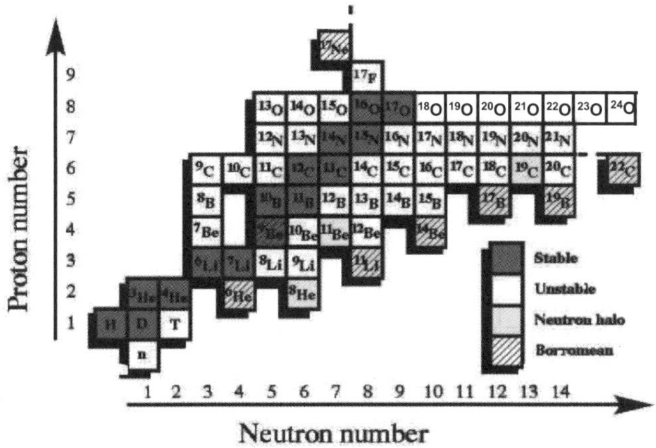

Fig. 2. The lower part of the nuclear chart.

While ^{22}C is highlighted as the heaviest Borromean nucleus discovered, ^{24}O is not Borromean, but a new *doubly* magic nucleus, $Z = 8$, $N = 16$. The experimental study[6] of ^{22}C by Tanaka et al. is very similar to Tanihata et al.'s 1985 study, both measuring interaction cross sections from which matter radii were deduced via a Glauber analysis. Figure 3 compares the findings in the two experiments, in time a quarter of a century apart. The inference that the outstanding size of ^{22}C is Borromean, like ^{6}He and ^{11}Li, is convincing. Not only is the two neutron separation energy low 420$\pm$ 940 keV, the ^{21}C is unbound and ^{20}C has a significant binding ($\sim$ 3.3 MeV), and the last two neutrons are likely to be preferentially in s motion with respect to the ^{20}C "core", if the $N = 16$ closure is similar to that of ^{24}O discussed next.

Thus last year the first measurements of the momentum distribution for one-neutron removal from ^{24}O were reported.[7] The results are explained with a nearly pure $s_{1/2}$ neutron spectroscopic factor of 1.74$\pm$0.19 within the eikonal model. We find pioneers Tanihata and Jonson in the quite long list of authors.

A recent *ab initio* calculation[8] within coupled-cluster theory reproduces the experimental spectroscopic finding and supports ^{24}O being a doubly closed nucleus. *Ab initio* calculations have already made it to the catwalk, such as the coupled-cluster calculation by Hagen et al[9] for the $\frac{1}{2}^{+}$ one-proton excited halo state in ^{17}F, highlighted in Science Daily.

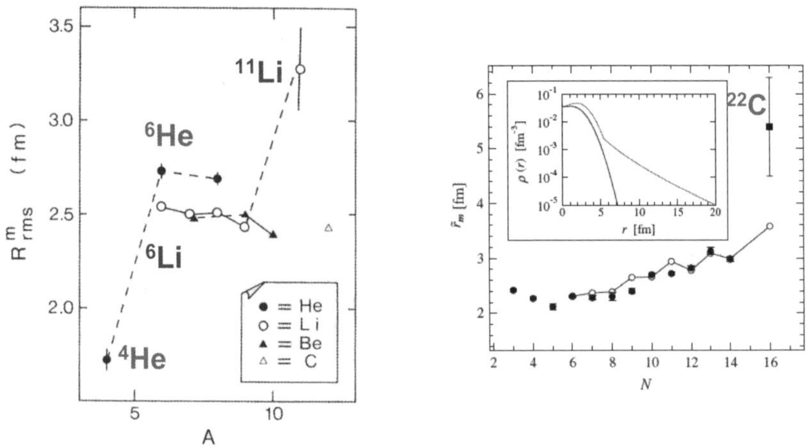

Fig. 3. Left: r.m.s. matter radii for helium and lithium isotopes from[2] , right: r.m.s. matter radii for carbon isotopes from[6] .

Conceptually new ways of understanding emergent structures, such as Borromean few-body models, continue to fascinate the public, and also help more ambitious theory efforts to find direction. This should be kept in mind also in years to come, in spite of the welcomed progress made by our (mostly young) *computational physicists* with *ab initio* approaches. It is gratifying to observe that also new theory work makes it to the catwalk.

3. Lessons from bound states

In the lower part of the Table of Isotopes (Fig. 2) ${}^{4}_{2}\mathrm{He}_2$ and ${}^{16}_{8}\mathrm{O}_8$ are the two tightly bound doubly closed nuclei, serving as "cores" for two chains of isotopes towards the neutron dripline. In the case of helium the chain is discontinues, containing only the Borromean nucleus ${}^{6}\mathrm{He}$ and the dripline nucleus ${}^{8}\mathrm{He}$. A three-body modelling ${}^{8}\mathrm{He} = {}^{6}\mathrm{He} + n + n$ gives a deceptive impression of a Borromean system: The two-neutron separation energy is small and there are no binary bindings between the constituents which move mainly in relative p motion. The core ${}^{6}\mathrm{He}$ is however very fragile, being itself Borromean. This rules out "sequentially Borromean" modelling. An extended few-body Borromean concept with respect to the tightly bound ${}^{4}\mathrm{He}$ may however be considered, and appears justified by the loose binding and large spatial extension of the p motion dominated ${}^{8}\mathrm{He}$.

The oxygen chain is quite different, starting up by being d dominated which prevents halo formation. The chain is continuous all the way to the

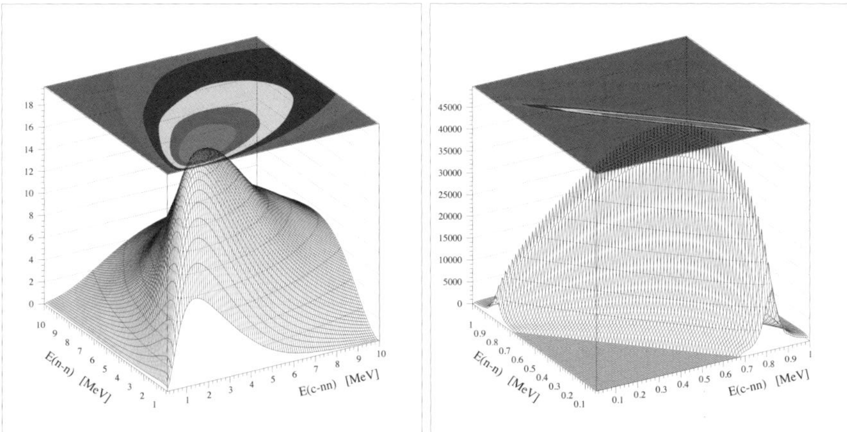

Fig. 4. Energy correlation plot for 2^+ states. Left: Without final state interaction. Right: With FSI.

dripline nucleus ^{24}O, which corresponds to $d_{5/2}$ and $s_{1/2}$ closure. The latter (s motion) might suggest a Borromean nature, but ^{23}O is bound.[10] The oxygen chain is much more complex than our test bench helium isotopes.

The character of the neutron chains is drastically amended by changing the proton number from shell closure. The carbon isotopes serve as a good recent example, and also lithium (see Fig. 2). In the latter case, by adding a proton to $Z = 2$ closure, the dripline nucleus becomes the Borromean ^{11}Li. In the case of carbon the neutron dripline is pulled in (to $N = 16$) while in lithium it is at $N = 8$, but with p^2, s^2 mixture.

4. Including the Borromean Continuum

The intensifying international "drang nach the driplines", has led to striking new discoveries concerning their location, new magic numbers, and also rich insight in the structure at the limits of nuclear existence. Going beyond the driplines by adding neutrons, to submerged rocks or distant envisioned exotic islands is already more than imagination. Ongoing exploration of the transition into the Borromean continuum, defined by the range of validity of the three-body cluster model ($\sim$ 12 MeV for ^{6}He i.e. below the $t + t$ threshold), has already taught us useful lessons, and made nuclear physics partner in a what today is called Many-Body Open Quantum Systems.

There are three types of correlations: The intrinsic correlated structure of the three-body continuum itself, the correlations in the ground state only (or No Final State Interactions), and the transition energy correlations,

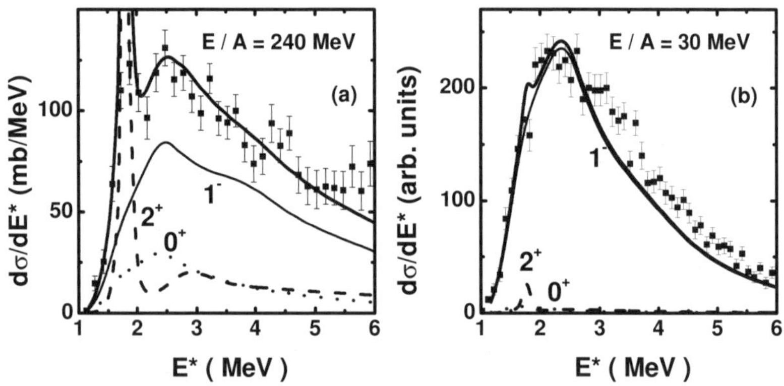

Fig. 5. Comparison of theoretical ^{6}He excitation spectra[12] (thick solid curves) for ^{6}He + ^{208}Pb Coulomb dominated breakup at high (panel a) and low (panel b) collision energies with experimental data.[13,14] The theoretical dipole 1^-, quadrupole 2^+ and monopole 0^+ contributions are shown for both energies.

filtered by the ground state, and with all final state interactions (FSI) included. The first would require $3 \to 3$ scattering. The second (NoFSI) is used in sudden approximation (or to some extent in the Serber break-up model), while the third (FSI) is probed by nuclear and electromagnetic responses which mix properties of the previous two and where the ground state still may play a very visible role. Only combined analysis of all three gives a complete understanding of the three-body continuum structure.

The clearest case of a three-body resonance is the well known 2^+ resonance in ^{6}He at 0.8 MeV above the three-body threshold. The correlation plots[11] in Fig. 4 show this now classic example of a three-body resonance. The most remarkable feature of the 'true' three-body resonance, exhibited by ^{6}He (Fig. 4(right)), is that it exists in lowest hypermoment configurations, which corresponds to the three particles interacting while all near each other. The main criterion for existence and defining property of any intrinsic resonant state, i.e. that it should not depend on the excitation mechanism (electromagnetic, strong or weak interaction etc.), serves as a signature of three-body resonant nature of observed resonant-like enhancement in cross-sections. Description of the excitation spectrum of a cluster nucleus in few-body breakup is a key objective for continuum spectroscopy. A practical way to test theoretical assumptions is to calculate the spectrum using the same structure model for different reaction mechanisms. Figure 5 shows comparison of the calculated ^{6}He excitation spectra[12] (thick solid curves) with experimental data for breakup reactions on lead target at collision energies 240[13] and 30 MeV/nucleon,[14] obtained at GSI and GANIL, respectively. Theory reproduces the shape of the low-lying excitation spectrum for both collision energies, though the breakup mechanisms are quite different. Since the target is a heavy nucleus, Coulomb dissociation dominates at both energies. At low collision energy, the soft dipole mode (suggested by cluster pioneer K. Ikeda[15]) dominates the spectrum, and only a small remnant of the 2^+ (1.8 MeV) three-body resonance is visible. At high collision energy, the dipole mode remains large but quadrupole and monopole excitations give considerable contributions to the cross section, in particular the quadrupole resonance.

5. Two-Proton Emitters

The theoretical studies of systems which undergo "spontaneous" three-body decay differ qualitatively from the nuclear physics methods used for studies of weakly bound three-body nuclear states. To understand, for example, the structure of the neutron halo in the ^{6}He ground state by nuclear reactions, we need to excite i.e. destroy this nucleus - induce its breakup. The reaction mechanism plays an important role in these processes and simultaneously introduces uncertainties in the theoretical treatment. However, if we turn from the ^{6}He ground state to its *isobaric* partner, the ^{6}Be ground state (which has to have the same nuclear structure within the paradigm of isobaric symmetry), we go from the Borromean halo situation to the case of true $2p$ decay. In this case the system decays *spontaneously* (without external influence) and as far as we will be able to calculate theoretically the process of three-body decay properly, the interpretation of experimental results becomes more certain. The first consistent quantum mechanical theory of two-proton radioactivity (three-body decay) was developed in 2000.[16,17] Predictions for lifetimes and correlations strongly differed from the predictions of the widespread quasi-classical diproton and simultaneous decay models. With subsequent experimental discovery of $2p$ radioactivity the basic predictions of the quantum mechanical model were confirmed.[18,19] Despite the advances in the experimental studies, the 'heavy' $2p$ emitters (e.g. ^{19}Mg, ^{45}Fe, ^{54}Zn) are only so far produced with relatively low statistics.

One might expect lighter two-proton emitters such as ^{6}Be to be more easily populated in various reactions, providing statistics sufficient for detailed comparison with theory. In the first paper[20] devoted to studies of ^{6}Be ground state decay to the ^{4}He + p + p channel, the ^{4}He spectrum was measured in the ^{6}Be center of mass. The authors could not fit the data using

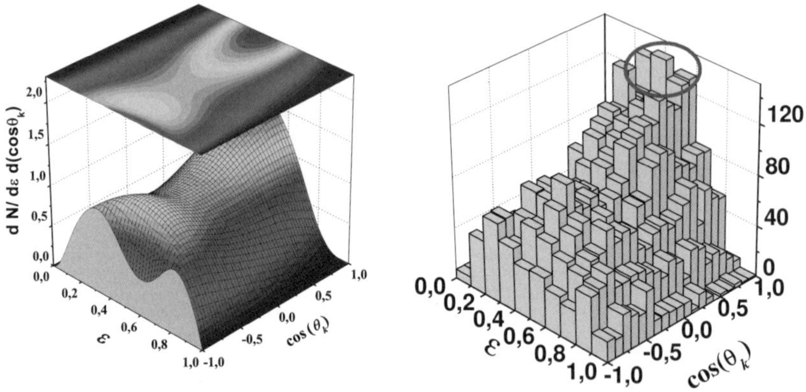

Fig. 6. Comparison of angular-energy correlations from $2p$ decay of ^{6}Be in the Y Jacobi coordinate system[21] : theoretically predicted (left panel), measured in the ^{10}B(p, n)^{10}C→^{6}Be + α reaction[22] (right panel).

simple decay scenarios (phase volume, p-p final state interaction, sequential decay through ^{5}Li) and concluded "... no incoherent sum of the processes considered here will fit the data. Perhaps a full three-body calculation is necessary to understand the energy spectrum."

The $2p$ decay of ^{6}Be was very recently measured[22] in the ^{10}B(p, n)^{10}C→^{6}Be + α reaction. Figure 6 shows energy-angular correlations for ^{6}Be. The theory[21] is in good agreement with experimental data. It is interesting to note that a peculiarity of the ^{6}Be decay spectrum, mentioned in the first paper[20] devoted to ^{6}Be studies, namely that the largest cross section occurs when two protons are emitted 180° from one another with nearly equal energies, is confirmed in the new experiment and reproduced well by theory. See the largest peak in the theoretical spectrum and the region marked by an ellipse in the experimental spectrum.

6. Conclusions and Outlook

The Borromean halo nuclei display universality in few-body systems with large scattering lengths, thus their properties have reduced dependence on the details of the binary interactions at short distances. Recent years have seen a rapidly growing literature from nano- to femtometer scale dealing with exotics such as Efimov states and Borromean halos and new structural insight linked to universality.[23] Possibilities on nano scale to tune the experimental conditions to produce the exotics have certainly brought this physics to the catwalk.

A quarter of a century has elapsed since the discovery by Tanihata and his team at Berkeley that led to halo physics. The field has been driven by remarkable experimental progress and discoveries, and more is to come with new and upgraded facilities, like FAIR in Europe. From finding adequate phenomenological ways to describe the physics of halo ground states, ambitions have grown to also understand the three-body halo continua. Theory has gradually become a useful partner for experiment, also concerning predictions, although reaction theory remains a challenge, in particular at lower energies. This has however not prevented theory from having ambitions to go further, to also understand the self-organization of halo systems in an *ab initio* manner, i.e. starting out from nucleon constituents and their interactions. This paper is a reminder on how the road, partly a catwalk, has been made - while we go.

References

1. H. Kragh, *Quantum Generations — A History of Physics in the Twentieth Century*, Princeton University Press, Princeton, New Jersey, 1999.
2. I. Tanihata et al., *Phys. Rev. Lett.* **55** (1985) 2676
3. P. G. Hansen and B. Jonson, *Europhysics Letters* **4** (1987) 409
4. M. V. Zhukov et al., *Phys. Rep.* **231** (1993) 151
5. Yu. Ts. Oganessian, V. I. Zagrebaev and J. S.Vaagen, *Phys. Rev. Lett.* **82** (1999) 4996
6. K. Tanaka et al., *Phys. Rev. Lett.* **104** (2010) 062701
7. R. Kanungo et al., *Phys. Rev. Lett.* **102** (2009) 152501
8. O. Jensen, *Spectroscopic Factors with Coupled-Cluster. Connecting ab initio Nuclear Structure to Reactions*, Thesis, Bergen University, 2010 (submitted)
9. G. Hagen et al., *Phys. Rev. Lett.* **104** (2010) 182501
10. C. R. Hoffman et al., *Phys. Rev. Lett.* **100** (2008) 152502
11. B. V. Danilin et al., *Phys. Rev. C* **73** (2006) 054002
12. S. N. Ershov, L. V. Grigorenko, J. S. Vaagen and M. V. Zhukov, *J. Phys. G: Nucl. Part. Phys.* **37** (2010) 064026
13. T. Aumann et al., *Phys. Rev. C* **59** (1999) 1252
14. N. Orr, *arXiv:nucl-ex* (2008) 0803.0886
15. K. Ikeda, *INS Report JHP-7* (1988) in Japanese
16. L. V. Grigorenko et al., *Phys. Rev. Lett.* **85** (2000) 22
17. L. V. Grigorenko et al., *Phys. Rev. C* **64** (2001) 054002
18. K. Miernik et al., *Phys. Rev. Lett.* **99** (2007) 192501
19. I. Mukha et al., *Phys. Rev. C* **77** (2008) 061303(R)
20. D. F. Geesaman et al., *Phys. Rev. C* **15** (1977) 1835
21. L. V. Grigorenko et al., *Phys. Rev. C* **80** (2009) 034602
22. K. Mercurio et al., *Phys. Rev. C* **80** (2009) 024306
23. E. Braaten and H.-W. Hammer, *Phys. Rep.* **428** (2006) 259

NUCLEAR MATTER, NUCLEI, AND NEUTRON STARS IN HADRON AND QUARK-HADRON MODELS

S. SCHRAMM

Frankfurt Institute for Advanced Studies, Goethe University,
Frankfurt, D-60438, Germany
E-mail: schramm@fias.uni-frankfurt.de

V. DEXHEIMER

Gettysburg College,
300 North Washington Street Gettysburg, PA 17325 USA
E-mail: vantoche@gettysburg.edu

R. NEGREIROS

Frankfurt Institute for Advanced Studies, Goethe University,
Frankfurt, D-60438, Germany
E-mail: negreiros@fias.uni-frankfurt.de

T. SCHÜRHOFF

Frankfurt Institute for Advanced Studies, Goethe University,
Frankfurt, D-60438, Germany
E-mail: schrhoff@fias.uni-frankfurt.de

We develop a unified model of hadrons and quarks. Within this approach we investigate the phase structure of the model as function of temperature and chemical potential. Computing the equation of state of cold matter we determine neutron and hybrid star masses and radii. In an extension of the investigation we consider the cooling behavior of the compact stars and derive a general relation between the star's mass and rotation and its cooling behavior. Finally we study the effect of Δ resonances for star matter, especially with respect to possible solutions of stars with small radii.

Keywords: Chiral models, hadronic matter, neutron stars, hybrid stars.

1. Introduction

A central topic in modern nuclear physics is the study of the general behavior of highly excited strongly interacting matter. One expects a transition from the effective degrees of freedom at low excitations, hadrons, to a phase of quarks and gluons at high temperatures and/or densities. This can be described by the restoration of chiral symmetry and the deconfinement transition. The actual structure of the transition is still unknown, perhaps with the exception in the case of vanishing chemical potential, where lattice results show a cross-over transition. Some lattice results also point to a critical end-point of a first-order transition line in temperature and chemical potential.[1] In order to describe the matter a main challenge for theory is to develop models that can cover all the ranges of temperatures and densities. This requires a model that is able to reproduce nuclear matter ground state properties and the quark and gluon phase beyond the phase transition. In the following we develop an effective theoretical description along these lines. Results for neutron star properties and their cooling behavior within this approach will be presented.In addition we study the influence of the Δ baryons on stellar properties.

2. Effective Chiral Model

In order to describe hadronic physics as well as quark degrees of freedom we couple an effective chiral flavor-SU(3) model for hadrons[2] with a so-called PNJL quark model. The model includes the lowest SU(3) multiplets of hadrons and quarks, where baryonic and constituent quark vacuum masses are generated via spontaneous symmetry breaking by the coupling of the particles to the scalar fields, which attain non-zero vacuum expectation values. Explicitly the Lagrangian reads:[4]

$$L = L_{Kin} + L_{Int} + L_{Self} + L_{SB} - U. \tag{1}$$

Here, the interaction term between baryons, quarks and mesons is given by

$$L_{Int} = -\sum_i \bar{\psi}_i[\gamma_0(g_{i\omega}\omega + g_{i\phi}\phi + g_{i\rho}\tau_3\rho) + M_i^*]\psi_i,$$

with the effective baryon and quark mass M_i^*

$$M_i^* = g_{i\sigma}\sigma + g_{i\delta}\tau_3\delta + g_{i\zeta}\zeta + \delta M_i \ . \tag{2}$$

δM_i is a small explicit mass term. The self interactions of the scalar fields, generating the spontaneous symmetry breaking, and of the vector fields are contained in

$$
\begin{aligned}
L_{Self} = & -\frac{1}{2}(m_\omega^2\omega^2 + m_\rho^2\rho^2 + m_\phi^2\phi^2) \\
& + g_4\left(\omega^4 + \frac{\phi^4}{4} + 3\omega^2\phi^2 + \frac{4\omega^3\phi}{\sqrt{2}} + \frac{2\omega\phi^3}{\sqrt{2}}\right) \\
& + k_0(\sigma^2 + \zeta^2 + \delta^2) + k_1(\sigma^2 + \zeta^2 + \delta^2)^2 \\
& + k_2\left(\frac{\sigma^4}{2} + \frac{\delta^4}{2} + 3\sigma^2\delta^2 + \zeta^4\right) + k_3(\sigma^2 - \delta^2)\zeta \\
& + k_4 \ \ln \frac{(\sigma^2 - \delta^2)\zeta}{\sigma_0^2\zeta_0} .
\end{aligned} \tag{3}
$$

Finally, the explicit symmetry breaking term L_{SB} has the form

$$
L_{SB} = m_\pi^2 f_\pi \sigma + \left(\sqrt{2}m_k^2 f_k - \frac{1}{\sqrt{2}} m_\pi^2 f_\pi\right)\zeta \ . \tag{4}
$$

The mesons included are the vector isoscalar fields ω and ϕ, the vector isovector ρ, the scalar isoscalars σ and ζ (non-strange and strange quark-antiquark state, respectively) and the scalar-isovector state δ. The isovector mesons affect isospin asymmetric matter, which makes them relevant for the study of compact stars. The various couplings are fitted to reproduce the vacuum masses of the baryons and mesons, nuclear saturation properties (density $\rho_0 = 0.15$ fm^{-3}, binding energy per nucleon $B/A = -16.00$ MeV, nucleon effective mass $M_N^* = 0,67\ M_N$, compressibility $K = 297.32$ MeV), asymmetry energy ($E_{sym} = 32.50$ MeV), and reasonable values for the hyperon potentials ($U_\Lambda = -28.00$ MeV, $U_\Sigma = 5.35$ MeV, $U_\Xi = -18.36$ MeV). The numerical values of the couplings can be found in[2–4] .

In order to be able to study the transition from hadronic to quark degrees of freedom the model incorporates a Polyakov loop field Φ in analogy to the PNL models for quarks.[5,6] The field has an effective potential of the form

$$
\begin{aligned}
U = & (a_0T^4 + a_1\mu^4 + a_2T^2\mu^2)\Phi^2 \\
& + a_3T_0^4 \log(1 - 6\Phi^2 + 8\Phi^3 - 3\Phi^4)
\end{aligned} \tag{5}
$$

depending on the temperature T and the baryochemical potential μ. The μ independent terms have been fitted, such that thermodynamical quantities of quenched lattice calculation agree well with the model results over a wide range of temperature. The other two parameters of the potential have been adjusted in order to have a phase diagram with a first-order phase transitions at high densities which ends in a critical end-point at a temperature T and chemical potential μ in agreement with projections by lattice calculations.[1] In order to suppress quarks at low densities and temperatures as well as suppress hadrons at high excitation energies an effective mass shift is introduced in the form:

$$\delta M_B^* = g_{B\Phi}\Phi^2 \quad , \quad \delta M_q^* = g_{q\Phi}(1-\Phi) \ . \tag{6}$$

3. Results

In an earlier investigation, where we fine-tuned the couplings to properties of finite nuclei, we computed nuclei in a two-dimensional calculation including nuclear deformation.[7] The corresponding equations of motion for the mesonic fields as well as the Dirac equation for the nucleons were solved on a spatial grid. Computing all nuclei with known binding energies an overall accuracy of the energies can be achieved with an average error for nuclei with 50 or more nucleons of

$$\epsilon_E(A > 50) \approx 0.21\% \ . \tag{7}$$

This results compares very well with standard relativistic and non-relativistic mean-field nuclear structure models. As can be seen in Fig. 1 magic numbers can be reproduced as well as the measured deformations of many nuclei. Thus this shows that a more general hadronic model approach can be used to perform competitive nuclear structure calculations.

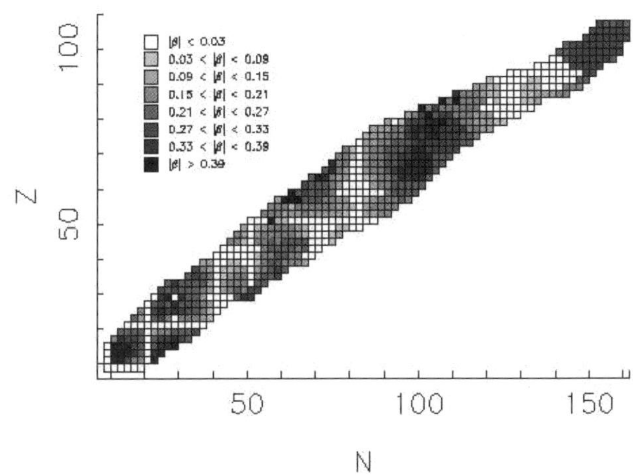

Fig. 1. Nuclear chart as calculated in the effective chiral model. The gray level encodes the deformation (absolute value) of the corresponding nucleus. One can clearly observe lines of spherical nuclei corresponding to magic numbers.

Investigating the full model and tuning the couplings in order to reproduce high-mass neutron stars as well as having reasonable nuclear ground state values[4] we minimize the grand canonical potential in mean-field approximation following from the Lagrangian discussed in the previous section.[4] Figure 2 shows the resulting phase transition curve for symmetric nuclear matter and star matter in β equilibrium. The critical end-point occurs at $T_c \approx 167$ MeV and $\mu_c \approx 354$ MeV in agreement with[1] . The phase transition to quarks occurs at 4 times ground state density for vanishing temperature. Since the model features a realistic nuclear ground state the figure also shows a quantitatively reasonable first-order liquid gas phase transition.

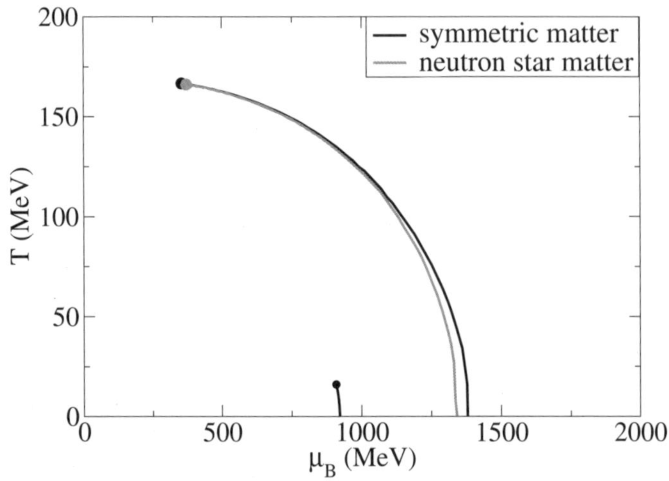

Fig. 2. Phase diagram as function of temperature and baryon chemical potential. The first order phase transition stops at a critical end point as suggested in.[1] At $T = 0$ the transition takes place at 4 times nuclear matter saturation density. The phase transition for symmetric and star matter is shown.

Computing the resulting field values one can determine energy density and pressure for stellar matter. Using the equation of state we solve the Tolman-Oppenheimer-Volkov equations for non-rotating, spherical neutron stars.[8,9] The resulting masses and radii of the stars are shown in Fig. 3. Excluding quarks the maximum attainable mass is about 2.1 solar masses. Taking quarks into account the maximum mass drops to ≈ 2 solar masses. Note that when quarks are populated the region of stable stars is cut off as the equation of state in this regime becomes softer and cannot support stable high-mass star solutions any more. This is true for a Maxwell construction of the phase transition assuming local charge neutrality of the matter. In a Gibbs construction, which only assumes global charge neutrality, allowing for a mixture of quark and baryonic phases with opposite charge, this sharp cut off is softened as can be seen in the inset of the figure. As a result at the highest star masses a core of about 2 km size develops that contains a mixture of quarks and baryons. This can also be observed in Fig. 4, which shows the densities of baryons and quarks as function of chemical potential. Around the phase transition the plot shows this mixture of states.

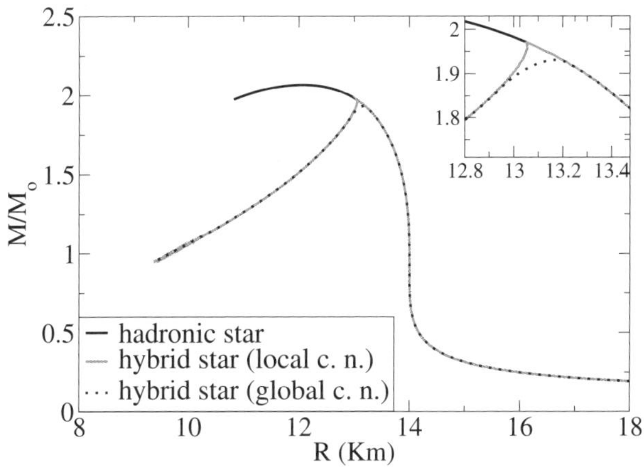

Fig. 3. Masses and radii of neutron stars. The results for purely baryonic stars and for stars including the quark phase are shown. The inset illustrates the difference between a Maxwell and a Gibbs construction of the hadron-quark phase transition.

One of the most interesting observables concerning neutron stars is their temperature compared to an age estimate of the star, defining the cooling behavior of the neutron star. In order to study this feature we determine the equation of state of the hadron-quark model for stellar matter at finite temperature. Using this result in a cooling simulation of the star we compute the cooling behavior stars with different masses.[10] Figure 5 shows the result of the calculation. Compared to the measurements also shown in the figure, one can conclude that a good agreement with data can be achieved if the corresponding (unknown) star masses are roughly below 1.2 solar masses. For higher masses the model predicts much faster cooling rates. The main reason for this behavior is the on-set of direct Urca processes for

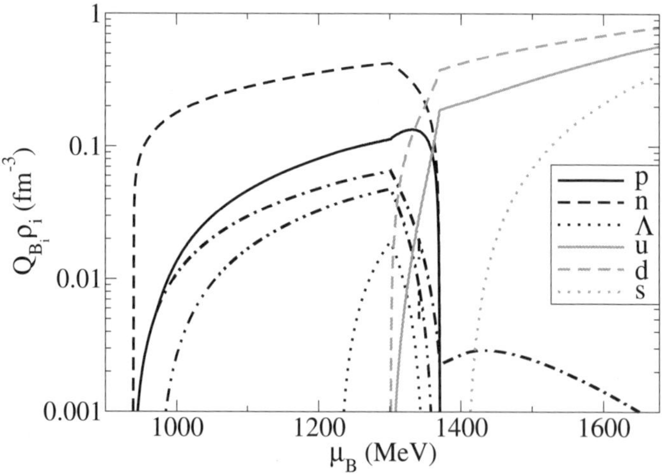

Fig. 4. Baryon number densities of particles for star matter as function of baryon chemical potential. The mixture of baryons and quarks results from the Gibbs construction of the first order phase transition assuming a mixed phase with global charge conservation.

heavy stars. The most important processes are

$$n \rightarrow p + e^- + \bar{\nu}_e \quad , \quad p + e^- \rightarrow n + \nu_e \quad . \tag{8}$$

These processes generate fast cooling via neutrino emission. In a more general analysis we take into account rotation of the star. As rotation reduces the central energy density of the star also the threshold of the direct Urca process shifts to higher masses. The result of this analysis is shown Fig. 6. Here one can read off the ranges of frequencies and masses of the stars that allow for the direct Urca process to take place and there leads to fast cooling of the star.

In a separate investigation concentrating on a purely baryonic star, for simplicity, we investigate the possibility of the existence of neutron stars with rather small radii around 10 km and below, as has been suggested in an analysis of masses and radii of three neutron stars discussed in.[11] Here we also include the spin-3/2 baryonic multiplet as degrees of freedom as has already been studied in.[3] As with other baryons the couplings of the particle to the scalar fields are fixed by reproducing baryonic vacuum masses. The vector interaction strength is less clear. Using vector dominance arguments a natural choice is to take the same value as in the case of the corresponding baryonic octet state with the same number of light and strange quarks. Starting from this assumption we allow for a small reduc-

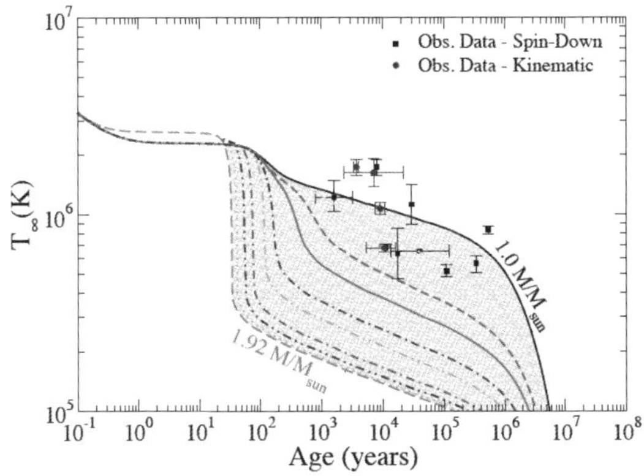

Fig. 5. Cooling behavior as function of time for compact stars as calculated in this model. Observational data are also shown. One can observe a good agreement with the data assuming low star masses below 1.2 solar masses (assuming non-rotating stars). For higher masses the direct URCA process leads to enhanced cooling.

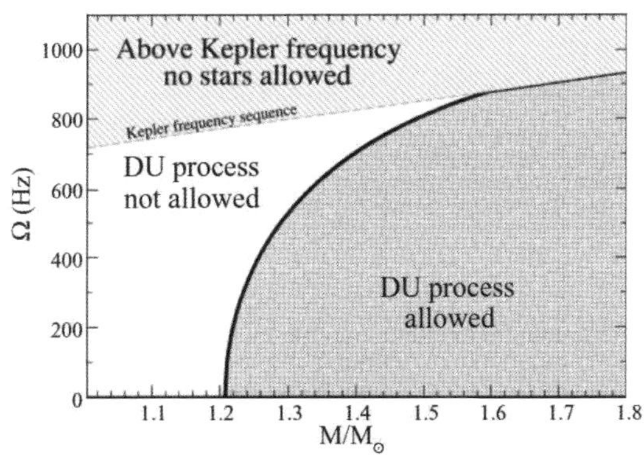

Fig. 6. The figure illustrates the range of star masses and rotation frequencies for which the Urca process takes place (DU process allowed). The shaded region at high frequencies is not physically realized as the stars would be instable with respect to mass shedding.

tion from this canonical value.[12] The resulting star masses and radii are shown in Figure 7. One can see that for a moderate change in the vector strength of about 10 percent ($r_V = 0.9$) one obtains good agreement with the values obtained in the analysis of the observational data. Thus it is possible to reproduce stars with relatively small radii in a purely baryonic model without introducing a more exotic solution like a quark star. The population densities of the various baryons as function of chemical potential are shown in Fig. 8. It is interesting to observe a very different structure compared to calculations not including the Δ resonances. In fact, beyond 4 times nuclear ground state densities the stellar matter is dominated by Δ's.

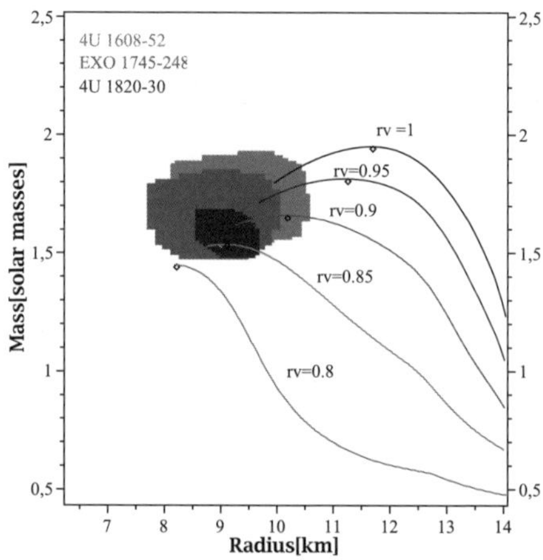

Fig. 7. Neutron star masses and corresponding radii. The strength of the vector-interaction of the Delta resonance is varied. Also shown are the results of an analysis of three observed neutron stars.[11] The model calculation can reproduce the observational data for a relative vector strength $r_v = 0.9$.

4. Summary

We developed a unified hadron-quark model with the correct degrees of freedom on both sides of the QCD phase transition. Within this approach we were able to reproduce a phase transition line with a critical end point for values of temperature and chemical potential as were suggested by lattice gauge calculations. Using these model parameters we studied neutron star properties and found heavy stars with maximum mass values of around 2 solar masses. The core of the star contained a mixture of baryons and quarks. Including a study of the cooling of the star we observed agreement with data for low-mass stars. In the case of high masses the direct Urca process led to fast cooling rates. We determine a general parameter region of stars with respect to rotational frequency and mass that allowed for the Urca process and thus for fast cooling.

In an investigation of purely baryonic but taking into account the baryonic spin-3/2 multiplet, in particular the Δ resonances, we show that one

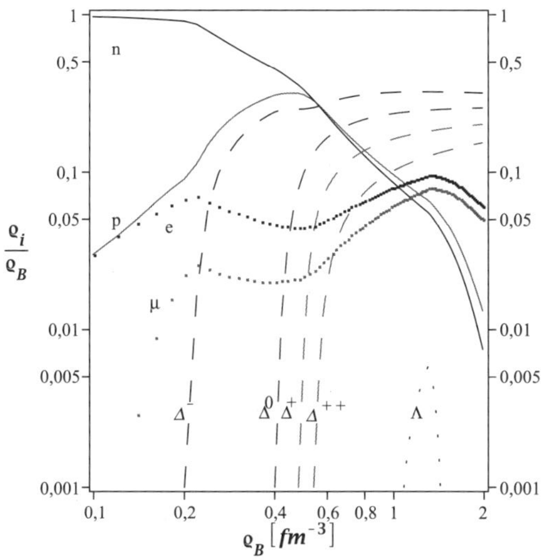

Fig. 8. Particle number densities as function of density including Δ resonances. One can observe that the Δs dominate other particles at densities $\rho > 4\rho_o$.

is able to reproduce stars with small radii, as suggested by the analysis of observational data, by a slight tuning of the vector coupling strength of the spin-3/2 particles without the need for exotic particles.

References

1. Z. Fodor and S. D. Katz, JHEP **0404**, 050 (2004) [arXiv:hep-lat/0402006].
2. P. Papazoglou, D. Zschiesche, S. Schramm *et al.*, Phys. Rev. **C59**, 411-427 (1999).
3. V. Dexheimer and S. Schramm, Astrophys. J. **683**, 943 (2008).
4. V. A. Dexheimer and S. Schramm, Phys. Rev. C **81**, 045201 (2010).
5. K. Fukushima, Phys. Lett. B **591**, 277 (2004) [arXiv:hep-ph/0310121].
6. C. Ratti, M. A. Thaler and W. Weise, Phys. Rev. D **73**, 014019 (2006).
7. S. Schramm, Phys. Rev. C. **66**, 064310 (2002).
8. R. C. Tolman, Phys. Rev. **55**, 364 (1939).
9. J. R. Oppenheimer and G. M. Volkoff, Phys. Rev. **55**, 374 (1939).
10. R. Negreiros, V. Dexheimer, and S. Schramm, Phys. Rev. C **82**, 035803 (2010).
11. F. Özel, G. Baym, and T. Güver, arXiv:1002.3153v1 (2010).
12. T. Schürhoff, S. Schramm, and V. Dexheimer, Ap. J. **724**, L74.

DETECTING QUARK MATTER IN THE EARLY UNIVERSE BY GRAVITATIONAL WAVES

S. SCHETTLER*, M. T. BOECKEL AND J. SCHAFFNER-BIELICH

Institute for Theoretical Physics, Philosophenweg 16, Heidelberg University, Germany
**E-mail: s.schettler@thphys.uni-heidelberg.de*

For large baryochemical potential μ_{B} strongly interacting matter might undergo a first order phase transition at temperatures $T \sim 100 - 200$ MeV. Within standard cosmology, however, μ_{B} is assumed to be very small leading to a crossover. We discuss implications of a first order QCD transition at high μ_{B} being consistent with current observations. In this contribution we concentrate on effects on the gravitational wave spectrum. There are other interesting cosmological signals as a modification of the power spectrum of dark matter, the production of stellar black holes, and the seeds for the extragalactic magnetic fields which we briefly address also.

Keywords: Cosmological QCD phase transition; gravitational waves.

1. Introduction

In the standard scenario of the hot big bang a period of inflation is assumed to have taken place in the very early universe. During inflation metric perturbations are created of which the tensorial part forms a spectrum of primordial gravitational waves (GWs) presumably propagating through spacetime until today. Not interacting with the medium they can retain information about the very early stages of the cosmos. Nevertheless, the evolution of the universe can leave imprints in the primordial GW signal because the GW spectrum depends on the behavior of the Hubble parameter H given by the Friedmann equation

$$H^2 \equiv \left(\frac{\dot{a}}{a}\right)^2 = \frac{8\pi G}{3}\rho \propto g(T)T^4.$$

The proportionality holds for a radiative medium with $g(T)$ effective relativistic degrees of freedom. A GW starts to decay when the wavelength λ becomes smaller than the Hubble radius H^{-1}, i.e. at horizon entry. H^{-1} depends on g and T whereas, by contrast, λ grows proportional to the scale

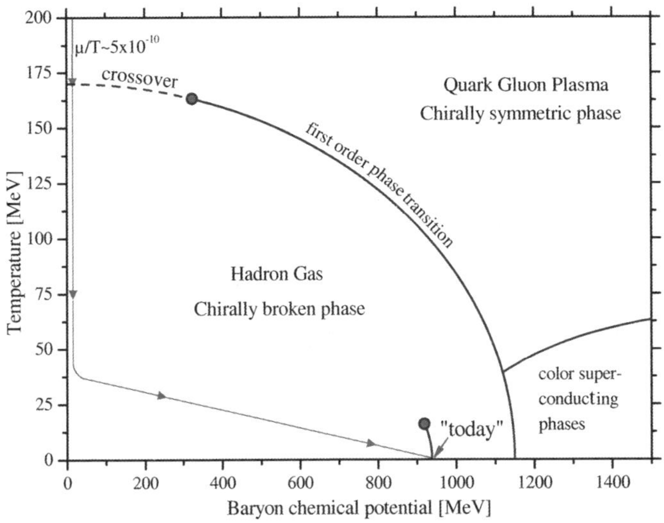

Fig. 1. Outline of a possible phase diagram for strongly interacting matter. In the conventional picture the early universe evolves at small baryochemical potential, where lattice QCD predicts a crossover. The evolution of the universe follows the arrow, see also.[1] (Figure taken from Ref. 2.)

factor a—independently of the properties of the medium. Thus, a drop in the relativistic degrees of freedom, as expected during the QCD phase transition, can modify the spectrum. In this contribution we present our results for the primordial GW spectrum after different types of QCD phase transitions including one at high μ_B featuring a short period of inflation.[2,3] The little inflation also modifies the power spectrum of large-scale structure up to an enclosed dark matter mass of $M \sim 10^6 M_\odot$ and dilutes the cold dark matter density up to a factor of 10^{-9}. This requires a reduced WIMP annihilation cross section or a larger WIMP mass which could be probed at the LHC. During a first order transition collisions of charged bubbles can generate magnetic fields which could be the seed fields for the galactic and extragalactic magnetic fields observed today. We point out that the concept of a short inflationary period during the QCD phase transition has also been introduced earlier by Kämpfer et al.[4,5] and Borghini et al.[6]

2. QCD Phase Transition with a Little Inflation

From lattice calculations we know that for small baryochemical potential strongly interacting matter undergoes a crossover from the hadronic phase to the quark-gluon plasma. As observations of the cosmic microwave background and measurements of the abundances of light elements suggest, the baryon to photon ratio in the early universe and equivalently the baryochemical potential are well within the crossover region of the QCD phase diagram: $n_\mathrm{B}/s \sim n_\mathrm{B}/n_\gamma \sim \mu_\mathrm{B}/T \sim 10^{-9}$, see Fig. 1. This ratio is conserved except at baryogenesis and during a first order phase transition with release of latent heat. At $\mu_\mathrm{B}/T \sim 1$ there are no reliable data from lattice QCD, and a first order phase transition is not excluded. As will be discussed in the following, in this case there may be observable signals from the transition. But how do we circumvent the bounds on μ_B set by big bang nucleosynthesis (BBN) data? A possible scenario is the following (see also Fig. 2): Before QCD time a strong baryon asymmetry is produced, e.g. by an Affleck-Dine baryogenesis. This can lead to $\mu_\mathrm{B}/T \sim 1$ and therefore allows for a first order transition during which the universe is trapped in a false metastable vacuum. The energy density is then dominated by the potential of this state. Since the latter is constant in time, it leads to an exponential growth of the scale factor and thus to an additional period of (de Sitter) inflation:

$$H^2 \propto \rho = \mathrm{const.} \quad \Longrightarrow \quad \frac{\dot{a}}{a} = \mathrm{const.} \quad \Longrightarrow \quad a \propto \exp(H \cdot t).$$

After the inflationary period the baryochemical potential should be at the value needed for BBN to produce the right amount of light elements. Tagging quantities before the inflation with a subscript i and those afterwards with f we can write down

$$\left(\frac{\mu_\mathrm{B}}{T}\right)_\mathrm{f} \approx \left(\frac{a_\mathrm{i}}{a_\mathrm{f}}\right)^3 \left(\frac{\mu_\mathrm{B}}{T}\right)_\mathrm{i} \overset{!}{\approx} 10^{-9}.$$

Hence, to provide the necessary dilution of μ_B only $N = \ln(a_\mathrm{f}/a_\mathrm{i}) \sim \ln(10^3) \sim 7$ e-folds are sufficient. This is few in comparison to the corresponding value $N \sim 50$ in standard inflation. Nevertheless, the production of hadronic bubbles within the supercooled quark-gluon plasma must be suppressed in order to delay the phase transition. The corresponding nucleation rate depends on the difference in the free energy of the two phases and on the surface tension.[7] For high densities this is discussed in Refs. 8,9. The attenuation of the fraction μ_B/T is accompanied by a strong supercooling which passes into a period of reheating after the little inflation. Thereby the transition temperature T_c is reached again and the evolution of the universe returns to its conventional path.

Finally we should mention that the scenario described above is only one possibility for a first order QCD phase transition in agreement with observation. Another one includes a large lepton asymmetry and is discussed in Ref. 10.

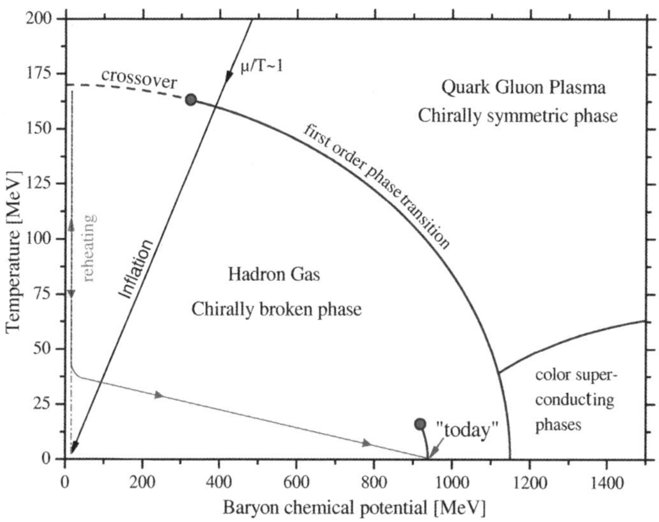

Fig. 2. Possible track of the universe through the phase diagram for strongly interacting matter. The system crosses the first order line starting from high chemical potential μ_B. It is supercooled during a short period of inflation followed by the actual phase transition when reheating sets in. (Figure taken from Ref. 2.)

3. Tensor Perturbations and the QCD Trace Anomaly

In the future, observation of gravitational waves may allow for distinction between different types of QCD transitions. To see this we need the (gauge invariant) equation of motion for the tensor perturbation amplitude $v_k = a \cdot h_k$ in Fourier space. Making use of conformal time η and denoting the corresponding derivatives with primes we get

$$v_k''(\eta) + \left(k^2 - \frac{a''}{a}\right) v_k(\eta) = 0$$

where

$$\frac{a''}{a} = \frac{4\pi G a^2}{3}\left(\rho - 3p\right).$$

The last expression is proportional to the trace anomaly to which GWs are sensitive. In order to calculate the evolution of the primordial GW spectrum around QCD time we use parameterizations corresponding to several lattice calculations, see Fig. 3. The lattice results were obtained with improved staggered fermion actions (asqtad, p4, and stout) with and without including a hadron resonance gas (HRG).[11–13] The resulting GW spectra show a step-like structure at a critical frequency ν^* which corresponds to the mode

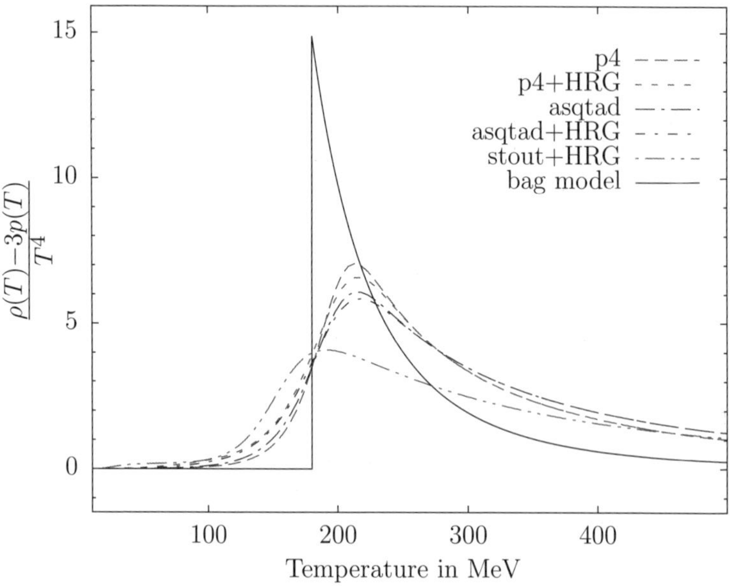

Fig. 3. The trace anomaly of strongly interacting matter. The dashed curves depict parameterizations of results of lattice QCD. In part also the hadron resonance gas (HRG) has been taken into account. (Figure taken from Ref. 2.)

that enters the Hubble horizon at the end of the phase transition. In Fig. 4 the resulting energy density in GWs per logarithmic frequency interval is shown:

$$\Omega_{\rm g}(k) = \frac{1}{\rho_{\rm c}} \frac{d\rho_{\rm g}}{d\ln k} \quad \text{with } \rho_{\rm c} \text{ being the critical density.}$$

The spectra in Fig. 4 are rather insensitive to details of the phase transition. This is because entropy is conserved during a phase transition described by the bag model or lattice data at vanishing $\mu_{\rm B}$. Entropy conservation in turn leads to

$$\frac{\Omega_{\rm g}(\nu \gg \nu^*)}{\Omega_{\rm g}(\nu \ll \nu^*)} = \left(\frac{g_{\rm f}}{g_{\rm i}}\right)^{1/3} \approx 0.7$$

if we choose the typical values $g_{\rm i} = 51.25$ and $g_{\rm f} = 17.25$ as in Ref. 14. However, in case of large entropy release, as it is expected at the end of an inflationary period, the resulting spectrum differs drastically: Because during inflation the successive entrance of modes into the horizon is reversed to a successive horizon exit, the high frequency modes in Fig. 5 are strongly suppressed with respect to the ones with low frequency. In addition to the

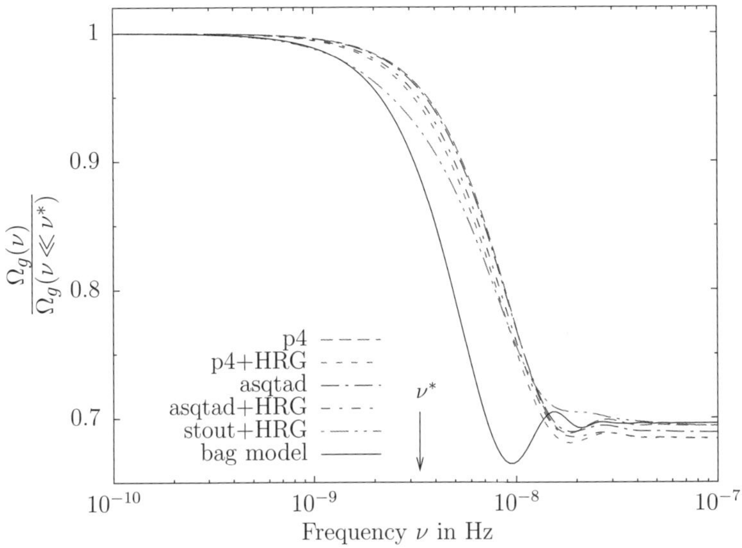

Fig. 4. The energy density of GWs per logarithmic frequency interval after the QCD phase transition assuming entropy conservation. High frequency modes are suppressed by a factor of ~ 0.7. The bag model calculation has been done with $T_c = 180$ MeV. The dashed curves represent calculations with lattice data. (Figure taken from Ref. 2.)

consequential drop $\Omega_g(\nu) \propto \nu^{-4}$ we observe strong oscillations within the spectrum. They emerge because each mode is at a different phase at horizon exit during inflation.

4. Observations of Gravitational Waves

In Fig. 6 different possible GW spectra after the QCD phase transition are displayed. In the upper panel the strain amplitude of the primordial GWs after a conventional QCD transition is compared with a spectrum resulting from the inflationary scenario (omitting the oscillations). The step frequency in the amplitude is close to the highest sensitivity of pulsar timing methods. We also display a spectrum which is expected after a phase transition which additionally includes a long period of domination of kinetic energy of a scalar field (kination). Kination can occur during the process of reheating after inflation and here serves as an example of how future measurements of the GW spectrum could shed light on the time evolution of the equation of state in the early universe. The lower panel shows possible spectra of GWs emanating from bubble collisions and turbulences during a first order QCD transition. They most strikingly differ at high frequencies:

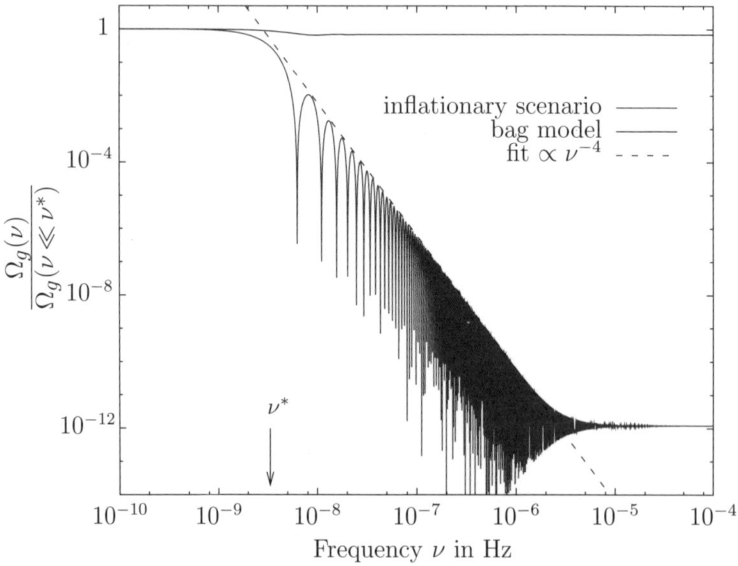

Fig. 5. The GW spectrum after a QCD transition with a little inflation. The inflationary scenario entails a much stronger dilution than a transition with conserved entropy. (Figure taken from Ref. 2.)

In this regime the slopes of the spectra deviate because in Ref. 15 multi-bubble collisions are included. Note that these analytic estimates were done for a general first order transition at QCD time and do not include a short period of inflation. For both we chose the maximal strength still being consistent with PPTA measurements. Recent analyses of data measured by the European Pulsar Timing Array EPTA[16] pose similar constraints on the spectra displayed in the lower panel. The Square Kilometre Array SKA is expected to push these limits further down. By contrast, a detection of the primordial GW spectrum seems impossible in the foreseeable future. However, comparison of Planck[17] measurements at $\nu \sim 10^{-18}$ Hz with those of the proposed Big Bang Observer BBO[18] at $\nu \sim 1$ Hz will potentially allow for a discrimination between an inflationary transition and the standard one.

5. Summary

We have presented an alternative scenario for the cosmological QCD phase transition which includes a short period of inflation. The scenario depends on a strong baryogenesis before the QCD scale allowing the universe to intersect the first order transition line which is suggested by effective models

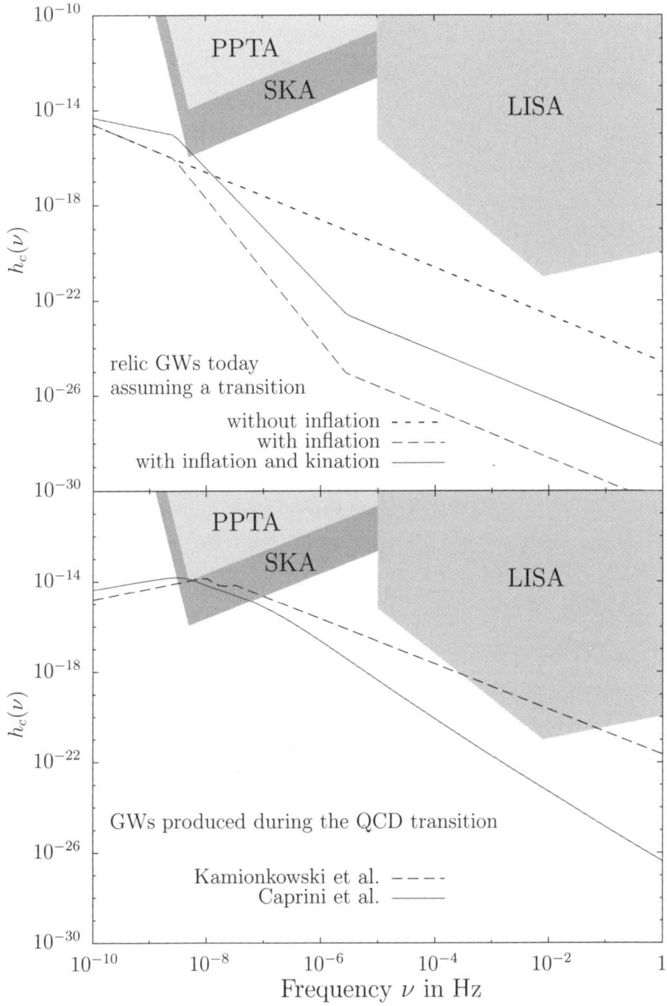

Fig. 6. Different GW spectra after the QCD transition. The sensitivities of PPTA (Parkes Pulsar Timing Array), SKA (Square Kilometre Array) and LISA (Laser Interferometer Space Antenna) are displayed. PPTA already constrains the GW production during the QCD phase transition. The spectra in the lower panel are based upon analytic estimates in Ref. 15 (Kamionkowski et al.) and Ref. 19 (Caprini et al.). See also Refs. 20–22. The displayed spectrum of relic GWs has the largest amplitude consistent with COBE.[23] (Figure taken from Ref. 2.)

of QCD. The system needs to be trapped in a metastable vacuum state such that the corresponding vacuum energy leads to an inflationary phase. Such a first order phase transition would generate signals which can potentially be observed: In this contribution we discussed the strong suppression of GWs with frequencies above 10^{-8} Hz and the production of GWs in bubble collisions and turbulences during the phase transition. We also mentioned effects on the dark matter power spectrum, different requirements on the WIMP annihilation cross section, and the production of primordial magnetic fields.

Acknowledgments

One of us (JSB) is indebted to Walter Greiner for his continuing and stimulating interest in this work. This work is supported by BMBF under grant FKZ 06HD9127, by DFG within the framework of the excellence initiative through the Heidelberg Graduate School of Fundamental Physics, the International Max Planck Research School for Precision Tests of Fundamental Symmetries (IMPRS-PTFS), the Gesellschaft für Schwerionenforschung GSI Darmstadt, the Helmholtz Graduate School for Heavy-Ion Research (HGS-HIRe), the Graduate Program for Hadron and Ion Research (GP-HIR), and the Helmholtz Alliance Program of the Helmholtz Association contract HA-216 "Extremes of Density and Temperature: Cosmic Matter in the Laboratory".

References

1. M. J. Fromerth and J. Rafelski, *arXiv:astro-ph/0211346* (2002).
2. S. Schettler, T. Boeckel and J. Schaffner-Bielich, *Phys. Rev.* **D83**, 064030 (2011).
3. T. Boeckel and J. Schaffner-Bielich, *Phys. Rev. Lett.* **105**, 041301 (2010).
4. V. G. Boyko, L. L. Jenkovszky, B. Kämpfer and V. M. Sysoev, *Astron. Nachr.* **311**, 265 (1990).
5. L. L. Jenkovszky, V. M. Sysoev and B. Kämpfer, *Z. Phys. C* **48**, 147 (1990).
6. N. Borghini, W. N. Cottingham and R. V. Mau, *J. Phys. G* **26**, 771 (2000).
7. L. P. Csernai and J. I. Kapusta, *Phys. Rev. D* **46**, 1379 (1992).
8. L. F. Palhares and E. S. Fraga, *Phys. Rev. D* **82**, 125018 (2010).
9. B. W. Mintz, E. S. Fraga, G. Pagliara and J. Schaffner-Bielich, *Phys. Rev. D* **81**, 123012 (2010).
10. D. J. Schwarz and M. Stuke, *J. Cosmol. Astropart. Phys.* **0911**, 25 (2009).
11. A. Bazavov et al., *Phys. Rev. D* **80**, 014504 (2009).
12. S. Borsányi, Z. Fodor, C. Hoelbling, S. D. Katz, S. Krieg, C. Ratti and K. K. Szabó, *J. High Energy Phys.* **9**, 73 (2010).
13. S. Borsányi, G. Endrődi, Z. Fodor, A. Jakovác, S. D. Katz, S. Krieg, C. Ratti and K. K. Szabó, *J. High Energy Phys.* **11**, 77 (2010).
14. D. J. Schwarz, *Mod. Phys. Lett. A* **13**, 2771 (1998).
15. M. Kamionkowski, A. Kosowsky and M. S. Turner, *Phys. Rev. D* **49**, 2837 (1994).

16. R. van Haasteren et al., *arXiv:1103.0576* (2011).
17. The Planck Collaboration, *arXiv:astro-ph/0604069* (2006).
18. V. Corbin and N. J. Cornish, *Class. Quant. Grav.* **23**, 2435 (2006).
19. C. Caprini, R. Durrer and X. Siemens, *Phys. Rev. D* **82**, 063511 (2010).
20. S. J. Huber and T. Konstandin, *J. Cosmol. Astropart. Phys.* **9**, 22 (2008).
21. M. Kramer, D. C. Backer, J. M. Cordes, T. J. W. Lazio, B. W. Stappers and S. Johnston, *New Astron. Rev.* **48**, 993 (2004).
22. T. Kahniashvili, A. Kosowsky, G. Gogoberidze and Y. Maravin, *Phys. Rev. D* **78**, 043003 (2008).
23. B. Allen, The Stochastic Gravity-Wave Background: Sources and Detection, in *Some Topics on General Relativity and Gravitational Radiation*, ed. J. A. Miralles, J. A. Morales, & D. Saez (Cambridge University Press, 1997).

DECONSTRUCTING THE INTERIORS OF COMPACT STARS

ARMEN SEDRAKIAN

Institute for Theoretical Physics, J. W. Goethe-University,
*D-60438 Frankfurt am Main, Germany * E-mail: sedrakian@th.physik.uni-frankfurt.de*
http://th.physik.uni-frankfurt.de/ sedrakian/

Massive neutron stars may harbor deconfined quark matter in their cores. I review some recent work on the microphysics and the phenomenology of compact stars with cores made of quark matter. This includes the equilibrium and stability of non-rotating and rapidly rotating stars, gravitational radiation from deformations in their quark cores, neutrino radiation and dichotomy of fast and slow cooling, and pulsar radio-timing anomalies.

Keywords: Neutron stars; equations of state of neutron-star matter; Relativity and gravitation in astrophysics.

1. Introduction

Deconstruction is an approach, introduced by French philosopher Jacques Derrida, which rigorously pursues the meaning of a text to the point of exposing the supposed contradictions and internal oppositions upon which it is founded - showing that those foundations are irreducibly complex, unstable, or impossible. It is an approach that may be deployed in philosophy, literary analysis, or other fields. The studies of compact star may well follow this philosophical approach to achieve a unique understanding of the state of dense matter in the interiors of compact stars, by exposing various mutually contradicting theories to the experimental/observational test.

Because of the quark substructure of nucleons predicted by quantum chromodynamics (QCD), nuclear matter will undergo a phase transition to quark matter if squeezed to sufficiently high densities. In the quark matter phase the "liberated" quarks occupy continuum states which, in the low-temperature and high-density regime, arrange themselves in a Fermi sphere. The Fermi-sphere determines the form of low-lying excitation spectrum of quark matter, which resembles that of less exotic low-temperature systems found in condensed matter (e.g. electron gas or ultracold atomic vapor) or hadronic physics (e.g. nuclear or neutron matter). In analogy to these system the attractive interaction between quarks, mediated by the gluon exchange (which is responsible for the bound state spectrum of QCD, e.g., the nucleon and mesons) leads to quark superconductivity and superfluidity (color superconductivity) via the Bardeen-Cooper-Schrieffer mechanism.[1] Furthermore, under stellar conditions the pairing between the two light flavors of quarks occurs at finite isospin chemical potential, i.e., when the Fermi surfaces of up and down quarks are shifted apart by an amount which is of the same order of magnitude as the gap in the quasiparticle spectrum. Under these conditions the pairing between the fermions persists, but the actual pairing pattern may be significantly different from that of the BCS and is likely to involve breaking of the spatial symmetries by the condensate order parameter.

During the last decade there has been a substantial progress in understanding of pairing in two-component asymmetric superconductors. Firstly, new developments on (variations of) the Larkin-Ovchinnikov-Fulde-Ferrell (LOFF) phase revealed novel lattice structures of the order parameter.[2] Secondly, it has been suggested that such systems may actually phase separate into normal and superconducting domains[3] or the pairing may require changes in the shapes of the Fermi surfaces.[4] Controlled experiments on two-component cold atomic vapors with mismatched Fermi surfaces show that the phase separation scenario is at work. Whether this is the case in the related systems such as the isospin asymmetric nuclear matter or deconfined quark matter is still unclear. Therefore, a major goal of astrophysics of dense matter is to find and to quantify the manifestations of dense phases in the observable properties of compact stars. Here we review some recent progress towards this goal.

2. Color Superconducting Phases with Broken Spatial Symmetries

The quark-quark interaction is strongest for pairing between up and down quarks which are antisymmetric in color.[1] At intermediate densities quark matter is composed of up and down quarks, while strange quarks can appear in substantial amounts at higher densities. The leading candidate phase at these densities is the so-called 2SC phase, which is characterized by the order parameter $\Delta \propto \langle \psi^T(x) C\gamma_5 \tau_2 \lambda_2 \psi(x) \rangle$, where τ_2 is the Pauli matrix in the isospin state, λ_2 is the Gell-Mann matrix in the color space, $C = i\gamma^2\gamma^0$ is the matrix of charge conjugation. The isospin chemical potential is determined by the β-equilibrium condition $\mu_d - \mu_u = \mu_e \sim 100$ MeV among d and u quarks and electrons, where μ_i with $i = d, u, e$ are the chemical potentials. The Fermi spheres of up and down quarks are shifted apart by this amount. Thus, the cross-flavor pairing must overcome the disruptive effect of mismatched Fermi surfaces. Furthermore, if the physical strange quark mass is close to its current mass ~ 100 MeV, electrons may be gradually replaced by strange quarks, which again will disfavor the cross-flavor pairing.

A superconducting phase with asymmetric cross-species pairing needs to optimize the overlap between the Fermi surfaces (which is perfect in the symmetric BCS state) to attain the maximum possible condensation energy. There are a number of ways to achieve this: (a) condensate that carries non-zero momentum with respect to some fixed frame. The penalty in the energy budget for the (always positive) kinetic energy of condensate motion is compensated by the gain in the (negative) condensation energy. In three-flavor quark matter the face-centered cubic (fcc) lattices were identified as having particularly low free energy[2] in the regime where the Ginzburg-Landau (GL) theory is applicable. The spatial form of the condensate beyond the GL regime is not known, therefore current treatments beyond this regime require some simplifying assumptions.[5,6] (b) A more symmetric superconducting phase exploits deformations of Fermi spheres as the mechanism of restoring the coherence needed for pairing at the cost of kinetic energy loss caused by these deformations. The deformed Fermi surface phase requires minimal breaking of spatial symmetry and is more symmetric than the lattice phases above.[4]

In this review we will be concerned mainly with the 2SC and the crystalline color superconducting (CCS) phases. We expect that the phenomenology of related phases with cross-flavor pairing is essentially the same.

3. Equilibrium and Stability

The central question, which we address in this section, is whether the equation of state of matter at high densities admits stable configurations of self-gravitating objects in General Relativity featuring deconfined quark matter. The deconfinement phase transition from baryonic to quark matter leads to a softening of the equation of state which could lead to an instability towards a collapse into a black hole. In Ref. 7 the quark phase was studied in the Nambu–Jona-Lasinio (NJL) model, which is a low-energy non-perturbative approximation to QCD, that is anchored in the low-energy phenomenology of the hadronic spectrum. While dynamical symme-

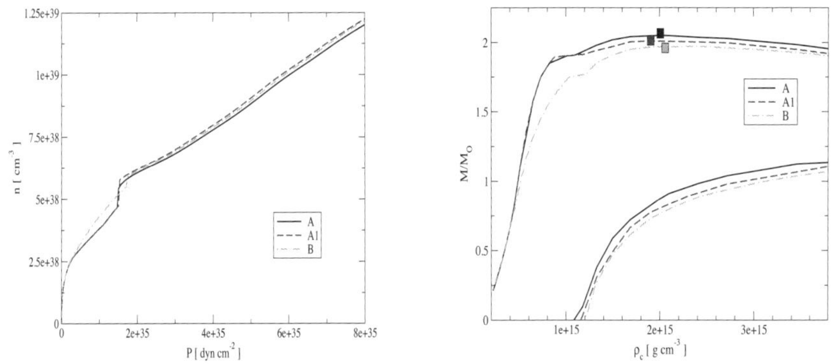

Fig. 1. *Left panel.* Number density versus pressure for three models. For the models A and A1 the nuclear (low density) equation of state is the same; for the models A and B the quark (high density) equation of state is the same. *Right panel.* Dependence of the total stellar mass and the mass of the quark core in units of solar mass $M_\odot$ on the central density for non-rotating configurations. The lower set of curves represents the masses of the CCS quark cores, the upper set - the total masses of the configurations. The maximal masses are marked with boxes.

try breaking, by which quarks acquire mass, is incorporated in this model, it lacks confinement. Physically, the true nuclear equation state must go over to some sort of quark equation of state at some density if the deconfinement has to take place in nature. If pressure vs chemical potential curves for chosen equations of state of nuclear and quark matter do not cross, then the models are incompatible in the sense that they can not describe the desired transition between the nuclear and quark matter. The low-density equation of state of nuclear matter and the high-density equation of state of CCS matter were matched in Ref. 7 at an interface via the Maxwell construction. The phase with largest pressure is the one that is realized at a given chemical potential. Thus, at the deconfinement phase transition there is a jump in the density at constant pressure as illustrated in Fig. 1, left panel. The low-density nuclear equations of state are based on the Dirac-Bruckner-Hartree-Fock approach and are the most hard equations of state in our collection.[8,9] It should be noted that the matching with softer equations of state can be enforced by varying the normalization of the pressure ("effective bag constant") of the quark matter, in which case the density of deconfinement is a free parameter.

The spherically symmetric solutions of Einstein's equations for self-gravitating fluids are given by the well-known Tolman-Oppenheimer-Volkoff equations.[8] A generic feature of these solutions is the existence of a maximum mass for any equation of state; as the central density is increased beyond the value corresponding to the maximum mass, the stars become unstable towards collapse to a black hole. A criterion for the stability of a sequence of configurations is the requirement that the derivative $dM/d\rho_c$ should be positive (the mass should be an increasing function of the central density).

For configurations constructed from a purely nuclear equation of state the stable sequence extends up to a maximum mass of the order $2\ M_\odot$ (Fig. 1), this large value being a consequence of hardness of the equation of state. The hybrid configurations branch off from the nuclear configurations when the central density reaches that of the deconfinement phase transition. It is seen that a stable branch of hybrid stars emerges in the range of central densities $1.3 \leq \rho_c \leq 2.5 \times 10^{15}$ g cm^{-3}. The masses of the CCS quark cores cover the range $0 \leq M_{\rm core}/M_\odot \leq 0.75 - 0.88$ for central densities $1.3 \times 10^{15} \leq \rho_c \leq 2 \times 10^{15}$. Thus, the quark core mass ranges from one third to about the half of the total stellar mass.

Millisecond neutron stars can rotate at frequencies which are close to the limiting orbital Keplerian frequency at which mass shedding from the equatorial plane starts. The Keplerian frequency sets an upper limit on the rotation frequency, since other (less certain) mechanisms, such as secular instabilities, could impose lower limits on the rotation frequency. The mass versus central density dependence of compact stars rotating at the Keplerian frequency is similar to that for non-rotating stars with the scales for mass shifted to larger values.[7] The increase in the maximum mass for stable hybrid configurations (in solar mass units) is $2.052 \rightarrow 2.462$ for the model A, $2.017 \rightarrow 2.428$ and 2.4174 for the model A1 (there are two maxima) and $1.981 \rightarrow 2.35$ for the model B.[7]

Thus, a new branch of *stable hybrid configurations* with CCS matter cores emerges within a broad range of central densities. The quark equation of state depends only on the NJL model parameters (i.e. does not contain an additional "bag constant"). A quite general conclusion of our analysis is that the nuclear equation of state needs to be *hard* to enable thermodynamical equilibrium with the NJL model quark matter.

4. Gravitational Radiation

Continuous gravitational waves emitted by non-axisymmetric rotating compact stars are expected to be in the bandwidth of current gravitational wave interferometric detectors. Upper limits on the strain of gravitational

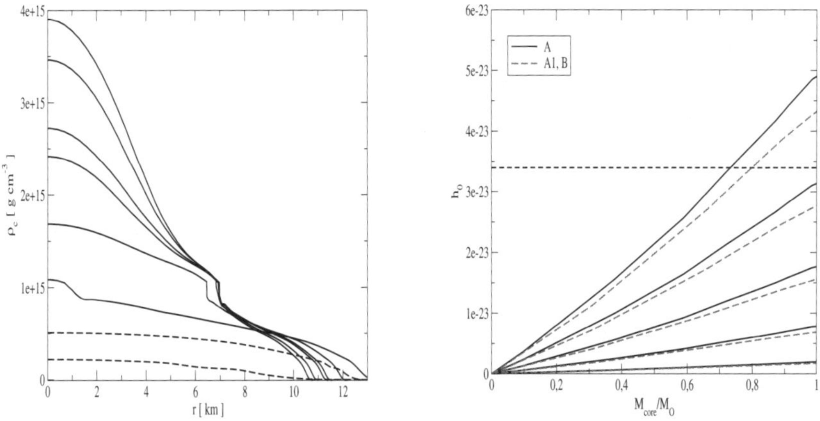

Fig. 2. *Upper panel.* Dependence of density on the internal radius (model A). The dashed lines correspond to purely nuclear, the solid lines - to hybrid configurations. Density jumps at the phase transition to quark matter. *Lower panel.* The strain of gravitational wave emission of the Crab pulsar. Each triple of curves corresponds to the gap value (from top to bottom) $\Delta = 50$, 40, 30, 20, and 10 MeV and the breaking strain $\bar{\sigma}_{\rm max} = 10^{-4}$. The horizontal line shows the current experimental upper limit.

waves from a selection of radio pulsars were set by the LIGO collaboration.[10] Gravity waves arise from time-dependent quadrupole deformations of masses. Therefore, rotating isolated neutron stars will emit gravitational radiation if their mass distribution is non-axisymmetric with respect to their rotation axis. The axial symmetry of the star's core can be broken by the solid deformations in CCS matter, whose shear modulus $\mu_{\rm shear}$ was computed in Ref. 11. Initially, the gravitational wave emission from a quark star made of uniform-density incompressible CCS matter was estimated by Lin.[12] Haskell et al.[12] estimated core deformations and associated with them constraints on the QCD parameters for sequences of 1.4 $M_\odot$ and $R = 10$ km stars composed of CCS incompressible quark cores and hadronic shells obeying $n = 1$ polytropic equation of state. Gravitational radiation of models based on realistic equations of state were considered in Ref. 13. One key difference to the previous models is that realistic hybrid configurations have masses that are close to the maximum sustainable mass $M_{\rm max} \simeq 2M_\odot$. Furthermore, the quadrupole moment of the quark core was computed from realistic density profile, which differs substantially from constant density profile of an incompressible fluid, see Fig. 2. The characteristic strain amplitude of gravitational waves emitted by a triaxial star rotating about its principal axis is $h_0 = 16\pi^2 G c^{-4} \epsilon I_{zz} \nu^2 r^{-1}$, where ν is the star's rotation frequency, r is the distance to the observer, $\epsilon = (I_{xx} - I_{yy})/I_{zz}$ is the equatorial ellipticity, I_{ij} is the tensor of moment of inertia, G is the gravitational constant, c is the speed of light. The elastic deformations are assumed to be small perturbation on the background equilibrium of the star. Fig. 2 displays the strain of gravitational wave emission as a function of the core mass $M_{\rm core}$ and the current upper limit for the Crab pulsar $h_0 < 3.4 \times 10^{-25}$, which is rotating at the frequency $\nu = 29.6$ Hz at the distance 2 kpc.[10] The value of the breaking strain of CCS core is $\bar{\sigma}_{\rm max} = 10^{-4}$. The dependence of the stain amplitude on the microscopic parameters follows from the chain $h_0 \propto Q_{\rm max} \propto \mu_{\rm shear} \propto \bar{\sigma}_{\rm max}\Delta^2$, whereby $Q_{\rm max}$ is the maximal quadrupole moment of the CCS core. The strain of gravitational wave emission is close to the upper limit for $\Delta = 50$ MeV. If, however, the gaps are small more "optimistic" values of $10^{-3} \leq \bar{\sigma}_{\rm max} \leq 10^{-2}$ will be needed to generate sizeable strain amplitude. The Crab pulsar's current limit implies $\bar{\sigma}_{\rm max}\Delta^2 \sim 0.25$ MeV^2, assuming maximally strained matter. The evolutionary avenues that may lead to such maximal deformations are not known. Pulsars may simply preserve their initial deformations as they cool down and solidify in the CCS state. We conclude that *massive compact stars with CCS quark cores are strong candidate sources of gravitational wave emission.* Improved upper limits on the strain amplitude, can narrow down the admissible range of parameters of the CCS matter in the future.

5. Neutrino Emission and Cooling

Neutrino emissivities control the cooling rate of a neutron star during the first $10^4 - 10^5$ yr of their evolution. At later times the photon emission from the surface and the heating in the interior are the main factors. Depending on the dominant neutrino emission process in the neutrino emission era $t \leq 10^5$ yr the cooling could be slow (standard) or fast (nonstandard).

The slow cooling scenario is based on neutrino cooling via the modified Urca and bremsstrahlung processes. The fast cooling requires unconventional processes such as decays in pion/kaon condensates, direct Urca process on nucleons, hyperons, or quarks. Phase-space arguments show that fast cooling neutrino processes have typically temperature dependences $\sim T^6$, whereas those leading to slow cooling require additional phase space for the spectator particle and their emissivities scale as T^8.

An inspection of the observational data on neutron star surface temperatures, which is commonly presented on a plot of photon-luminosity (or surface temperature) vs age, shows that the data cannot be described by a single cooling track; a regulator is needed that will cool some stars faster than the others.[8,9]

Thus, we are led to examine the potential fast cooling processes in the quark cores of compact stars. The key process is the Urca processes (quark β-decay) $d \to u+e+\bar{\nu}_e$ and $u+e \to d+\nu_e$, which is permitted in *interacting* quark matter for any asymmetry between u and d quarks. The effect of asymmetric pairing is well demonstrated on the example of 2SC phase.[14,15] The result for the emissivity depends on whether the parameter $z = \Delta/\delta\mu$ is larger or smaller than unity, where Δ is the gap in the 2SC phase and $\delta\mu = \mu_d - \mu_u$ is the shift in the chemical potentials of the up and down quarks. For $z > 1$ all the particle modes are "gaped", therefore, as the temperature is lowered, the emissivity is suppressed (for asymptotically low temperatures exponentially). When $z < 1$ there are gapless modes in the quasiparticle spectrum; this implies that the neutrino production is not affected by color superconductivity. As a result, the superconducting quark matter cools at a rate comparable to the unpaired matter. More generally, we may conclude that *the existence of ungaped segments on the Fermi surfaces of quarks will lead to an enhanced cooling of these stars.* A density dependent z parameter may resolve the dichotomy of the fast and slow cooling by providing a smooth transition between these extremes.

6. Radio-Timing and Rotational Anomalies

Pulsars are nearly prefect clocks, whose pulsed emissions, with a typical periodicity of seconds or less, are locked to the rotation period of the star. Their periods increase gradually over time, due to a secular loss of rotational energy. Some pulsars show deviations from this regularity. The pulsar timing anomalies divide roughly into three types. (i) *Glitches.* These are distinguished by abrupt increases in the rotation and spin-down rates of pulsars by amounts $\Delta\Omega/\Omega \sim 10^{-6} - 10^{-8}$ and $\Delta\dot{\Omega}/\Omega \sim 10^{-3}$. After a glitch, $\Delta\Omega/\Omega$ and $\Delta\dot{\Omega}/\Omega$ slowly relax toward their pre-glitch values, on a time scale of order weeks to years, in some cases with permanent hysteresis effects. Such behavior is attributed to a component within the star that is only weakly coupled to the rigidly rotating normal component responsible for the emission of pulsed radiation. (ii) *Timing Noise.* These represent irregular, stochastic deviations in the spin and spin-down rates that are superimposed on the near-perfect periodic rotation of the star. Whether the superfluids are involved in the generation of timing noise remains unclear. (iii) *Long-Term Periodic Variabilities.* Observed in the timing of few pulsars, most notably PSR B1828-11, these deviations strongly constrain theories of superfluid friction, if their periodicities are interpreted in terms of precession. (Note however that the Tkachenko modes of the vortex lattice are a viable alternative to the precession interpretation.[17]) The importance of the inferred precession mode stems from the fact that it involves non-axisymmetric perturbations of the rotational state, removing the degeneracy with respect to ζ ($\zeta' \simeq 0$) that is inherent in the interpretation of post-glitch dynamics.

The macroscopic physics of neutron-star rotation and its anomalies observed in the timing of pulsars can be described within the hydrodynamic theory of superfluids. At the local hydrodynamical scale, the rate of angular momentum transfer between the superfluid and normal components is determined by the equation of motion of a vortex line in a neutral superfluid

$$\omega_S(\boldsymbol{v}_S - \boldsymbol{v}_L) \times \boldsymbol{\nu} + \zeta(\boldsymbol{v}_L - \boldsymbol{v}_N) + \zeta'(\boldsymbol{v}_L - \boldsymbol{v}_N) \times \boldsymbol{\nu} = 0\,, \tag{1}$$

where $\boldsymbol{v}_S$ and $\boldsymbol{v}_N$ are the superfluid and normal fluid velocities, $\boldsymbol{v}_L$ is the velocity of the vortex, $\boldsymbol{\nu}$ is a unit vector along the vortex line, ω_S is the unit of circulation, and ζ, ζ' are (dimensionless) friction coefficients, also known as the drag-to-lift ratios. These coefficients encode the essential information on the microscopic processes of interaction of vortices with the ambient unpaired fluid.

The potential role of quark matter in the rotational dynamics of compact stars is not understood yet. Some of the candidate phases, such as the color-flavor-locked (CFL) phase, rotate by creating vortices,[18] which interact with the phonon gas within this phase.[19] However, phonons do not interact electromagnetically with the rest of the star, therefore it is unclear on which timescales the CFL quark core will couple to the observable crust. Such a coupling is manifest for the 2SC and related phases, where the magnetic field (partially) penetrates into the superfluid region.

This work was in part supported by the Deutsche Forschungsgemeinschaft (Grant SE 1836/1-1).

References

1. D. Bailin and A. Love, Phys. Rept. **107**, 325 (1984); M. Alford and K. Rajagopal, in *Pairing in Fermionic Systems*, edited by A. Sedrakian, J. W. Clark and M. Alford, (World Scientific, Singapore, 2006), p. 1, arXiv:hep-ph/0606157; S. B. Rüster, V. Werth, M. Buballa, I. A. Shovkovy and D. H. Rischke, *ibid.*, p. 63, arXiv:nucl-th/0602018; T. Schafer, *ibid.*, p. 109, arXiv:nucl-th/0602067; M. G. Alford, arXiv:0907.0200 [nucl-th].
2. J. A. Bowers and K. Rajagopal, Phys. Rev. D **66**, 065002 (2002) [arXiv:hep-ph/0204079]; K. Rajagopal and R. Sharma, Phys. Rev. D **74**, 094019 (2006)

[arXiv:hep-ph/0605316]; J. Phys. G **32**, S483 (2006) [arXiv:hep-ph/0606066]; N. D. Ippolito, G. Nardulli and M. Ruggieri, JHEP **0704**, 036 (2007) [arXiv:hep-ph/0701113].

3. P. F. Bedaque, H. Caldas and G. Rupak Phys. Rev. Lett. **91**, 247002 (2003); H. Caldas, Phys. Rev. Rev. A **69**, 063602 (2004).
4. H. Müther and A. Sedrakian, Phys. Rev. Lett. **88**, 252503 (2002) [arXiv:cond-mat/0202409]; Phys. Rev. C **67**, 015802 (2003) [arXiv:nucl-th/0209061]; Phys. Rev. D **67**, 085024 (2003) [arXiv:hep-ph/0212317]; A. Sedrakian, in *Superdense QCD Matter and Compact Stars*, edited by D. Blaschke and D. Sedrakian, (Springer, Dordrecht, 2006), p. 209, arXiv:nucl-th/0312053.
5. D. Nickel and M. Buballa, Phys. Rev. D **79**, 054009 (2009) [arXiv:0811.2400 [hep-ph]].
6. A. Sedrakian and D. H. Rischke, Phys. Rev. D **80**, 074022 (2009) arXiv:0907.1260 [nucl-th].
7. N. Ippolito, M. Ruggieri, D. H. Rischke, A. Sedrakian and F. Weber, Phys. Rev. D **77**, 023004 (2008) [arXiv:0710.3874 [astro-ph]].
8. F. Weber, *Pulsars as astrophysical laboratories for nuclear and particle physics*, Bristol, U.K.: Institute of Physics, 1999;
9. A. Sedrakian, Prog. Part. Nucl. Phys. **58**, 168 (2007) [arXiv:nucl-th/0601086].
10. B. Abbott, *et al.* (LIGO Scientific Collaboration) Astrophys. J. Letters **683**, 45 (2008).
11. M. Mannarelli, K. Rajagopal and R. Sharma, Prog. Theor. Phys. Suppl. **174**, 39 (2008).
12. L.-M. Lin, Phys. Rev. D **76**, 081502(R) (2007); B. Haskell, N. Andersson, D. I. Jones, and L. Samuelsson, Phys. Rev. Lett. **99**, 231101 (2007).
13. B. Knippel and A. Sedrakian, Phys. Rev. D **79**, 083007 (2009) [arXiv:0901.4637 [astro-ph.SR]].
14. M. Alford, P. Jotwani, C. Kouvaris, J. Kundu and K. Rajagopal, Phys. Rev. D **71**, 114011 (2005) [arXiv:astro-ph/0411560]; R. Anglani, G. Nardulli, M. Ruggieri and M. Mannarelli, Phys. Rev. D **74**, 074005 (2006) [arXiv:hep-ph/0607341].
15. P. Jaikumar, C. D. Roberts and A. Sedrakian, Phys. Rev. C **73**, 042801(R) (2006) [arXiv:nucl-th/0509093].
16. A. Sedrakian, I. Wasserman and J. M. Cordes, Astrophys. J. **524**, 341 (1999), arXiv:astro-ph/9801188; B. Link, Astrophys. Space Sci. **308**, 435 (2007); T. Akgun, B. Link and I. Wasserman, Mon. Not. Roy. Astron. Soc. **365**, 653 (2006) [arXiv:astro-ph/0506606]. B. Link, arXiv:astro-ph/0608319.
17. J. Noronha and A. Sedrakian, Phys. Rev. D **77**, 023008 (2008) [arXiv:0708.2876 [astro-ph]].
18. K. Iida and G. Baym, Phys. Rev. D **66**, 014015 (2002).
19. M. Mannarelli, C. Manuel and B. A. Sa'd, Phys. Rev. Lett. **101**, 241101 (2008) [arXiv:0807.3264 [hep-ph]].

A ZEPTO-SECOND ATOMIC CLOCK FOR NUCLEAR CONTACT TIME MEASUREMENTS AT SUPERHEAVY COLLISION SYSTEMS

H. BACKE

Institute of Nuclear Physics, Joh. Gutenberg-University, D-55099 Mainz, Germany,
E-mail: backe@kph.uni-mainz.de
www.kph.uni-mainz.de

The distance between the centers of gravity of two heavy colliding nuclei $R(t)$ as function of time can be employed to describe the dynamical behavior of the di-nuclear system during the contact phase. In order to get information on this quantity, it was suggested by G. Soff, J. Reinhard, B. Müller and W. Greiner[1,2] many years ago to investigate the spectral shape of electrons emitted in dissipative heavy ion collisions. It has been shown that in the limit of lowest order adiabatic perturbation theory the δ-electron spectrum just reflects the Fourier transform of the function $\dot{R}(t)/R(t)$. With this tool nuclear contact times in the order of 1 zs $= 1 \cdot 10^{-21}$ s have been determined for the U+Au collision system at an uranium beam energy of 8.65 MeV/u.[9] The δ electron spectra observed for the U+Pd ($Z_u = 138$) collision system at 6.1 MeV/u at impact parameters for which nuclear forces become important, most likely cannot anymore be analyzed within the framework of such a model because of its breakdown if time changes of $\dot{R}(t)/R(t)$ become too rapid.

Keywords: Heavy ion collisions at the Coulomb barrier; nuclear contact time measurements; δ electron spectroscopy.

1. Introduction

I would like to start my talk by going back nearly 30 years ago. It was the time when the UNILAC at GSI in Darmstadt overcame his childhood deceases and started to produce first internationally recognized results. Also organizationally childhood deceases were overcome. I remember that the run on beam time was tremendous and beam was distributed in the beginning in quantities of hours. Anyway, it was the time when newspapers reported in connection with top-level science in Germany also from accomplishments at GSI: "Positrons from nowhere" was the title of an article in the newspaper "Frankfurter Allgemeine" from June 09, 1982, with the sub-

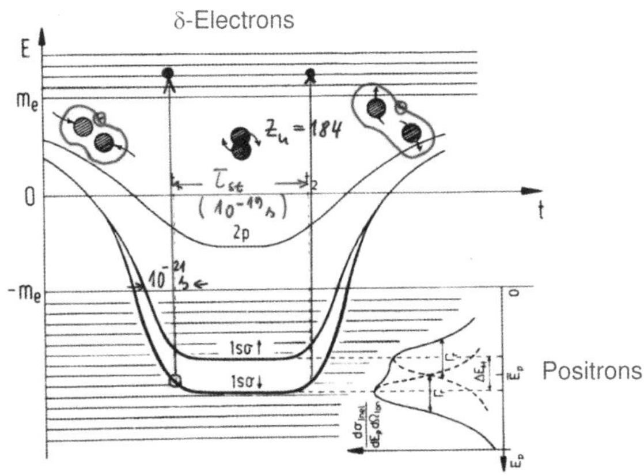

Fig. 1. Schematic representation of the correlation diagram for the inner electronic shells in an inelastic U+U-collision as function of time. The resonances in the positron continuum are indicated together with the relevant widths.[3]

title "Breakdown of the vacuum in 'supercritical' fields. Super-atoms in the collision of Uranium atoms". In this article it was pointed out "One of the difficulties is that the collision of the uranium nuclei lasts only 10^{-21} s, but that the creation of spontaneous positrons requires about a factor of 100 more time. ... Therefore, the Frankfurt theoreticians proposed already some years ago to delay the process between the colliding partners by means of nuclear reactions."

What is meant will be explained by means of Fig. 1. Electrons from the 1sσ or 2pσ shell of the quasi-molecule are excited due to the strong dynamics in the approaching phase of the nuclei and the remaining hole may decay by positron emission. Since the collision lasts only about 1 zs $= 1 \cdot 10^{-21}$ s, the spectrum levels off exponentially with a characteristic energy of about 660 keV. Those positron spectra were already observed at that time. However, the proposal of Rafelski, Müller and Greiner[3] was to take advantage of a possible sticking time of the two nuclei to prolong the diving of the mentioned levels into the negative energy continuum. If the holes have time enough to decay spontaneously, positron lines with a width of about 10 keV should be observable. I must admit that, when I first saw this picture, I was absolutely fascinated from this idea. We had just performed an experiment in which we looked to the positron spectrum in coincidence to fission fragments which signal a nuclear contact. I remember the tension when inspecting statistically insignificant peaks which always appeared. Experts already told us that one cannot expect in a deep inelastic collision such long sticking times. But we thought that clever experiments might reveal them. Of course, the best possibility was to directly search for the spontaneous positron peak. This direction was pursued by three experimental groups at GSI. I favored from the beginning another approach, the measurement of nuclear contact times by means of δ-electron spectroscopy.

In this contribution a short overview will be given what has been accomplished in the field of studying the nuclear reaction dynamics, and in particular the interaction time of two colliding nuclei, by means of δ-electron spectroscopy. This part will be short since most of the results have been published. I will dwell a little bit more in detail on a striking experimental result observed for the U+Pd collision system at a beam energy of 6.1 MeV/u which could not be explained up to now.

2. Contact Time Measurements for the Deep Inelastic ^{238}U+^{197}Au Collision System

If the two colliding nuclei come in close contact in a deep inelastic collision, the repulsive nuclear forces cause a rotation by an angle θ of the di-nuclear system before the nuclei separate again, see Fig. 21 of Schröder et al.[4] If this angle would be known, the nuclear contact time could simply be determined by the formula $T = J\theta/L$, with J the moment of inertia of the di-nuclear system, and $L = Mbv_0$ the angular momentum. The latter depends on the mass M of the projectile, its velocity v_0 at infinity and the impact parameter b. Unfortunately, only the angle in the exit channel can be measured. The impact parameter is not an observable. Experimental evidence on fragment-mass distributions suggests that during the time the two constituents of the intermediate double-nucleus system interact with each other, a mass equilibration process proceeds which is accompanied by a damping of the relative kinetic energy into other degrees of freedom. It is generally accepted that the total kinetic energy loss T_{KEL} signifies the total interaction time T, i.e. $L = L(T_{KEL})$. This view is supported by classical dynamical calculations[5–7] which show that the energy loss is a monotonic function of the initial angular momentum and the total interaction time.

The consequence of this picture of a deep inelastic collision for the design of an experimental setup is that it must be able to determine T_{KEL} and the mass transfer. The setup which fulfills this requirements is shown in Fig. 2. It allows the measurement of the complete kinematics for binary, ternary and quadruple reactions. In the latter, one or both colliding particles undergo fission, respectively.

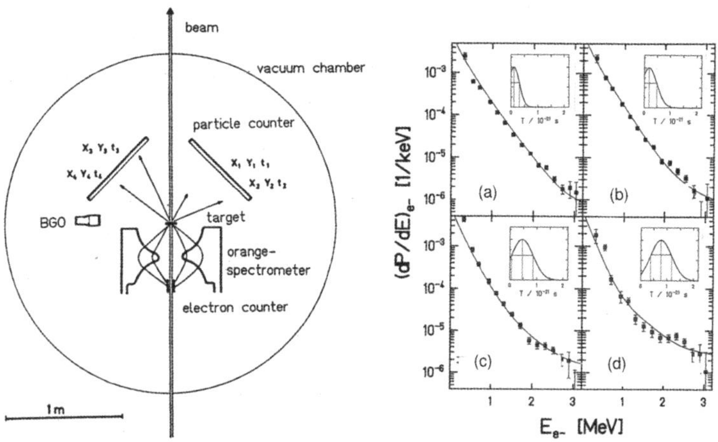

Fig. 2. Left panel: Experimental Setup.[9] The parallel plate particle counters with a size of 91×91 cm^2 and a resolution $\delta x = \delta y = 2$ mm (FWHM) allow a measurement of the velocity vectors of up to four particles in the exit channel. For that purpose advantage is taken of the beam time structure which is delivered in bunches with a duration of typically 0.4 ns. These numbers allow a measurement of the total kinetic energy loss with an accuracy $\delta T_{KEL} = 40$ MeV, a mass resolution $\delta m = 5$ u, and the scattering angle in the CM system with an accuracy $\delta\theta_{CM} = 0.5°$. Electrons are measured with an orange spectrometer[10] located in backward direction. The BGO detector measures the γ-ray spectrum from which the conversion electron background can be determined. Right panel: δ-electron spectra for the ^{238}U+^{197}Au ternary reaction at a beam energy of 8.65 MeV/u for various T_{KEL}: $\overline{T}_{KEL} = 100$ MeV (a), $\overline{T}_{KEL} = 190$ MeV (b), $\overline{T}_{KEL} = 280$ MeV (c), and $\overline{T}_{KEL} = 350$ MeV (d). The insets show the contact time distributions as used for the best fits (full lines).[11]

With this experimental setup the δ-electron spectra for the ^{238}U+^{197}Au ternary reaction at a beam energy of 8.65 MeV/u have been measured as function of the T_{KEL}, see Fig. 2, right panel.[11] The insets show the contact time distributions which were used to reproduce the measurement as good as possible. For further details see the papers of P. Senger et al.[11] and J. Stroth et al.[9] It can be concluded that long contact times of up to 10 zs as predicted by the analysis of deep inelastic collisions with the schematic model of Schröder at al.[4] could not be verified by the analysis of the experimental δ-electron spectra.

3. The ^{238}U+^{108}Pd Collision System at the Coulomb Barrier

Two typical δ-electron spectra for the ^{238}U+^{108}Pd collision system at 6.1 MeV/u are shown in Fig. 3. The data were taken from the doctoral

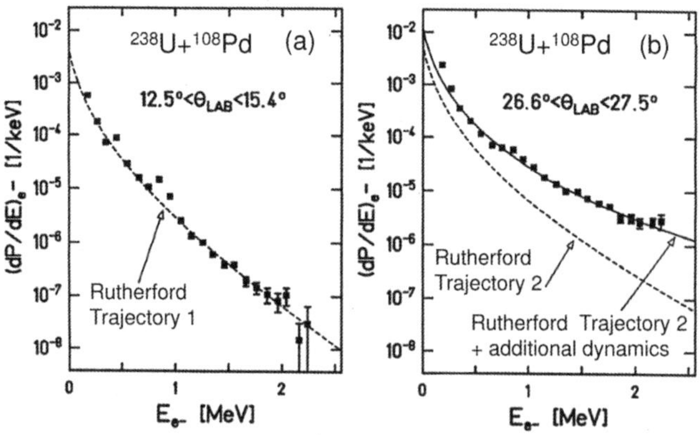

Fig. 3. Two typical δ-electron spectra for the $^{238}U + ^{108}Pd$ scattering system at an uranium beam energy of 6.1 MeV/u. The left panel (a) shows a spectrum for a mean impact parameter $< b >_1 = 15.5$ fm ($\varepsilon_1 = 2.51$), the right panel (b) for $< b >_2 =$ 4.6 fm ($\varepsilon_2 = 1.21$). The curves assigned with "Rutherford Trajectory" are calculated with Eqn. (1) with $\hat{t}_1 = 0.841$ zs, left panel, and with $\hat{t}_2 = 0.624$ zs, right panel. The curve "Rutherford Trajectory 2 + additional dynamics" includes an additional 44 % contribution with $\hat{t}_{add} = 0.1$ zs.

theses of J. Stroth[12] and S. Mojumder.[13] The left panel of Fig. 3 (a) shows a δ-electron spectrum for which the two colliding particles follow pure Rutherford trajectories. This can be concluded from the selected scattering angle region which corresponds to a mean impact parameter $< b >_1 = 15.5$ fm or a distance of closest approach $R_{min} = 23.6$ fm far above the safe distance $R_{safe} = 1.5\,(A_1^{1/3} + A_2^{1/3}) + 3$ fm $= 19.4$ fm for which nuclear reactions can be neglected. For the right panel (b) the mean impact parameter is $< b >_2$ $= 4.6$ fm which corresponds to $R_{min} = 14.9$ fm, a distance which is below the interaction radius $R_{int} = 1.4\,(A_1^{1/3} + A_2^{1/3})$ fm $= 15.4$ fm. In this regime nuclear levels may be excited not only by Coulomb excitation but also by nuclear reactions like the transfer of nucleons. The data were fitted with a scaling model developed by E. Kankeleit[8] which describes the δ-electron spectrum as

$$\frac{dP_\delta}{dE_{e^-}} = P_0 \frac{(1 + 0.174/\varepsilon)^2}{(E_{e^-} - E_B)^2} \exp\Big[-2\frac{(E_{e^-} - E_B)}{\hbar}\hat{t}(\varepsilon)\Big] \tag{1}$$

with

$$\hat{t}(\varepsilon) = \frac{a}{v_0}\Big(\varepsilon + 1.6 + \frac{0.449}{\varepsilon}\Big) \tag{2}$$

a characteristic time which depends on the eccentricity $\varepsilon =$ $1/\sin(\theta_{CM}/2) = \sqrt{(b/a)^2 + 1}$, $E_B \approx 0.1$ MeV the mean binding energy of the emitted electron in the initial state, θ_{CM} the scattering angle in the CM system, b the impact parameter, and $2a = 13.5$ fm the distance of closest approach in a head-on collision.

The striking feature of this evaluation is that a big excess yield of electrons is observed for the spectrum shown in Fig. 3 (b) in comparison with the scaling model which fits very well the data shown in Fig. 3 (a). It has been carefully investigated whether this additional yield may originate from the de-excitation of nuclear levels excited by the mentioned nuclear reactions. It turned out[12] that the only conceivable source of electrons would be a de-excitation via E0 transitions the contribution of which cannot be estimated from the γ-ray spectra as taken with the BGO detector shown in Fig. 2. However, E0 transitions of levels with an energy above about 1.1 MeV may also decay by pair conversion and, if this happens, such a contribution would be observable in a positron spectrum. A positron spectrum has been recorded under rather similar conditions as for the spectrum shown in Fig. 3 (b).[14] The spectrum is shown in Fig. 4 together with the hypothetical assumption that the excess electron yield in Fig. 3 (b) originates from E0 transitions. Since no indications of this extra yield can be found in the positron spectrum, the observed yield can be explained by pair conversion derived from the γ-ray spectrum, it is rather unlikely that E0 transitions are responsible for the observed excess yield in the electron spectrum of Fig. 3 (b). In the following it will be assumed, therefore, that the latter is of atomic origin.

The underlying dynamics of the curve assigned as "Rutherford Trajectory 1 (2)" in Fig. 3 is a function of type $t/[t^2 + \hat{t}(\varepsilon)^2]$. The factor $\exp[-(E_{e^-} - E_B)\hat{t}(\varepsilon)/\hbar]$ of Eqn. (1) is just the Fourier-transform of this

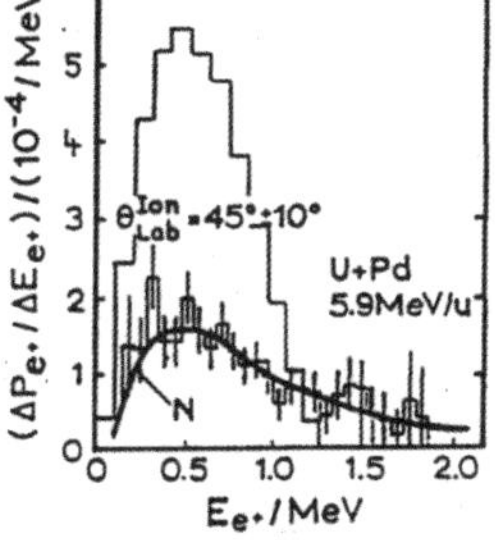

Fig. 4. The error bars show a positron spectrum[14] taken under similar conditions as the δ electron spectrum depicted in Fig. 3 (b). The full curve corresponds to the hypothetical assumption that the excess electron yield originates from E0 transitions, see Ref.[12]

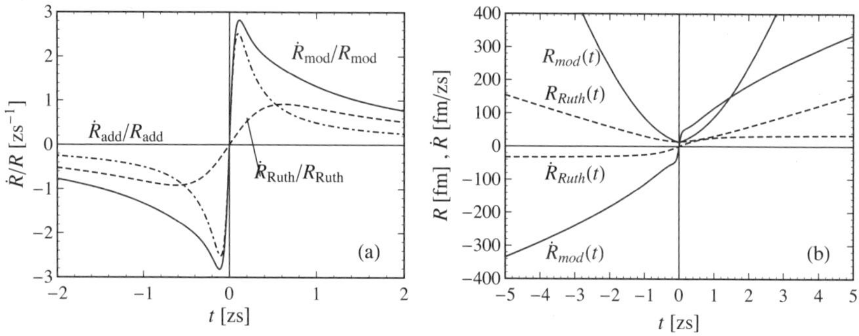

Fig. 5. (a) Illustration of the dynamics as given by Eqn. (3). The dashed curve represents the dynamics of a pure Rutherford trajectory with $\hat{t} = 0.624$ zs, the dotted-dash curve the additional term with $\hat{t} = 0.1$ zs. (b) Solutions of the differential equation Eqn. (3). The dashed curves depict $R(t)$ and $\dot{R}(t)$ for a pure Rutherford trajectory.

function. The best fit curve in Fig. 3 (b) labelled by "Rutherford Trajectory 2 + additional dynamics" is the Fourier transform of the function

$$\frac{\dot{R}_{mod}}{R_{mod}}(t) = (1 + 0.174/\varepsilon_2)\Big(\frac{t}{t^2 + \hat{t}_2^2} + 0.44\frac{t}{t^2 + \hat{t}_{add}^2}\Big). \tag{3}$$

It is depicted in Fig. 5 (a). It consists of a part originating from a Rutherford trajectory and an additional part which had to be added to fit the observed δ-electron spectrum.[12]

Eqn. (3) is just a differential equation for the relative distance between the colliding nuclei $R_{mod}(t)$ which can be solved numerically. The result is shown in Fig. 5 (b). It can immediately be recognized that the solution has no physical meaning since the relative velocity $\dot{R}_{mod}(t)$ is a monotonically increasing function and does not approach for $\mid t \mid \rightarrow \infty$ the Rutherford trajectory $\dot{R}_{Ruth}(t)$ which is shown in Fig. 5 (b) as a dashed curve. In other words, the energy is not conserved. The reason for this behavior can be found in an un-physical assumption for the additional dynamics $\dot{R}_{add}/R_{add}(t)$ which implies long range forces.

To avoid long range forces, the model has been refined taking into account the well known fact that nuclear forces are negligibly small at distances larger than the safe distance $R_{safe} = 19.4$ fm. However, all attempts to find an additional dynamics of the type $\dot{R}_{add}/R_{add}(t) = \text{sign}(t) \mid t \mid^n \exp[- \mid t \mid /\tau]$ without long tails failed. Also inclusion of the finite size of the nuclei by replacing in the differential equation (3) the distance R by $R = 2\sqrt{< r^2 >}$ [15] did not improve the results. However, quite satisfactory fit results were obtained if a symmetric term was added to the dynamics of a Rutherford trajectory resulting in

$$\frac{\dot{R}_{mod}}{R_{mod}}(t) = \frac{\dot{R}_{Ruth}}{R_{Ruth}}(t) + c_1 \exp\Big[-\Big(\frac{t}{\tau}\Big)^2\Big] \tag{4}$$

with $\tau = 0.27$ zs. Notice that the distance of the nuclei amounts for a Rutherford trajectory at the time τ to 16.4 fm which is close to $R_{int} = 15.4$ fm where nuclear forces get of importance.

The additional symmetric term in Eqn. (4) has most likely also no physical meaning. Instead, it might be conjectured that the first order adiabatic perturbation approach underlying all scaling lows[16] cannot anymore be applied. In other words, if nuclear forces come into play close to the distance of closest approach of the colliding nuclei, the system Hamiltonian may change so rapidly that the first order adiabatic perturbation theory approach breaks down.[17] It is well known that acceleration results in radiation. However, it is an open question how strongly the virtual electromagnetic field couples to the electrons surrounding the di-nuclear system.

4. Discussion and Conclusion

With the zs-atomic clock, which is based on δ-electron spectroscopy, sticking times in the dissipative nuclear collisions of uranium and gold nuclei at 8.65 MeV/u not longer than $1.5 \cdot 10^{-21}$ s have been measured.[9] The about a factor 10 longer sticking times derived with a pure kinematical model are not related to observables on which experimentally can be triggered. The preparation of well defined kinematical conditions as, e.g., T_{KEL} or a very asymmetric mass transfer suffers from the fact that diffusion processes, once the nuclei have formed a neck, do not allow to determine the impact parameter.

The experimentally observed excess electron yield for the U+Pd ($Z_u = 138$) collision system at conditions when nuclear forces become important is most likely of dynamical origin. However, the dynamics under the influence of the nuclear forces cannot anymore be analyzed with scaling laws derived on the basis of first order adiabatic perturbation theory. The reason may be found in its break down if the system Hamiltonian, characterized by $\dot{R}/R$, changes too rapidly.

Other nice results have been obtained by the tool of δ-electron spectroscopy. In particular the work of the TORI group on the elastoplasticity in dissipative Pb+Pb collision at 12 MeV/u[18] must be mentioned here. However, it would be worthwhile to re-investigate whether the first order adiabatic perturbation theory approach holds for the analysis of this collision system, or not.

Unfortunately, even now about 30 years later after the very suggestive Fig. 1 has been published neither the spontaneous positron peak has been observed nor the model dependent predictions of long sticking times in dissipative nuclear collisions have been verified by an atomic clock experiment. The nice positron peaks which were claimed to be observed by not only one group at GSI in the 80th and early 90th finally disappeared. It remains a challenge to observe the mentioned effects at GSI, GANIL, FAIR or elsewhere. Experimentalists feel confident to address again the subject of spontaneous positron emission.[19] However, a long way must be gone to attain again the maturity of the experimental setups already reached at GSI more than 20 years ago.

5. Closing Remarks

I would like to close my talk with a worldly wisdom of Arthur Schopenhauer, one famous German Philosopher [a]. He died 150 years ago in Frankfurt. "From the point of view of youth, life is an infinitely long future; from the point of view of old age, a very short past". This sounds for many of us not very comforting. Therefore, I would like to interpret Schopenhauer: What holds for yourself is a question of your own perception since whether we feel young or old is not a question of our age assigned in terms of years.

Dear Professor Greiner, you became now 75, and I also would like to congratulate you. But with respect to your creativity, originality, and productivity in physics you are as young as ever, as young as when I met you for the first time in Frankfurt about 43 years ago. I wish you with all my heart many fruitful years to come. Feel young, the community needs your enthusiasm, however, please also take care of you.

Acknowledgments

Recent fruitful discussions with H. Feldmeier and G. Wolschin on the subject of nuclear contact times as determined via δ-electron spectroscopy are gratefully acknowledged.

[a]Arthur Schopenhauer (* 22. Februar 1788 in Danzig; 21. September 1860 in Frankfurt am Main) "Aphorismen zur Lebensweisheit", Reclam S. 238, English translation: From the point of view of youth, life is an infinitely long future; from the point of view of old age, a very short past, so that it is represented to us at the beginning as things when we place opera glasses to our eyes and finally then the eyepiece. One must become old, that is, to have lived a long time, in order to recognize how short life is.

I thank J. Stroth for inviting me to give a talk at GSI at October 12, 2010 on the occasion of P. Senger's 60th birthday which prompted me to address once more the puzzle of the δ-electron excess yield in the U+Pd collision system at 6.1 MeV/u.

References

1. J. Reinhard, B. Müller, W. Greiner and G. Soff, *Z. Phys. A* **292**, 211 (1979).
2. G. Soff, J. Reinhardt, B. Müller and W. Greiner, *Phys. Rev. Lett.* **43**, 1981 (1979).
3. J. Rafelski, B. Müller and W. Greiner, *Z. Physik A* **285**, 49 (1978).
4. W.U. Schröder, J.R. Birkelund, J.R. Huizenga, K.L. Wolf and J.E. Viola Jr., *Phys. Rep.* **45**, 301-345 (1978).
5. K. Siwek-Wilczyńska and J. Wilczyński *Nucl. Phys. A***264**, 115 (1976).
6. E. Gross, H. Kalinowski and J.N. De, in *Symposium on Classical and Quantum Mechanical Aspects of Heavy Ion Collisions, Heidelberg, 1974* (Springer-Verlag Lecture Notes in Physics)**33**, 194 (1975).
7. C. Riedel, G. Wolschin and W. Nörenberg *Z. Physik A***290**, 47-55 (1979).
8. E. Kankeleit, *Nukleonika* **25**, 253 (1980).
9. J. Stroth, H. Backe, M. Begemann-Blaich, K. Bethge, H. Bokemeyer, M. Dahlinger, W. Konen, P. Kosmadakis, S. Mojumder, P. Senger and K. Stiebing, *Z.Phys. A* **357**, 441-455 (1997).
10. E. Moll and E. Kankeleit,*Nukleonika* **7**, 180 (1965).
11. P. Senger, H. Backe, M. Begemann-Blaich, H. Bokemeyer, P. Glssel, D. v. Harrach, M. Klver, W. Konen, K. Poppensieker, K. Stiebig, J. Stroth and K. Wallenstein, in *Physics of Strong Fields*, Maratea (Italy) 1986, ed.: W. Greiner, *NATO Adv. Study Inst. Ser. B* **153**, 423-439 (1987).
12. J. Stroth, *Untersuchungen der Reaktionsdynamik von Schwerionenreaktionen via δ-Elektronenspektroskopie*, doctoral thesis, University of Mainz 1990, GSI-90-28 Report November 1990, ISSN 0171-4546.
13. S. Mojumder, *Untersuchungen der Emission von Delta-Elektronen in binären Schwerionenstössen an der Coulomb-Barriere*, doctoral thesis, University of Frankfurt 1990, GSI-90-12 Report Juni 1990, ISSN 0171-4546.
14. H. Backe, in *Present Status and Aims of Quantum Electrodynamics (Mainz 1980)*, eds. G. Gräff, E. Klempt, G. Werth, *Lecture Notes in Physics **143**, Springer Verlag Berlin, Heidelberg, New York 1981*, p. 277
15. E. Kankeleit, in *Physics of Strong Fields*, Maratea (Italy) 1986, ed.: W. Greiner, *NATO Adv. Study Inst. Ser. B* **153**, 359-371 (1987).
16. B. Müller, J. Reinhardt, W. Greiner and G. Soff *Z. Phys. A* **311**, 151-155 (1983).
17. A.B. Migdal, in *Qualitative Methods in Quantum Theory*, Advanced Book Program, Adison-Wesley Publishing Co., New York 1977, p. 115
18. M. Rhein, R. Barth, E. Ditzel, H. Feldmeier, E. Kankeleit, V. Lips, Ch. Müntz, W. Noerenberg, H. Oeschler, A. Piechaczek, W. Polai and I. Schall, *Phys. Rev. Lett.* **69**, 1340 (1992).
19. C. Golabek, S. Heinz, W. Mittig et al., *Eur. Phys. J. A* **43**, 251-259 (2010).

Nuclear Heavyweights

Yuri Oganessian
Flerov Laboratory of Nuclear Reactions (Joint Institute for Nuclear Research)
141980 Dubna, Moscow region, Russia
oganessian@jinr.ru

Abstract. The observation of atomic numbers Z that are 40% larger than that of Bi, the heaviest stable element, is an impressive extension of nuclear survival. Although the super heavy nuclei (SHN) are at the limits of Coulomb stability, shell stabilization lowers the ground-state energy, creates a fission barrier, and thereby enables the SHE to exist. The fundamentals of the modern theory concerning the mass limits of nuclear matter have been experimentally verified.

We are going to consider how big a nucleus may be, what is the limit of atomic nuclei mass and how it is determined.

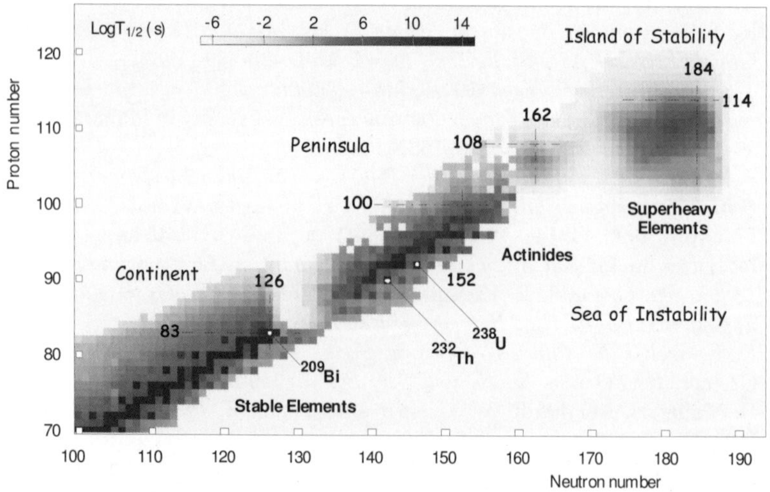

Figure. 1. Map of the nuclides in the region of heavy nuclei. Black squares - isotopes of stable elements. The half-lives of all artificial elements are denoted by different degrees of darkening (upper scale). New nuclear shells Z=108, N=162 and Z=114, N=184 are predicted by the macro-microscopic nuclear theory.

The nuclei shown in Fig.1 having different numbers of protons and neutrons have different binding energy and, therefore, different decay probability (half-life).

In the region of transuranium elements, a sharp decrease in the stability of nuclei with increasing atomic number is observed. Indeed adding 10 protons and 4 neutrons to ^{238}U lead to decrease in the period of α-decay of ^{252}No by a factor of 10^{16} with respect to uranium. An even stronger decrease is expected for another type of decay - the spontaneous fission. Here the drop is 23 orders of magnitude! A simple extrapolation into the region of heavier nuclei shows that the limits of existence of elements will be determined by spontaneous fission. When the half-life is reduced to 10^{-14} s, the nuclei will decay long before they obtain its atomic structure. According to a simple extrapolation, this situation arises even for nuclei with atomic numbers 106-108.

Such a limit of atomic nuclei was predicted in by N. Bohr and J. A. Wheeler [1], long before the first artificial elements were obtained. The fission of heavy nuclei was simulated by the process of deformation of a drop of charged liquid (the nuclear liquid drop model). According to Bohr and Wheeler, in the case of uranium a potential barrier about 6 MeV high prevents the fission of the nucleus (Fig.2 upper panel). If the energy more than 6 MeV is imported into the uranium nucleus, it will split into two fragments within about 10^{-19} sec. However, fission can occur spontaneously, by tunneling through the fission barrier. The half-life of uranium relative to fission is 10^{16} years. Increasing the number of protons in the nucleus (advancing to the transuranium elements) reduces the fission barrier and greatly increases the probability of spontaneous fission. When fission barrier is drop out, the nucleus will split into two fragments for about 10^{-19} s. According to the calculations of the liquid drop model, this situation occurs immediately after Z > 100. It should be noted that when neutron-rich

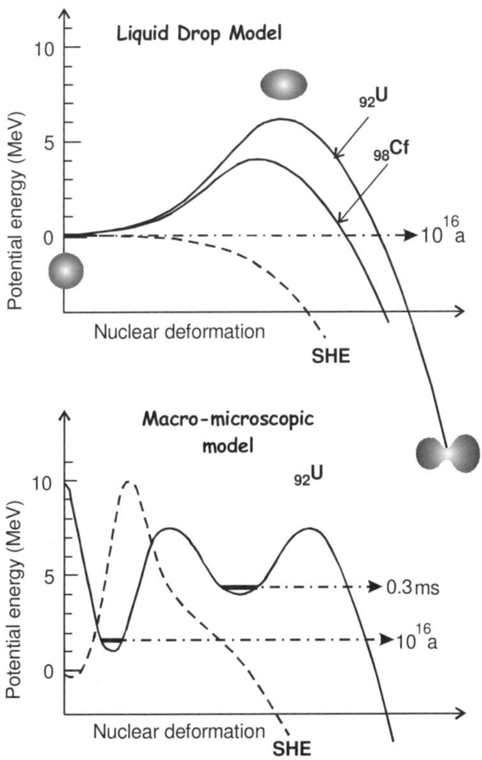

Figure 2. Potential energy of heavy nuclei (Z ≥ 92) *vs.* deformation. Upper graph show the calculation in the liquid drop model; lower graph – with shell corrections. Dotted curve - the fission barrier of the nuclei with Z=108, N= 162.

isotope of artificial elements up to Fermium (Z = 100) were synthesized using high flux reactors, it seemed that the theoretical predictions have been confirmed.

However, in 1962, in our Laboratory, an unexpected effect was observed, which caused great doubts in the analogy of nuclear fission and a liquid drop. The isotope of ^{242}Am, obtained in a nuclear reaction, experienced spontaneous fission with two, very different half-lives: 10^{14}a, and 0.014 s [2]. Subsequently, a similar phenomenon was observed in more than 31 nuclei with Z = 92-97 [3]. Definitely, fission in these nuclei occurs from two states - the ground and isomeric states (Fig.2 lower panel). But the shape isomerism of nuclei is completely incompatible with describing the fission process as a split of a liquid drop. In subsequent experiments to measure the heights of fission barriers of nuclei with Z = 80-100, significant differences from the predictions of the liquid drop model of fission were also found.

It is noteworthy that by this time it was already known that the nuclei in their ground states differ in form and energy. At certain «magic» numbers of protons and/or neutrons (closed shells) the nuclear binding energy considerably increases. But, it was always assumed that reaching strong deformation on the way to splitting into two fragments, the shell effect disappears and the fission process can still be considered in the liquid drop approximation. In fact, as was shown in [4], nuclear structure does not disappear with increasing deformation, but is modified and continues to play a significant role in the process of nuclear fission. Calculation of the potential energy of the nucleus at large deformations, taking into account structural effects, explained the experimentally observed high fission barrier of the «doubly magic» nucleus ^{208}Pb (Z = 82, N = 126), as well as the shape isomerism of uranium and transuranic nuclei, as a consequence of the double-humped fission barrier of many isotopes of the actinides.

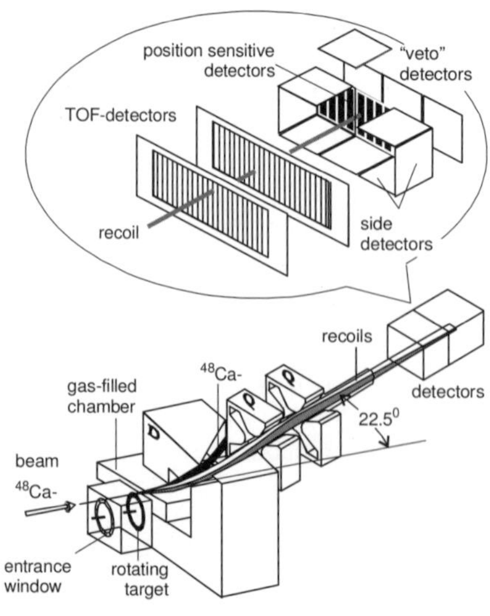

Figure 3. A schematic view of gas-filled recoil-separator, which was used in the experiments on the synthesis of superheavy elements.

It was natural to extend this approach to the region of nuclei with Z ≥ 104, where, according to the droplet model, nuclei may not exist. A formal calculation led to unexpected results. It turned out that in the deformed nucleus with mass 270 (Z = 108 and N = 162) a fission barrier occurs, which leads to an increase in its half-life to seconds (instead of 10^{-19} sec). But even higher stability appears in heavier (superheavy) nuclei with proton number Z = 114 and a large number of neutrons N = 184. Here the strong effect of new nuclear shells works, similarly to the shells Z = 82 and N = 126 in the «doubly magic» spherical nucleus ^{208}Pb. Some new calculations indicate that on the map of nuclei-heavyweights form a large enough area, called the «Island of Stability» of superheavy elements (see Fig.1). In the region where the liquid-drop model predicted that the nuclei should decay within 10^{-19} sec, now the half-lives of «long-lived» nuclei at the peak of the island of stability is expected to reach thousands and even millions of years! (see, for example, review [5]).

Hypothetical superheavy elements radically change our ideas about the limits of the material world, and they indicate very significant expansion of the boundaries of existence of atomic nuclei and chemical elements. Walter Greiner, from his earliest years in science, has been deeply involved in this progress and he made significant contribution in microscopic theory of the atomic nuclei.

Here we have a problem how to check this theoretical prediction experimentally. The known method of the production of elements heavier than uranium in successive neutron capture (with high flux reactor and even in nuclear explosions) confined only to the observation of nuclei only up to ^{257}Fm. In reactions induced by light ions (Z_i = 6-10) it was advance to elements with Z = 106-108 only. Entry into the area of Z>106 occurred after the discovery the so-called «cold fusion» reactions of the target nuclei - ^{208}Pb or ^{209}Bi and massive projectiles with Z_i = 22-30 [6] were used (see talk of Dr. Sigurd Hofmann at this meeting). In the reactions of «cold fusion» the formed compound nuclei are slightly heated, they are cooled by the emission of one neutron, and therefore have a relatively high survival rate. This was their main advantage, which allowed in the years 1978-1998 to synthesize 6 new elements up to 112 [7].
Unfortunately, as will follow later, this method cannot be used for the synthesis of nuclei located on the «island of stability» of the heaviest elements. The problem is that with increasing the mass and charge of the projectile-nuclei (having fixed target-nuclei: ^{208}Pb or ^{209}Bi) the Coulomb repulsion grows also. It greatly decreases the probability of formation of the compound nuclei with the growth of its charge number.

There is another fundamental limitation also. In cold fusion reactions, due to lack of neutrons, the produced heavy nuclei are outside the predicted «island of stability». Even for the heaviest stable isotopes of elements 110, 111, and 112 obtained in this way, the half-lives are estimated to be only tens or hundreds of microseconds. It seemed that using more asymmetric masses of the interacting nuclei (the target nucleus

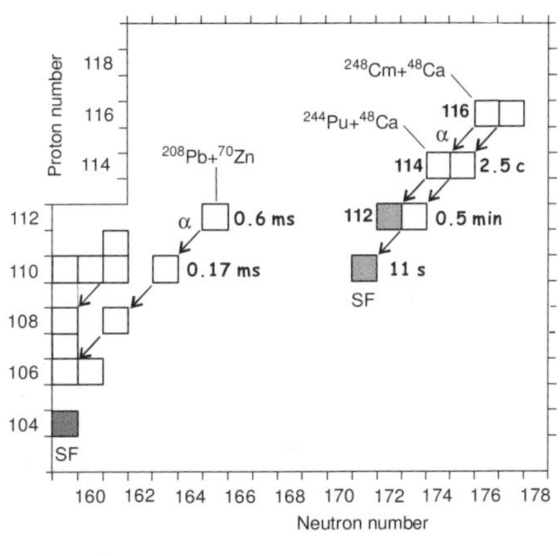

Figure 4. Decay chains of isotopes with Z=114 and 116, synthesized in the fusion reactions with calcium-48 projectiles

is heavier than the mass of projectile) might reduce the strength of the Coulomb repulsion and thus raise the estimated probability of fusion by 4-5 orders of magnitude. But the compound nuclei formed in these reactions are very hot; the losses during their cooling are even greater. Unsuccessful attempts to synthesize superheavy elements in reactions of «hot fusion», undertaken in various laboratories in the years 1977-1985, led to the pessimistic view that the elements heavier than 112 might be quite stable, but it would be practically impossible to produce them.

In this conclusion, in our opinion, there was an inherent contradiction. If the theoretical predictions were correct, the nuclei near the closed shells should have high fission barriers. In fact, it is this circumstance that causes the enormous stability of superheavy nuclei against spontaneous fission. But the high fission barrier must also prevent *the fission of the heated compound nucleus* and enhance its survival in the process of neutron emission during its cooling. In other words, the nuclei lying in the «island of stability» should fission less strongly and should survive easier. Moreover, as we approach the boundaries of the „Island of Stability», the rise of the formation cross sections of superheavy nuclei would be a direct indication of the inclusion of the effect of the new nuclear shells with Z = 114 (probably 120-122) and N = 184. Based on these assumptions, we chose a different reaction of synthesis of superheavy elements. Neutron-rich nuclei of SHE can in principle be synthesized if the heaviest actinides: ^{244}Pu, ^{243}Am, ^{248}Cm, ^{249}Bk and ^{251}Cf, obtained in powerful nuclear reactors, were used as targets. As the projectile, we have chosen the rare isotope of calcium with mass number 48. As a result of two years work using an ECR-source and the U-400 accelerator, a stable beam of ^{48}Ca ions was obtained with high intensity: 5.10^{12} -1.10^{13} pps. This allowed conducting long-term experiments (6000 hours/year) over the past 10 years. The registration of rare events of formation and decay of superheavy atoms was performed by the gas-filled recoil separator (Fig.3). In this setup, the nuclei emitted from the target, enter a hydrogen atmosphere (pressure about 1 Torre) and are separated by mass and ionic charge in a magnetic field and focus on the focal plane detector, located at a distance of 4 m from the target. The reaction products pass the distance between the target and the detector for 1 microsecond. The separator has a transmission for evaporation residues about 35-40%; the by-products of the reaction are suppressed by factor of 10^4-10^7 [8]. The focal plane detector registers the energy, position and time when the recoil nucleus stops in it. Then similar information follows concerning the charged particles (alpha particles or fission fragments) emitted in the decay of the implanted nucleus.

We recall that the theoretical predictions of the nuclei within the «Island of Stability» are stable against spontaneous fission. In the process of successive alpha-decay, the daughter nuclei will move away from the closed shells and will approach the boundaries of the island.

The likelihood of spontaneous fission increases rapidly and eventually the chain will be interrupted by spontaneous fission. In other words, we expect that the decay of a superheavy nucleus will have the form of a chain (radioactive family), consisting of successive alpha transitions,

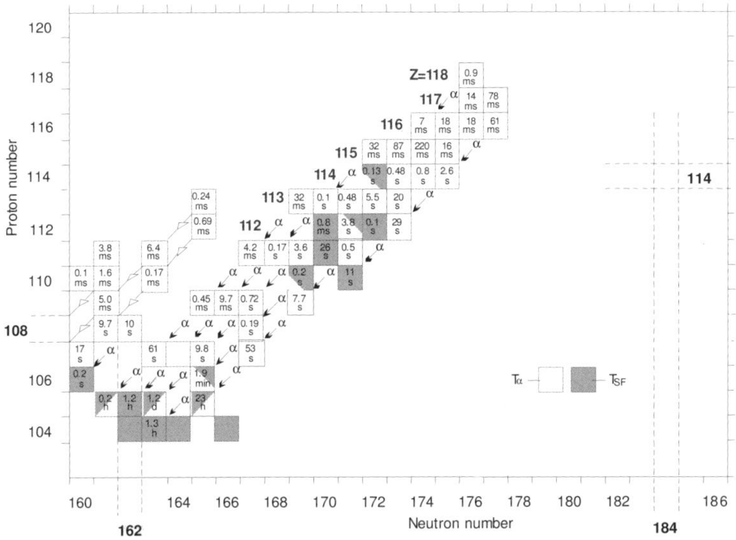

Figure 5. Nuclei with Z=112-118 and products of their radioactive decay - the isotopes of elements 111, obtained in the fusion reactions of the Actinides and ^{48}Ca. The experimental value half-lives are shown in the squares.

relatively long in time, terminated by spontaneous fission. Our focal plane detectors have efficiency of 87% for the registration of alpha particles, almost 100% - for a single spontaneous fission fragment, 42% - for both fragments. The location of the stopped recoil nucleus at the working surface of the detector is determined with an accuracy of better than 0. 2%.

The coordinates of the follow-up signals can be used to establish the genetic link between the recoil nucleus and its decay. The probability of random coincidences of signals, simulating a correlated decay, can be significantly reduced by switching off the accelerator beam between the registration of the recoil nucleus and the following first α-particle of its decay. For example, the probability of random coincidences in the registration of one event of formation and decay of the nucleus of the isotope of element 117 by the chain ^{293}117-α→^{289}115-α→^{285}113-α→^{281}Rg (SF), is 3.10^{-11} [9].

Just with this setup were synthesized isotopes of superheavy elements with atomic numbers 114 and 116 for the first time [10]. They showed a huge increase in nuclear stability by increasing the number of neutrons in the heavy nucleus (Fig.4). Adding to the previously known nuclei ^{273}110 and ^{275}112 8 neutrons led to an increase in their half-lives by almost 100 000 times!

Our data on the decay characteristics of the nuclei with Z = 112, 114 and 116 have been reproduced in 2007-2010 in a number of independent experiments [11-15]. Later, the heaviest element with atomic number 118 [16] was synthesized, followed by the synthesis of elements with odd atomic numbers: 113, 115 and 117 [17,9] in the reactions ^{237}Np, ^{243}Am and ^{249}Bk + ^{48}Ca. The data on the decay characteristics of the isotopes of 6 new superheavy elements, obtained up till now, together with those of the products of their radioactive decay (altogether 47 new nuclei) are shown in Fig.5. They can now be compared with the predictions of the microscopic theory [18].

Such a comparison is shown in Fig.6 for even-Z nuclei, for which the calculation can be done with the greatest accuracy. On the basis of the entire set of experimental data on the properties of superheavy nuclei produced in reactions induced by ^{48}Ca some general conclusions can be made.

1. The largest part of the nuclei undergoes alpha decay. The experimental energy of the alpha decay of superheavy nuclei differs from the calculated values by no more than 5-7%. This is a good agreement with the calculations of the nuclear macro-microscopic model. The periods of α-decay of the new nuclides were, on the average, up to 10 times higher than predicted by the theory.

2. Spontaneous fission was observed only in 14 isotopes (see Fig.5). They belong to two groups: the isotopes of elements 110-114 with N = 169-172 and the isotopes of lighter elements with Z = 104-106 and N = 160-165. Spontaneously- fissioning nuclei of the first group are located between the closed neutron shells N = 162 and N = 184; they are away from them by 7-12 neutrons. The smallest value of $T_{SF} \approx 10^{-3}$ sec is observed for the even-even nucleus of element 112 with N = 172. As we move away from this area towards an excess or deficit of neutrons, the partial spontaneous fission half-life increases (top graph of Fig.6). For nuclei with Z ≥ 114 and N ≥ 174 the spontaneous fission was not observed. In the second group, the nuclei are located around the deformed shell N = 162; they have a relatively high stability. The even-odd nucleus of element 104 with N = 163 has a $T_{SF} \sim 5.10^3$ s. The long-lived odd-Z nuclei, the isotopes of element 105 ($T_{1/2} \approx 10^5$ sec), most probably undergo electron capture. Then

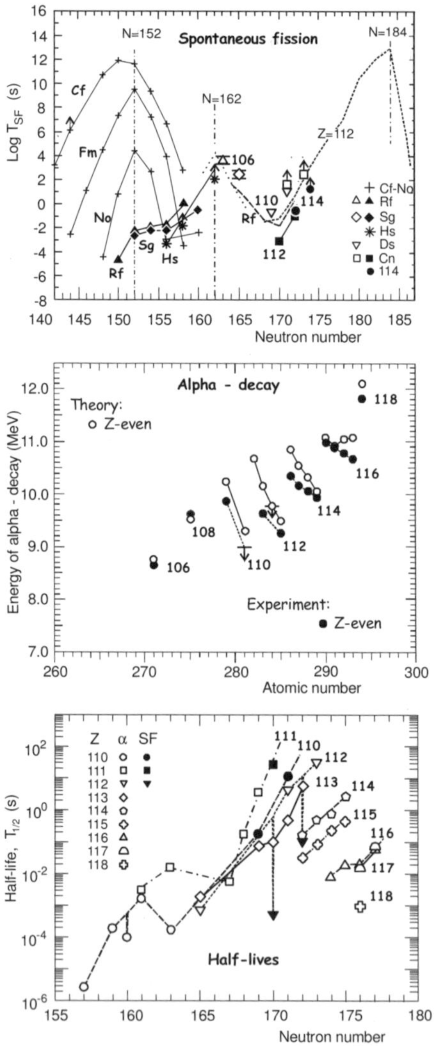

Figure 6. The experimental values of the spontaneous fission half-lives and alpha-decay energies are compared with theoretical expectations. The bottom graph shows the half-lives of superheavy nuclei vs N.

the experimentally observed spontaneous fission refers to the decay of even-even isotopes of element 104 with N = 164 and 166. For these nuclei, the so-called fast mode of symmetric fission is expected due to the effect of the nuclear shells Z = 50 and N = 82 in the fission fragments.

Conclusion

Increasing the atomic number of the last stable element - ^{209}Bi ($T_{1/2}$ = 1. 5.10^{19} years) by 40%, we are witnessing an impressive picture of the vitality of atomic nuclei. In the region of extreme Coulomb forces, due to the effect of new shells, the nuclear binding energy in the ground state increases, a fission barrier arises, and strongly increased stability of the heaviest nuclei leads to the existence of a large region of superheavy elements.
The fundamental predictions of the modern theories concerning the mass limits of the atomic nuclei have received their first experimental confirmation.

The experiments were carried out at the U-400 heavy ion cyclotron of the Flerov Laboratory of Nuclear Reactions (FLNR, JINR) in collaboration with LLNL (Livermore, USA), ORNL (Oak Ridge, USA), Research Institute of Atomic Reactors (Dimitrovgrad, Russia) and Vanderbilt University (Nashville, USA).

Dear Walter,
The above mentioned heavy element research collaboration is asking you to accept congratulations on your outstanding achievement in heavy ion physics!

References

[1] Bohr N and Wheller J A 1939 *Phys. Rev.* **56** 426
[2] Polikanov S M *et al.* 1962 *Soviet Phys.-JETF* **15** 1016
[3] Bjørnholm S and Linn J 1980 *Rev. Mod. Phys.* **52** 725
[4] Brack M *et al.* 1972 *Rev. Mod. Phys.* **44** 320
[5] Sobiczewski A and Pomorski K. 2007 *Prog. Part. Nucl. Phys.* **58**
[6] Oganessian Yu Ts 1975 *Lecture Notes in Physics,* Edt. J. Echlers et al Springer **33** 222
[7] Hofmann S and Munzenberg G 2000 *Rev. Mod. Phys.* **72** 733
[8] Oganessian Yu Ts *et al.* 2000 in Proceedings of the Fourth International Conference on DynamicalAspects of Nuclear Fission (World Scientific, Singapore 334
[9] Oganessian Yu Ts *et al.* 2010 *Phys. Rev.Lett.* **104** 142502
[10] Oganessian Yu Ts *et al.* 2000 *Phys. Rev.*C **62** 041604(R); C**63** 011301(R), (2004) C**70** 064609
[11] Hofmann S *et al.* 2007 *Eur. Phys. J.* A **32** 251
[12] Eichler R *et al. Nature* 2007 447 **72**
[13] Stavsetra L *et al.* 2009 *Phys. Rev. Lett.* **103** 132502
[14] Dullmann Ch E *et al.* 2010 *Phys. Rev. Lett.* **104** 252701

[15] Hofmann S *et al.* 2011 *Eur. Phys. J.* to be published
[16] Oganessian Yu Ts *et al.* 2006 *Phys. Rev.* C **74** 044602
[17] Oganessian Yu Ts *et al.* 2004 *Phys. Rev.* C **69** 021601(R) and 2005 C **72** 034611
[18] Muntian I Patyk Z and Sobiczewski 2003 A *Phys. At. Nucl.* **66** 1015

*This talk was presented also at the 24th Nuclear Physics Divisional Conference of the European Physical Society in Eilat, Israel, 2011

Memoriam : Irina Oganessian (1932 - 2010)

Irina was born in 1932 in Erevan. She graduated with excellence from Tchaikovsky Moscow State Conservatory. Being a young girl she came to the small city of Dubna (Moscow region) along with her husband, Yuri Oganessian, future famous nuclear physicist. For many years she worked here as a leading teacher at Dubna music school and as a head of the violinist ensemble. The ensemble has twice got the highest award of the International contest of instrumental music in Belgium. During her last years Irina Levonovna kept on her concert activity appearing on stage with the Dubna symphonic orchestra and giving solo concerts. She was a unique person who delighted her colleagues and friends.

DARK MATTER, GRAVITY WAVES, HIGGS BOSONS AND OTHER ELUSIVE ENTITIES

B.G. SIDHARTH

International Institute for Applicable Mathematics & Information Sciences Hyderabad,(India) & Udine (Italy)
B.M. Birla Science Centre, Adarsh Nagar, Hyderabad - 500 063 (India)
E-mail: iiamisbgs@yahoo.co.in

The twentieth century has thrown up exotic concepts – dark matter, gravity waves, Higgs Bosons, Magnetic Monopoles and so on. The sad truth is that even after several decades, these remain elusive to observation and experiment. Some are now questioning these conjectures. Their verification has become a matter more of hope than conviction. We will examine some alternatives in the light of the above pointing out that, on the other hand, an extra neutrino, recently predicted by the author may have just been discovered.

Keywords: Dark matter; gravity waves; Higgs bosons.

1. Dark Matter

It is well known that F. Zwicky introduced the concept of dark matter more than seventy five years ago to account for the anomalous rotation curves of the galaxies.[1,2] The problem was that according to the usual Newtonian Dynamics the velocities of the stars at the edges of galaxies should fall with distance as in Keplarian orbits, roughly according to

$$v \approx \sqrt{\frac{GM}{r}} \tag{1}$$

where M is the mass of the galaxy, r the distance from the centre of the galaxy of the outlying star and v the tangential velocity of the star. Observations however indicated that the velocity curves flatten out, rather than follow the law (1). This necessitated the introduction of the concept of dark matter which would take care of the discrepancy without modifying Newtonian dynamics. However even after nearly eight decades, dark matter has not been detected, even though there have been any number of candidates proposed for this, for example SUSY particles, massive neutrinos, undetectable brown dwarf stars, even black holes and so on.

Very recent developments are even more startling. These concern the rotating dwarf galaxies, which are satellites of the Milky Way.[3,4] These studies throw up a big puzzle. On the one hand these dwarf satellites cannot contain any dark matter and on the other hand the stars in the satellite galaxies are observed to be moving much faster than predicted by Newtonian dynamics, exactly as in the case of the galaxies themselves. Metz, Kroupa, Theis, Hensler and Jerjen conclude that the only explanation lies in rejecting dark matter and Newtonian gravitation. Indeed a well known Astrophysicist, R. Sanders from the University of Groningen commenting on these studies notes,[5] "The authors of this paper make a strong argument. Their result is entirely consistent with the expectations of modified Newtonian dynamics (MOND), but completely opposite to the predictions of the dark matter hypothesis. Rarely is an observational test so definite." Even more recently, dark matter has been ruled out by the observational studies of the Kavli Institute in California, of the interaction of electrons at the galactic edge with star light.[6] Finally, even more recent studies conclude that dark matter content has been vastly over estimated.[7] Sawangwit and Shanks note that, apart from the fact that candidates for dark matter like weakly interacting massive particles have not been detected so far, the dark matter problem in the Coma cluster is a factor of a hundred less than when Zwicky first proposed it. This is due to the discovery of hot gas in rich galaxy clusters. This makes the dark matter content only a factor of between four and five as a discrepancy, rather than the original six hundred times.

In this note we point out that this could indeed be so, though not via Milgrom's ad hoc modified dynamics,[2,8,9] according to which a test particle at a distance r from a large mass M is subject to the acceleration a given by

$$a^2/a_0 = MGr^{-2}, \tag{2}$$

where a_0 is an acceleration such that standard Newtonian dynamics is a good approximation only for accelerations much larger than a_0. The above equation however would be true when a is much less than a_0. Both the statements in (2) can be combined in the heuristic relation

$$\mu(a/a_0)a = MGr^{-2}. \tag{3}$$

In (3) $\mu(x) \approx 1$ when $x >> 1$, and $\mu(x) \approx x$ when $x << 1$. It must be stressed that (2) or (3) are not deduced from any theory, but rather are an ad hoc prescription to explain observations. Interestingly it must be mentioned that most of the implications of Modified Newtonian Dynamics or MOND do not depend strongly on the exact form of μ.

It can then be shown that the problem of galactic velocities is now solved.[8–12] Nevertheless, most physicists are not comfortable with MOND because of the ad hoc nature of (2) and (3).

We now come to the cosmological model described by the author in 1997 (Cf. Ref. 2,13 and several references therein), in which the universe, under the influence of dark energy would be accelerating with a small acceleration. Several other astrophysical relations, some of them hitherto inexplicable such as the Weinberg formula giving the pion mass in terms of the Hubble constant were also deduced in this model (Cf. also Ref. 14 and references therein). While all this was exactly opposite to the then established theory, it is well known that the picture was observationally confirmed soon thereafter through the work of Perlmutter and others (Cf. Ref. 14). Interestingly, in this model Newton's gravitational constant varied inversely with time.

Cosmologies with time varying G have been considered in the past, for example in the Brans-Dicke theory or in the Dirac large number theory or by Hoyle.[15–19] In the case of the Dirac cosmology, the motivation was Dirac's observation that the supposedly large number coincidences involving $N \sim 10^{80}$, the number of elementary particles in the universe had an underlying message if it is recognized that

$$\sqrt{N} \propto T \tag{4}$$

where T is the age of the universe. Equation (4) too leads to a G decreasing inversely with time as we will now show. We follow a route slightly different from that of Dirac.

From (4) it can easily be seen that

$$T = \sqrt{N}\tau \tag{5}$$

where τ is a typical Compton time of an elementary particle $\sim 10^{-23} secs$, because T, the present age of the universe is $\sim 10^{17} secs$. We also use the following well known relation which has been obtained a long time ago through different routes[20–22]

$$R = \frac{2GM}{c^2}. \tag{6}$$

Further multiplying both sides of (5) by c we get the famous Weyl-Eddington relation

$$R = \sqrt{N}l \tag{7}$$

where $l = c\tau$ is a typical Compton length $\sim 10^{-13} cms$. We will also use another well known relation (Cf. above references),

$$M = Nm,$$

where m is a typical elementary particle mass, like that of the pion, $10^{-25} gm$.

Use of (7) and the above in (6) now leads to

$$G = \frac{c^2 l}{2\sqrt{N}m} = \left(\frac{c^2 l\tau}{2m}\right) \cdot \frac{1}{T} \equiv \frac{G_0}{T}. \tag{8}$$

Equation (8) gives the above stated inverse dependence of the gravitational constant G on time, which Dirac obtained. On the other hand this same relation was obtained by a different route in the author's dark energy – fluctuations cosmology in 1997. This work, particularly in the context of the Planck scale has been there for many years in the literature (Cf.[2,14,24] and references therein). Suffice to say that all the supposedly so called accidental Large Number Relations like (7) as also the inexplicable Weinberg formula which relates the Hubble constant to the mass of a pion, follow as deductions in this cosmology. The above references give a comprehensive picture.

In any case, our starting point is, equation (8) where T is time (the age of the universe) and G_0 is a constant. Furthermore, other routine effects like the precession of the perihelion of Mercury and the bending of light and so on are also explained with (8) as will be briefly discussed below. We will also see that there is observational evidence for (8) (Cf. also[25] which described various observational evidences for the variation of G, for example from solar system observations, from cosmological observations and even from the palaeontological studies point of view).

With this background, we now mention some further tests for equation (8).

This could explain the other General Relativistic effects like the shortening of the period of binary pulsars and so on (Cf. Ref. 2,14,26,27 and other references therein). Moreover, we could now also explain, the otherwise inexplicable anomalous acceleration of the Pioneer space crafts (Cf. Ref. 2 for details). We will briefly revisit some of these effects later.

We now come to the problem of galactic rotational curves mentioned earlier (cf. Ref. 1). We would expect, on the basis of straightforward dynamics that the rotational velocities at the edges of galaxies would fall off according to

$$v^2 \approx \frac{GM}{r} \tag{9}$$

which is (1). However it is found that the velocities tend to a constant value,

$$v \sim 300km/sec \tag{10}$$

This, as noted, has lead to the postulation of the as yet undetected additional matter alluded to, the so called dark matter.(However for an alternative view point Cf.[28]). We observe that (8) can be written for an increase t,in time, small compared to the age of the universe, now written as t_0

$$G = \frac{G_0}{t_0 + t} = \frac{G_0}{t_0}\left(1 - \frac{t}{t_0}\right). \tag{11}$$

Using (11), let us consider the gravitational potential energy V between two masses, m_1 and m_2 by:

$$V = \frac{Gm_1m_2}{r_0} = \frac{G_0}{t_0} \cdot \frac{m_1m_2}{r_0}. \tag{12}$$

After a time t this would be, by (11),

$$V = \frac{G_0}{t_0}\left(1 - \frac{t}{t_0}\right)\frac{m_1m_2}{r}. \tag{13}$$

Equating (12) and (13) we get,

$$r = r_0\left(\frac{t_0}{t_0 + t}\right)\left(= r_0\left(1 - \frac{t}{t_0}\right)\right). \tag{14}$$

The relation (14) has been deduced by a different route by Narlikar.[1] From (14) it easily follows that,

$$a \equiv (\ddot{r}_o - \ddot{r}) \approx \frac{1}{t_o}(t\dddot{r}_o + 2\dot{r}_o) \approx -2\frac{r_o}{t_o^2} \tag{15}$$

as we are considering intervals t small compared to the age of the universe and nearly circular orbits. In (15), a or the left side of (15) gives the new extra effect due to (11) and (14), this being a departure from the usual Newtonian gravitation. Equation (15) shows (Cf. Ref. 26 also) that there is an anomalous inward acceleration, as if there is an extra attractive force, or an additional central mass.

So, introducing the extra acceleration (15), we get,

$$\frac{GMm}{r^2} + \frac{2mr}{t_o^2} \approx \frac{mv^2}{r}. \tag{16}$$

From (16) it follows that

$$v \approx \left(\frac{2r^2}{t_o^2} + \frac{GM}{r}\right)^{1/2}. \tag{17}$$

So (17) replaces (1) in this model. This shows that as long as

$$\frac{2r^2}{t_0^2} << \frac{GM}{r}, \tag{18}$$

Newtonian dynamics holds. But when the first term on the left side of (18) becomes of the order of the second (or greater), the new dynamical effects come in.

For example from (17) it is easily seen that at distances well within the edge of a typical galaxy, that is $r < 10^{23} cms$ the usual equation (9) holds but as we reach the edge and beyond, that is for $r \geq 10^{24} cms$ we have $v \sim 10^7 cms$ per second, in agreement with (10). In fact as can be seen from (17), the first term in the square root has an extra contribution (due to the varying G) which exceeds the second term as we approach the galactic edge, as if there is an extra mass, that much more.

We would like to stress that the same conclusions will apply to the latest observations of the satellite galaxies (without requiring any dark matter).

Let us for example consider the Megallanic clouds.[29] In this case, as we approach their edges, the first term within the square root on the right side of (17) or the left term of (18) already becomes of the order of the second term, leading to the new non Newtonian effects.

The point is that the above varying G scheme described in (8) or (17) reproduces all the effects of General Relativity as noted above, as also the anomalous acceleration of the Pioneer space crafts in addition to the conclusions of MOND regarding an alternative for dark matter, and is applicable to the latest observations of satellite galaxies. The satellite galaxy rotation puzzle is thus resolved.

2. General Relativity

It is now accepted that the tests which General Relativity proposed have all been observed, except for gravitational waves which have eluded detection for nearly a century. We would like to observe that the variation of the given constant mentioned earlier can also explain the various tests of General Relativity. Let us see how. We first observe that our starting point is the precession of the perihelion of mercury. We observe that from (8) or (11) it follows that[1]

$$G = G_o \left(1 - \frac{t}{t_o}\right) \tag{19}$$

where G_o is the present value of G and t_o is the present age of the Universe and t the (relatively small) time elapsed from the present epoch. Similarly one could deduce that (cf.Ref. 1),

$$r = r_o \left(\frac{t_o}{t_o + t} \right). \tag{20}$$

We next use Kepler's Third law:[30]

$$\tau = \frac{2\pi a^{3/2}}{\sqrt{GM}} \tag{21}$$

τ is the period of revolution, a is the orbit's semi major axis, and M is the mass of the sun. Denoting the average angular velocity of the planet by

$$\dot{\Theta} \equiv \frac{2\pi}{\tau},$$

it follows from (19), (20) and (21) that

$$\dot{\Theta} - \dot{\Theta}_o = \dot{\Theta}_0 \frac{t}{t_o},$$

where the subscript o refers to the present epoch.

Whence,

$$\omega(t) \equiv \Theta - \Theta_o = \frac{\pi}{\tau_o t_o} t^2. \tag{22}$$

Equation (22) gives the average perihelion precession at time 't'. Specializing to the case of Mercury, where $\tau_o = 0.25$ year, it follows from (22) that the average precession per year at time 't' is given by

$$\omega(t) = \frac{4\pi t^2}{t_0}. \tag{23}$$

Whence, considering $\omega(t)$ for years $t = 1, 2, \ldots, 100$, we can obtain from (23), the correct perihelion precession per century as,[26]

$$\omega = \sum_{n=1}^{100} \omega(n) \approx 43'',$$

if the age of the universe is taken to be $\approx 2 \times 10^{10}$ years.

Conversely, if we use the observed value of the precession in (23), we can get back the above age of the universe.

Interestingly it can be seen from (23), that the precession depends on the epoch.

We can similarly demonstrate that orbiting objects will have an anamolous inward radial acceleration.

For $r \sim 10^{14} cm$, as is the case of the space crafts Pioneer 10 or Pioneer 11, this gives, $a_r \geq 10^{-11} cm/sec^2$ This can be further refined to $a_\gamma \leq 10^{-10} cm$. Interestingly Anderson et al.,[31] claim to have observed an anomalous inward acceleration of $\sim 10^{-8} cm/sec^2$ for more than a decade.

We could also explain the correct gravitational bending of light in the same vein. The inexplicable anomalous accelerations of the Pioneer spacecrafts already alluded to, which have been observed by J.D. Anderson and coworkers at the Jet Propulsion Laboratory for well over a decade[31,32] have posed a puzzle. This can be explained, in a simple way as follows: In fact from the usual orbital equations we have[33]

$$v\dot{v} \approx -\frac{GM}{2tr}(1 + ecos\Theta) - \frac{GM}{r^2}\dot{r}(1 + ecos\Theta)$$

v being the velocity of the spacecraft and t is the time in general. It must be observed that the first term on the right side is the new effect due to (19). There is now an anomalous acceleration given by

$$a_r = \langle \dot{v} \rangle \text{anom} = \frac{-GM}{2trv}(1 + ecos\Theta)$$

$$\approx -\frac{GM}{2t\lambda}(1 + e)^3$$

where

$$\lambda = r^4\dot{\Theta}^2.$$

If we insert the values for the Pioneer spacecrafts we get

$$a_r \sim -10^{-7} cm/sec^2.$$

This is the anomalous acceleration reported by Anderson and co-workers.

We will next deduce that this case also explains correctly the observed decrease in the orbital period of the binary pulsar $PSR\,1913 + 16$, which has also been attributed to as yet undetected gravitational waves.[34]

It should also be remarked that in the case of gravitational radiation, there are some objections relevant to the calculation (Cf.Ref. 34).

Finally, we may point out that a similar shrinking in size with time can be expected of galaxies themselves, and in general, gravitationally bound systems. We will see a special case for the solar system.

To consider the above result in a more general context, we come back to the well known orbital equation[33]

$$d^2u/d\Theta^2 + u = \mu_0/h^2 \tag{24}$$

where $\mu_0 = GM$ and u is the usual inverse of radial distance.

M is the mass of the central object and $h = r^2 d\Theta/dt$ - a constant. The solution of (24) is well known,

$$lu = 1 + ecos\Theta$$

where $l = h^2/\mu_0$.

It must be mentioned that in the above purely classical analysis, there is no precession of the perihelion.

We now replace μ_0 by μ and also assume μ to be varying slowly because G itself varies slowly and uniformly, as noted earlier:

$$\dot{\mu} = d\mu/dt = K, \tag{25}$$

a constant remembering that $\dot{K} \sim 0(1/T^2)$ and so can be neglected.

Using (25) in (24) and solving the orbital equation (24), the solution can now be obtained as

$$u = 1/l + (e/l)cos\Theta + Kl^2\Theta/h^3 + Kl^2e\Theta cos\Theta/h^3. \tag{26}$$

Keeping terms up to the power of 'e' and $(K/\mu_0)^2$, the time period 'τ' for one revolution is given to this order of approximation by

$$\tau = 2\pi L^2/h. \tag{27}$$

From (26)

$$L = l - \frac{Kl^4\Theta}{h^3}. \tag{28}$$

Substituting (28) in (27) we have

$$\tau = \frac{2\pi}{h}\left(l^2 - \frac{2Kl^5\Theta}{h^3}\right). \tag{29}$$

The second term in (29) represents the change in time period for one revolution. The decrease of time period is given by

$$\delta\tau = 8\pi^2 l^3 K/\mu_0^2. \tag{30}$$

The second term in (28) indicates the decrease in latus-rectum.

For one revolution the change of latus-rectum is given by

$$\delta l = 2\pi K l^{2.5}/\mu_0^{1.5}. \tag{31}$$

In the solar system, we have,

$$K = 898800\, cm\, gm.$$

Using K and μ_0 to find the change in time period and the latus rectum in the varying G case by substituting in (30) and (31) respectively for Mercury we get

$$\delta T = 1.37 \times 10^{-5} sec/rev$$

$$\delta l = 4.54 cm/rev. \tag{32}$$

We observe that the equations (30), (31) or (32) show a decrease in distance and in the time of revolution. If we use for the planetary motion, the General Relativistic analogue of (24), viz.,

$$\frac{d^2u}{d\Theta^2} + u = \frac{\mu_0}{h^2}(1 + 3h^2u^2),$$

then while we recover the precession of the perihelion of Mercury, for example, there is no effect similar to (30), (31) or (32). On the other hand this effect is very minute– just a few centimeters per year in the case of the earth– and only protracted careful observations can detect it. Moreover these changes could also be masked at least partly, by gravitational and other perturbations.

However as noted, the decrease of the period in (30) has been observed in the case of Binary Pulsars.

Finally, it has been pointed out that (19) itself can be shown to yield an effect that simulates gravitational waves without invoking the full General Relativity.[2]

3. Higgs Bosons

We now come to another aspect. According to the standard model of particle physics we need the Higgs Bosons to provide mass to the various particles in the universe. However these Bosons, first suggested by Peter Higgs in 1965 have eluded detection for decades. More recently, hopes that these so called God particles would be detected at the LHC have received a segment, because, the Tevatron rules out the existence of Higgs particles in the expected range below $200GeV$. Physicists however are still hoping that the LHC would throw up the Higgs, though some are now veering to the view that physics beyond the standard model would be more interesting without the Higgs.

In this latter context, we will now argue that it is possible for both electromagnetism and gravitation to emerge from a gauge like formulation.[35] In Gauge Theory, which as we saw in Chapter 2, is a Quantum Mechanical generalization of Weyl's original geometry, we generalize, as is well known, the original phase transformations, which are global with the phase λ being a constant, to local phase transformations with λ being a function of the coordinates.[36] As is well known this leads to a covariant gauge derivative. For example, the transformation arising from $(x^\mu) \to (x^\mu + dx^\mu)$,

$$\psi \to \psi e^{-\imath\lambda} \qquad (33)$$

leads to the familiar electromagnetic potential gauge,

$$A_\mu \to A_\mu - \partial_\mu \lambda. \tag{34}$$

The above transformation, ofcourse, is a symmetry transformation. In the transition from (33) to (34), we expand the exponential, retaining terms only to the first order in coordinate differentials.

Let us now consider the case where there is a minimum cut off in the spactime intervals. As we saw this leads to a noncommutative geometry (Cf. Ref. 37)

$$[dx_\mu, dx_\nu] = O(l^2) \tag{35}$$

where l is the minimum scale. From (35) it can be seen that if $O(l^2)$ is neglected, we are back with the familiar commutative spacetime. The new effects of fuzzy spacetime arise when the right side of (35) is not neglected. Based on this we had argued in Chapter 5 that it is possible to reconcile electromagnetism and gravitation.[38–41] If in the transition from (33 to (34) we retain, in view of (35), squares of differentials, in the expansion of the function λ we will get terms like

$$\{\partial_\mu \lambda\} dx^\mu + (\partial_\mu \partial_\nu + \partial_\nu \partial_\mu) \lambda \cdot dx^\mu dx^\nu \tag{36}$$

where we should remember that in view of (35), the derivatives (or the product of coordinate differentials) do not commute as indeed we saw in Chapter 5. As in the usual theory the coefficient of dx^μ in the first term of (36) represents now, not the gauge term but the electromagnetic potential itself: Infact, in this noncommutative geometry, it can be shown that this electromagnetic potential reduces to the potential in Weyl's original gauge theory.[35,38]

Without the noncommutativity, the potential $\partial_\mu \lambda$ would lead to a vanishing electromagnetic field. However as we saw Dirac pointed out in his famous monopole paper in 1930 that a non integrable phase $\lambda(x, y, z)$ leads as above directly to the electromagnetic potential, and moreover this was an alternative formulation of the original Weyl theory.[42,43]
Returning to (36) we identify the next coefficient with the metric tensor giving the gravitational field:

$$ds^2 = g_{\mu\nu} dx^\mu dx^\nu = (\partial_\mu \partial_\nu + \partial_\nu \partial_\mu) \lambda dx^\mu dx^\nu. \tag{37}$$

Infact one can easily verify that ds^2 of (37) is an invariant. We now specialize to the case of the linear theory in which squares and higher powers of the deviation from the Minkowski metric, $h^{\alpha\beta}$ can be neglected. In this case it can easily be shown that

$$2\Gamma^{\beta}_{\mu\nu} = h_{\beta\mu,\nu} + h_{\nu\beta,\mu} - h_{\mu\nu,\beta} \tag{38}$$

where in (38), the Γs denote Christofell symbols. From (38) by a contraction we have

$$2\Gamma^{\mu}_{\mu\nu} = h_{\mu\nu,\mu} = h_{\mu\mu,\nu}. \tag{39}$$

If we use the well known gauge condition[44]

$$\partial_\mu \left(h^{\mu\nu} - \frac{1}{2}\eta^{\mu\nu} h_{\mu\nu} \right) = 0, \text{ where } h = h^{\mu}_{\mu}$$

then we get

$$\partial_\mu h_{\mu\nu} = \partial_\nu h^{\mu}_{\mu} \equiv \partial_\nu h \tag{40}$$

(40) shows that we can take the λ in (36) as $\lambda =$, both for the electromagnetic potential A_μ and the metric tensor $h_{\mu\nu}$. (39) further shows that the A_μ so defined becomes identical to Weyl's gauge invariant potential.[45]

However it is worth reiterating that in the present formulation, we have a noncommutative geometry, that is the derivatives do not commute and moreover we are working to the order where l^2 cannot be neglected. Given this condition both the electromagnetic potential and the gravitational potential are seen to follow from the gauge like theory. By retaining coordinate differential squares, we are even able to accommodate apart from the usual spin 1 gauge particles, also the spin 2 graviton which otherwise cannot be accommodated in the usual gauge theory. If however $O(l^2) = 0$, then we are back with commutative spacetime, that is a usual point spacetime and the usual gauge theory describing spin 1 particles.

We had reached this conclusion in Chapter 5 (Cf. Ref. 37), though from a different, non gauge point of view. The advantage of the present formulation is that it provides a transparent link with conventional theory on the one hand, and shows how the other interactions described by non Abelian gauge theories smoothly fit into the picture.

Finally it may be pointed out that we had already argued that a fuzzy spacetime input explains why the purely classical Kerr-Newman metric gives the purely Quantum Mechanical anomalous gyromagnetic ratio of the electron,[46,47] thus providing a link between General Relativity and electromagnetism. This provides further support to the above considerations.

Let us now return to the gauge field itself. As is well known, this could be obtained as a generalization of the above phase function λ to include fields with internal degrees of freedom. For example λ could be replaced by A_μ given by[48]

$$A_\mu = \sum_\imath A^\imath_\mu(x) L_\imath. \tag{41}$$

The gauge field itself would be obtained by using Stoke's Theorem and (41). This is a very well known procedure: considering a circuit, which for simplicity we can take to be a parallelogram of side dx and dy in two dimensions, we can easily deduce the equation for the field, viz.,

$$F_{\mu\nu} = \partial_\mu A_\nu - \partial_\nu A_\mu - \imath q[A_\mu, A_\nu], \tag{42}$$

q being the gauge field coupling constant.

In (42), the second term on the right side is typical of a non Abelian gauge field. In the case of the(U(1) electromagnetic field, this latter term vanishes.

Further as is well known, in a typical Lagrangian like

$$L = \imath\bar{\psi}\gamma^\mu D_\mu \psi - \frac{1}{4}F^{\mu\nu}F_{\mu\nu} - m\bar{\psi}\psi \tag{43}$$

D denoting the Gauge covariant derivative, there is no mass term for the field Bosons. Such a mass term in (43) must have the form $m^2 A^\mu A_\mu$ which unfortunately is not Gauge invariant.

This as we saw in Chapter 2, was the shortcoming of the original Yang-Mills Gauge Theory: The Gauge Bosons would be massless and hence the need for a symmetry breaking, mass generating mechanism.

The well known remedy for the above situation has been to consider, in analogy with superconductivity theory, an extra phase of a self coherent system (Cf. Ref. 48 for a simple and elegant treatment and also Refs. 36 and[49]). Thus instead of the gauge field A_μ, we consider a new phase adjusted gauge field after the symmetry is broken

$$W_\mu = A_\mu - \frac{1}{q}\partial_\mu \phi. \tag{44}$$

The field W_μ now generates the mass in a self consistent manner via a Higgs mechanism. Infact the kinetic energy term

$$\frac{1}{2}|D_\mu \phi|^2 \quad , \tag{45}$$

where D_μ in (45) denotes the Gauge , now becomes

$$|D_\mu \phi_0|^2 = q^2 |W_\mu|^2 |\phi_0|^2 , \tag{46}$$

Equation (46) gives the mass in terms of the ground state ϕ_0.

The whole point is as follows: The symmetry breaking of the gauge field manifests itself only at short length scales signifying the fact that the field is mediated by particles with large mass. Further the internal symmetry space of the gauge field is broken by an external constraint: the wave function has an intrinsic relative phase factor which is a different function of spacetime coordinates compared to the phase change necessitated by the minimum coupling requirement for a free particle with the gauge potential. This cannot be achieved for an ordinary point like particle, but a new type of a physical system, like the self coherent system of superconductivity theory now interacts with the gauge field. The second or extra term in (44) is effectively an external field, though (46) manifests itself only in a relatively small spatial interval. The ϕ of the Higgs field in (44), in analogy with the phase function of Cooper pairs of superconductivity theory comes with a Landau-Ginzburg potential $V(\phi)$.

Let us now consider in the gauge field transformation, an additional phase term, $f(x)$, this being a scalar. In the usual theory such a term can always be gauged away in the U(1) electromagnetic group. However we now consider the new situation of a noncommutative geometry referred to above,

$$[dx^\mu, dx^\nu] = \Theta^{\mu\nu}\beta, \beta \sim 0(l^2) \tag{47}$$

where l denotes the minimum spacetime cut off. Equation (47) is infact Lorentz covariant. Then the f phase factor gives a contribution to the second order in coordinate differentials,

$$\frac{1}{2}\left[\partial_\mu B_\nu - \partial_\nu B_\mu\right]\left[dx^\mu, dx^\nu\right]$$

$$+\frac{1}{2}\left[\partial_\mu B_\nu + \partial_\nu B_\mu\right]\left[dx^\mu dx^\nu + dx^\nu dx^\mu\right] \tag{48}$$

where $B_\mu \equiv \partial_\mu f$.

As can be seen from (48) and (47), the new contribution is in the term which contains the commutator of the coordinate differentials, and not in the symmetric second term. Effectively, remembering that B_μ arises from the scalar phase factor, and not from the non-Abelian gauge field, in equation (42) A_μ is replaced by

$$A_\mu \to A_\mu + B_\mu = A_\mu + \partial_\mu f. \tag{49}$$

Comparing (49) with (44) we can immediately see that the effect of noncommutativity is precisely that of providing a new symmetry breaking term

to the gauge field, instead of the ϕ term, (Cf. Refs. 50,51) a term not belonging to the gauge field itself.
On the other hand if we neglect in (47) terms $\sim l^2$, then there is no extra contribution coming from (48) or (49), so that we are in the usual non-Abelian gauge field theory, requiring a broken symmetry to obtain an equation like (49). This is not surprising because as noted several times if we neglect the term $\sim l^2$ in (47) then we are back with the usual commutative theory and the usual Quantum Mechanics.

The matters have been dealt with more recently too.[52]

4. Miscellaneous Remarks

The above consideration of non-commutative spacetime have also been shown to lead to the conclusion that the magnetic monopole is redundant, as indeed Dirac himself had suggested.[2,43] On the other hand, based on recent work with ultra high energy fermions, the author had suggested recently[53] that there would be an extra neutrino (rather like a sterile neutrino). It is quite remarkable that researches at Fermi Lab have just confirmed that indeed such a fourth flavour neutrino exists.[54]

Following the same train, the Hamiltonian gets a new Lorentz symmetry violating term, which makes it non Hermitian at very high energies. So we can expect a decay of very high energy particles, and as argued elsewhere, pions will be one of the products. Further, Einstein's General Relativity deals with gravitation and its unification with Quantum Mechanics or electromagnetism is still eluding us, even after a century. In this connection, We would first like to briefly touch upon the author's Planck oscillator model which, over the years, successfully describe phenomena from an elementary particle to the universe itself.

We can easily verify that the Planck scale l plays the role of the Compton length and the Schwarzchild radius of a black hole of the mass m[55]

$$l = \frac{\hbar}{2mc}, \, l = \frac{2Gm}{c^2} m \sim 10^{-5} gm \, l \sim 10^{-33} cm. \tag{50}$$

Today in various Quantum gravity approaches the Planck length l is considered to be the fundamental minimum length, and so also the time interval $t = l/c$. Spacetime intervals smaller than given in (50) are meaningless both classically and Quantum mechanically. Classically because we cannot penetrate the Schwarzchild radius, and Quantum mechanically because we encounter unphysical phenomena inside a typical Compton scale. We will return to this point but all this has been discussed in greater detail by the author and others (Cf. Ref. 2 and several references therein).

At another level, it may be mentioned that the author's 1997 cosmological model invoked a background dark energy and fluctuations therein to deduce a model of the universe that was accelerating with a small cosmological constant, together with several other relations completely consistent with Astrophysics and Cosmology (Cf. Ref. 14 and several references therein). The observations of distant supernovae by Perlmutter and others confirmed in 1998 the dark energy driven accelerating universe of the author. All this is well known.

It is against this backdrop that the author had put forward his model of Planck oscillators in the dark energy driven Quantum vacuum, several years ago (Cf. Ref. 24 and several references therein,[56]). Let us consider an array of N particles, spaced a distance Δx apart, which behave like oscillators that are connected by springs. As is known we then have[24,57]

$$r = \sqrt{N \Delta x^2} ka^2 \equiv k \Delta x^2 = \frac{1}{2} k_B T \tag{51}$$

where k_B is the Boltzmann constant, T the temperature, r the total extension and k is the spring constant given by

$$\omega_0^2 = \frac{k}{m} \tag{52}$$

$$\omega = \left(\frac{k}{m} a^2 \right)^{\frac{1}{2}} \frac{1}{2} = \omega_0 \frac{a}{r}. \tag{53}$$

In (52) ω_0 is the frequency of the individual oscillator, while in (53) ω is the frequency of the array of N oscillators, N given in (51).

We can easily show from the above theory of oscillators that an oscillator with the Planck mass and with a spatial extent at the Planck scale has the same temperature as the Beckenstein temperature of a Schwarzchild Black Hole of mass given by the Planck mass. The above results can be obtained by a different route as described in Ref. 58.

It has also been shown that, given the well known effect that the universe consists of $N \sim 10^{80}$ elementary particles like the pion, it is possible to deduce that a typical elementary particle consists of $n \sim 10^{40}$ Planck oscillators. These form a coherent array unlike in string theory, where we deal with single oscillators. Briefly, to recapitulate the known theory, using $N = n \sim 10^{40}$ in (51) we get

$$r = \sqrt{n} l \equiv L \sim 10^{-13} cm$$

r now being a typical elementary particle, Compton length L. Similarly, we can get from (53), M being the mass of an elementary particle,

$$M = \frac{m}{\sqrt{n}} \sim 10^{-25} gm.$$

It must be mentioned that in this theory, furthermore, $n \sim \sqrt{N}$, where $N \sim 10^{80}$ is the number of elementary particles in the universe (Cf. Ref. 58). It has also been shown that in the above approach there is a pleasing correspondence with the usual Hawking-Beckenstein theory of Black Hole Thermodynamics.[59]

We can push the above consideration further. So far we have considered only a coherent array. This is necessary for meaningful physics and leads to the elementary particle masses and their other parameters as seen above. Cercignani[60] had used Quantum oscillations, though just before the dark energy era – these were the usual Zero Point oscillations, which had also been invoked by the author in his model. Invoking gravitation, what he proved was, in his own words, "Because of the equivalence of mass and energy, we can estimate that this (i.e. chaotic oscillations) will occur when the former will be of the order of $G[(\hbar\omega)c^{-2}]^2[\omega^{-1}c]^{-1} = G\hbar^2\omega^3c^{-5}$, where G is the constant of gravitational attraction and we have used as distance the wavelength. This must be less than the typical electromagnetic energy $\hbar\omega$. Hence ω must be less than $(G\hbar)^{-1/2}c^{5/2}$, which gives a gravitational cut off for the frequency in the zero-point energy."

In other words he deduced that there has to be a maximum frequency of oscillators given by

$$G\hbar\omega_{max}^2 = c^5 \tag{54}$$

for the very existence of coherent oscillations (and so a coherent universe). We would like to point out that if we use the above in equation (54) we get the well known relation

$$Gm_P^2 \approx \hbar c \tag{55}$$

which shows that at the Planck scale the gravitational and electromagnetic strengths are of the same order. This is not surprising because it was the very basis of Cercignani's derivation – if indeed the gravitational energy is greater than that given in (55), that is greater than the electromagnetic energy, then the Zero Point oscillators, which we have called the Planck oscillators would become chaotic and incoherent – there would be no physics. Let us now speak in terms of the background dark energy. We also use the fact that there is a fundamental minimum spacetime interval, namely at the

Planck scale. Then we can argue that (55) is the necessary and sufficient condition for coherent Planck oscillators to exist, in order that there be elementary particles which as noted above has been shown to be the number of $n \sim 10^{40}$ coherent Planck oscillators, and the rest of the requirements for the meaningful physical universe. In other words gravitational energy represented by the gravitation constant G given in (55) is a measure of the energy from the background dark energy that allows a physically meaningful universe – in this sense it is not a separate fundamental interaction.

It is interesting that (55) also arises in Sakharov's treatment of gravitation where it is a residual type of an energy.[2,61]

To proceed if we use the expression for the elementary particle mass M seen above in terms of the Planck mass in (55), we can easily deduce

$$GM^2 \approx \frac{e^2}{n} = \frac{e^2}{\sqrt{N}} \tag{56}$$

where now $N \sim 10^{80}$, the number of particles in the universe.

Equation (56) has been known for a long time as an empirical accident, without any fundamental explanation. Here we have deduced it on the basis of the Planck oscillator model. Equation (56) too brings out the relation between gravitation and the background Zero Point Field or Quantum vacuum or dark energy. It shows that the gravitational energy has the same origin as the electromagnetic energy but is in a sense a smeared out effect over the N particles of the universe. In the context of the above considerations that (56) is deduced and not empirical as in the past, we can now claim that (56) gives the desired unified description of electromagnetism and gravitation.

Finally the following may be pointed out: The work stemming from the cosmology briefly referred to above, leads to a universal acceleration, which is of the order of the pioneer anomaly, as has been pointed out by the author. Indeed such a universal acceleration is now being factored in as a reality.[62]

Acknowledgment

I would like to compliment my friend Prof.Walter Greiner on his seventy-fifth birthday and the many decades of fruitful and important contributions to physics.

References

1. Narlikar, J.V. (1993). *Introduction to Cosmology* (Cambridge University Press, Cambridge), p.57.

2. Sidharth, B.G. (2008). *The Thermodynamic Universe* (World Scientific), Singapore.
3. Metz, M, et al. (2009). *The Astrophysical Journal* 10.1088/0004-637X/697/1/269, 2009.
4. Metz, M. et al. (2009). *Monthly Notices of the Royal Astronomical Society* 10.1111/j.1365-2966.2009.14489.x, 2009.
5. http://physorg.com
6. Petrosian, V. (2010). *Astrophysical Journal* February 10, 2010.
7. Sawangwit, U. and Shanks, T. (2010). *Astronomy and Geophysics* (Royal Astronomical Society) Vol.51 (5), October 2010, pg.14ff.
8. Milgrom, M. (1983). *APJ* 270, pp.371.
9. Milgrom, M. (1986). *APJ* 302, pp.617.
10. Milgrom, M. (1989). *Comm. Astrophys.* 13:4, pp.215.
11. Milgrom, M. (1994). *Ann.Phys.* 229, pp.384.
12. Milgrom, M. (1997). *Phys.Rev.E* 56, pp.1148.
13. Sidharth, B.G. (1998). *Int.J.Mod.Phys.A*, 13 (15), 1998, p.2599ff.
14. Sidharth, B.G. (2001). *Chaotic Universe: From the Planck to the Hubble Scale* (Nova Science, New York).
15. Barrow, J.D. and Parsons, P. (1997). *Phys.Rev.D.* Vol.55, No.4, 15 February 1997, pp.1906ff.
16. Narlikar, J.V. (1983). *Foundations of Physics* Vol.13, No.3, pp.311–323.
17. Narlikar, J.V. (1989). *Did the Universe Originate in a Big Bang?* in *Cosmic Perspectives* Biswas, S.K., Mallik, D.C.V. and Vishveshwara, C.V. (eds.) (Cambrdige University Press, Cambridge), pp.109ff.
18. Sidharth, B.G. (2008). *Ether, Space-Time and Gravity* **Vol.3** Michael Duffy (ed.) (Apeiron Press, USA).
19. Sidharth, B.G. (2003). *Chaos, Solitons and Fractals* 18, (1), pp.197–201.
20. Ruffini, R. and Zang, L.Z. (1983). *Basic Concepts in Relativistic Astrophysics* (World Scientific, Singapore), p.38.
21. Hayakawa, S. (1961). *Suppl of Progressive Theoretical Physics Commemmorative Issue* 1961, pp.858-860.
22. Nottale, L. (1993). *Fractal Space-Time and Microphysics: Towards a Theory of Scale Relativity* (World Scientific, Singapore), pp.312.
23. Weinberg, S. (1972). *Gravitation and Cosmology* (John Wiley & Sons, New York), p.61ff.
24. Sidharth, B.G. (2005). *The Universe of Fluctuations* (Springer, Netherlands).
25. Uzan, J.P. (2003). *Rev.Mod.Phys.* 75, April 2003, pp.403–455.
26. Sidharth, B.G. (2000). *Nuovo Cimento* **115B** (12), (2), pp.151ff.
27. Sidharth, B.G. (2006). *Foundations of Physics Letters* 19 (6), 2006, pp.611-617.
28. Sivaram, C. and Sabbata, V. de. (1993). *Foundations of Physics Letters* 6, (6).
29. Staveley-Smith, L. et al. (2002). *Astronomical Journal* Vol.24 (4), 2002, pp.1954-1974.
30. Goldstein, H. (1966). *Classical Mechanics* (Addison-Wesley, Reading, Mass.), pp.76ff.

31. Anderson, J.L. et al. *xxx.lanl.gov/gr-qc/9808081.*
32. Anderson, J.D. et al. (2002). *Phys.Rev.D* 65, pp.082004ff.
33. Sidharth, B.G. (2006). *Found.of Phys.Lett.* 19, (6), pp.611–617.
34. Davies, P. (1989). *The New Physics* Davies, P. (ed.) (Cambridge University Press, Cambridge), pp.446ff.- New Physics
35. Sidharth, B.G., *Ann.Fond.L.De Broglie* 30 (2), 2005, pp.151-156.
36. Jacob, M., (Ed.), "Gauge Theory and Neutrino Physics", North Holland, Amsterdam, 1978.
37. Sidharth, B.G., Annales de la Fondation Lois de Broglie, Volume 27 No.2, 2002, 333-342.
38. Sidharth, B.G., Nuovo Cimento, 116B (6), 2001, p.4ff.
39. Sidharth, B.G., Nuovo Cimento, 117B, (6), 2002, pp.703ff.
40. Sidharth, B.G., Found.Phys.Lett., August 2002.
41. Sidharth, B.G., Found.Phys.Lett., 15 (6), 2002, pp.577-583.
42. Dirac, P.A.M., in "Monopoles in Quantum Field Theory", Eds. N.S. Craigie, P. Goddard and W. Nahm, World Scientific, Singapore, 1982, p.iii.
43. Sidharth, B.G., Nuovo Cimento, (6), 118B, 2002, 703ff.
44. Ohanian, H.C., Rufkin, R., and Ruffini, R. (1994). *Gravitation and Spacetime* (W.W. Norton & Company New York) 1994, pp.64ff.
45. Bergmann, P.G. (1969). *Introduction to the Theory of Relativity* (Prentice-Hall, New Delhi), p248ff.
46. Sidharth, B.G., Gravitation and Cosmology, 4 (2) (14), 1998, p.158ff.
47. Sidharth, B.G., Found.Phys.Lett., 16 (1), 2003, pp.91-97.
48. Moriyasu, K., "An Elementary Primer for Gauge Theory", World Scientific, Singapore, 1983.
49. Taylor, J.C., "Gauge Theories of Weak Interactions", Cambridge University Press, Cambridge, 1978.
50. Sidharth, B.G., Proceedings of the Fifth International Symposium on "Frontiers of Fundamental Physics", Universities Press, Hyderabad, 2004 (In Press).
51. Sidharth, B.G. (2005). *Int.J.Mod.Phys.E.* 14 (2), 2005, 215ff.
52. Sidharth, B.G. (2010). *Int.J.Mod.Phys.E.* 19 (1), 2010, 79-91.
53. Sidharth, B.G. *arxiv 1008.2491.*
54. Roe, B. *http://physlinks.com*; to appear in *Phys.Rev.Letters.*
55. Kiefer, C. (2004). *Quantum Gravity* (Clarendon Press, Oxford).
56. Sidharth, B.G. (2004). *Found.Phys.Lett.* **17**, (5), pp.503-506.
57. Sidharth, B.G. (2002). *Found.Phys.Lett.* 15 (6), 2002, pp.577-583.
58. B.G. Sidharth. (2006). *Int.J.Mod.Phys.A.* 21, (31), December 2006, pp.6315.
59. Sidharth, B.G. (2009). *Int.J.Th.Phys.* 48 (8), pp.2427-2431.
60. Cercignani, C. (1998). *Found.Phys.Lett.* Vol.11, No.2, pp.189-199.
61. Sakharov, A.D. (1968). *Soviet Physics - Doklady* Vol.12, No.11, pp.1040–1041.
62. Smolin, L. (2006). *The Trouble with Physics*, (Houghton Mifflin Company, New York).

ONSET OF DECONFINEMENT IN NUCLEUS-NUCLEUS COLLISIONS

MARK I. GORENSTEIN

Bogolyubov Institute for Theoretical Physics, Kiev, Ukraine

The energy dependence of hadron production in relativistic nucleus-nucleus collisions reveals the anomalies – *kink, horn*, and *step*. They were predicted as the signals of the deconfinement phase transition and observed by NA49 collaboration in Pb+Pb collisions at the CERN SPS. This indicates the onset of the deconfinement in nucleus-nucleus collisions at about 30 AGeV.

1. Introduction

The data on nucleus-nucleus (A+A) collisions obtained up to 2000 admitted a possibility of a significant change in the energy dependence of the pion and strangeness yields which is located between the top AGS (11 AGeV) and SPS (158 AGeV) energies. A simplified quantitative model was developed, the Statistical Model of the Early Stage (SMES), which predicted a rapid change of the energy dependence of numerous experimentally detectable properties of the collisions, all appearing in a common energy domain and related to the onset of deconfinement at the early stage of A+A collisions.[1] This prediction triggered an experimental programme at the SPS – the energy scan programme – in which the NA49 experiment recorded central Pb+Pb collisions at several energies: 20, 30, 40, 80 and 158 AGeV.[2] The energy scan program at the CERN SPS results an observation of the predicted anomalies in the energy dependence of hadron production in the same domain of the collision energy. These 'anomalies' are interpreted as signals of the deconfinement phase transition in A+A collisions.

2. The Kink

It is natural to expect that with increasing of collision energy the energy density at the early stage of A+A collisions also increases. We may hope that with increasing of collision energy we can detect anomalies in the

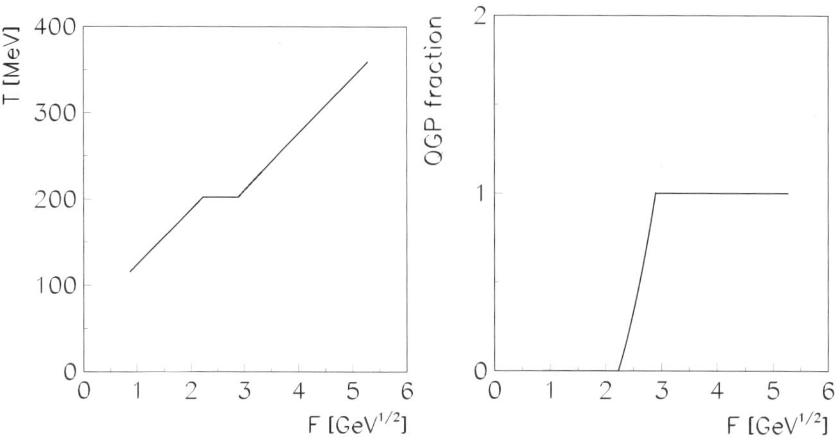

Fig. 1. *Left:* The early stage (initial) temperature of the fireball as a function of F. *Right:* The fraction of volume occupied by the QGP as a function of F.

energy dependence of hadron production properties and thus discover successive transitions between various phases of strongly interacting matter created at the early stage of collisions. At sufficiently high collision energy the matter may reach the quark-gluon plasma (QGP) phase. The created matter quickly expands, cools down and finally decays into hadrons. These decay products are measured in detectors surrounding the collision point.

The dependence of the early stage temperature T on the collision energy in the SMES is shown in Fig. 1 *left.* Outside the transition region T increases with F. The Fermi's energy measure F is defined in Eq. (1). Inside the transition region T is constant. The fraction of the volume occupied by the QGP at the early stage of A+A central collision increases rapidly in the transition region, as shown in Fig. 1 *right.*

The majority of all particles produced in high energy interactions are pions. Thus, pions carry basic information on the entropy created in the collisions. On the other hand, the entropy production should depend on the form of matter present at the early stage of collisions. A deconfined matter is expected to lead to the final state with higher entropy than that created by a confined matter. Consequently, it is natural to expect that the onset of creation of a deconfined matter should be signalled by an enhancement of the pion production. The simple intuitive argumentation can be further quantified within SMES assuming the generalized Fermi–Landau initial conditions: the initial volume is Lorenz-contracted, $V \propto \langle N_W \rangle \cdot (\sqrt{s_{NN}})^{-1}$, where $\sqrt{s_{NN}}$ is the c.m.s. energy of the nucleon pair and

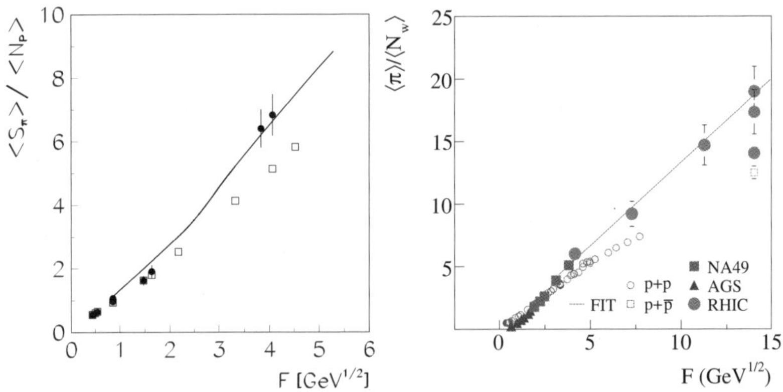

Fig. 2. *Left*: The $\langle S_\pi \rangle / \langle N_P \rangle$ ratio as a function F. Experimental data on central collisions of two identical nuclei are indicated by closed circles. The data correspond the compilations existed at 1998. They are compared with the SMES[1] shown by the solid line. The open boxes show results obtained for nucleon–nucleon interactions. *Right:* Energy dependence of the mean pion multiplicity per wounded nucleon measured in central Pb+Pb and Au+Au collisions (full symbols), compared to the corresponding results from $p + p(\bar{p})$ reactions (open symbols). The compilation of data is from Ref. 2.

$\langle N_W \rangle$ is the mean number of wounded nucleons (i.e. the number of nucleons participating in A+A inelastic reaction). A trivial dependence of the pion multiplicity on the size of colliding nuclei should be removed and thus a relevant observable is the ratio of the mean pion multiplicity $\langle \pi \rangle$ to $\langle N_W \rangle$. The initial energy density is given by $\varepsilon \propto gT^4 \propto (\sqrt{s_{NN}} - 2m_N) \cdot \sqrt{s_{NN}}$, where g is an effective number of internal degrees of freedom at the early stage and m_N is the nucleon mass. The pion multiplicity is proportional to the initial entropy, and the $\langle \pi \rangle / \langle N_W \rangle$ ratio can be thus calculated outside the transition region as:

$$\frac{\langle \pi \rangle}{\langle N_W \rangle} \propto \frac{V\ g\ T^3}{\langle N_W \rangle} \propto g^{1/4} (\sqrt{s_{NN}} - m_N)^{3/4} (\sqrt{s_{NN}})^{-1/4} \equiv g^{1/4}\ F\ . \quad (1)$$

Therefore, the $\langle \pi \rangle / \langle N_W \rangle$ ratio increases linearly with Fermi's energy measure F outside the transition region, and the slope parameter is proportional to $g^{1/4}$.[3] In the transition region, a steepening of the pion energy dependence is predicted, because of activation of partonic degrees of freedom, i.e. an effective number of internal degrees of freedom in QGP is larger than in the hadron gas (HG): $g_{QGP} > g_{HG}$.

The compilation of data on the pion multiplicity in central Pb+Pb (Au+Au) collisions and $p + p(\bar{p})$ interactions is shown in Fig. 2: the *left* figure presents the data available at 1998, and the *right* one shows presently available NA49 results.[2] In Fig. 2, *right*, the mean pion multiplicity $\langle\pi\rangle = 1.5\,(\langle\pi^-\rangle + \langle\pi^-\rangle)$ per wounded nucleon is shown as a function of F. The results from $p + p(\bar{p})$ interactions are shown by the open symbols. Up to the top SPS energy the mean pion multiplicity in $p+p$ interactions is approximately proportional to F. A fit of $\langle\pi\rangle/\langle N_W\rangle = bF$ yields a value of $b \cong 1.063$ GeV$^{-1/2}$. For central Pb+Pb and Au+Au collisions the energy dependence is more complicated. Below 40 AGeV the ratio $\langle\pi\rangle/\langle N_W\rangle$ is lower in A+A collisions than in $p + p(\bar{p})$ interactions (pion suppression) while at higher energies this ratio is larger in A+A collisions than in $p+p(\bar{p})$ interactions (pion enhancement). A linear fit, $\langle\pi\rangle/\langle N_W\rangle = a + bF$ for $F < 1.85$ GeV$^{1/2}$, gives $a \cong -0.45$ and $b \cong 1.03$ GeV$^{-1/2}$. The slope parameter fitted in the range $F > 3.5$ GeV$^{1/2}$ is $b \cong 1.33$. This is shown by the solid line in Fig. 2 (the lowest point at the top RHIC energy was excluded from the fit). Thus, in the region 15–40 AGeV between the highest AGS and the lowest SPS energy the slope increases by a factor of about 1.3. This increase of the slope for A+A collisions is interpreted within the SMES as due to an increase of the effective number of the internal degrees of freedom by a factor of $g_{QGP}/g_{HG} = (1.3)^4 \cong 3$.[1,3] It is caused by the creation of a transient state of deconfined matter at collision energies higher than 30 AGeV.

3. The Horn

The energy dependence of the strangeness to entropy ratio is a crucial signal of the deconfinement. The temperature dependence of a particle multiplicity is strongly dependent on the particle mass,

$$\langle N_i\rangle = \frac{g_i V}{2\pi^2}\int_0^\infty p^2 dp\, \exp\left(-\frac{\sqrt{p^2+m_i^2}}{T}\right) = \frac{g_i V}{2\pi^2}\, m_i^2 T\, K_2\left(\frac{m_i}{T}\right), \quad (2)$$

where g_i and m_i are respectively the degeneracy factor and particle mass, and K_2 is the modified Hankel function. For light hadrons ($m_l/T \ll 1$) one finds from Eq. (2), $\langle N_l\rangle \propto T^3$, whereas for heavy hadrons ($m_h/T \gg 1$) Eq. (2) leads to $\langle N_h\rangle \propto T^{3/2}\exp(-\,m_h/T)$. Within SMES at low collision energies, when the HG matter is produced, the strangeness to entropy ratio steeply increases with collision energy. This is due to a low temperature at the early stage and the high mass of the carriers of strangeness (the kaon mass), i.e. $m_K \gg T$ and the total strangeness is proportional to $T^{3/2}\exp(-\,m_K/T)$. On the other hand, the total entropy is approximately proportional to T^3. Therefore, the strangeness to pion ratio is ap-

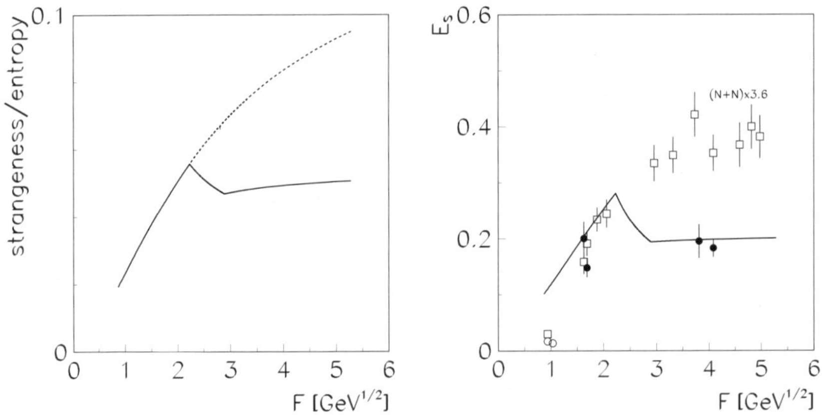

Fig. 3. *Left*: The ratio of the total number of s and $\bar{s}$ quarks and antiquarks to the entropy (solid line) as a function of F.[1] The dashed line indicates the corresponding ratio calculated assuming absence of the phase transition to the QGP. *Right*: The ratio E_S as a function of F. Experimental data on central collisions of two identical nuclei existed at 1998 are indicated by closed circles. These data should be compared with the SMES predictions[1] shown by the solid line. The open boxes show results obtained for nucleon–nucleon interaction, scaled be a factor 3.6 to match A+A data at AGS energy.

proximately $T^{-3/2}\exp(-m_K/T)$ in the HG and strongly increases with the collision energy. When the transition to a deconfined matter is crossed, the mass of the strangeness carriers is significantly reduced ($m_s \cong 150$ MeV, the strange quark mass). Due to the low mass ($m_s < T$) the strangeness yield becomes approximately proportional to the entropy (both are proportional to T^3), and the strangeness to entropy (or pion) ratio becomes independent of energy in the QGP. This leads to a "jump" in the energy dependence from the larger value for confined matter to the value for deconfined matter. Thus, within the SMES, the non-monotonic energy dependence of the strangeness to entropy ratio is followed by a saturation at the deconfined value which is a direct consequence of the onset of deconfinement taking place at about 30 AGeV.[1] This is illustrated by Fig. 3.

One can argue that the strangeness to entropy ratio is closely proportional to the two ratios directly measured in experiments: the $\langle K^+\rangle/\langle\pi^+\rangle$ ratio and the $E_s = (\langle\Lambda\rangle + \langle K + \overline{K}\rangle)/\langle\pi\rangle$ ratio. The energy dependence of these full phase space ratios is plotted in Fig. 4 for central Pb+Pb (Au+Au) collisions and $p+p$ interactions

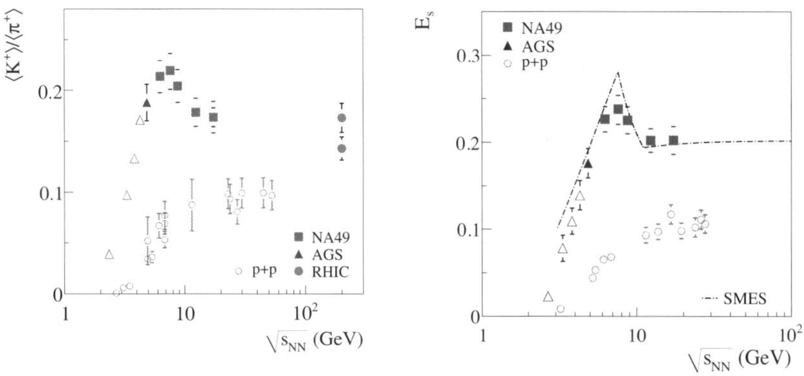

Fig. 4. *Left*: Energy dependence of the $\langle K^+\rangle/\langle\pi^+\rangle$ ratio measured in central Pb+Pb and Au+Au collisions (full symbols) compared to the corresponding results from $p+p$ reactions (open circles). *Right*: Energy dependence of the relative strangeness production as measured by the E_s ratio (see text) in central Pb+Pb and Au+Au collisions (full symbols) compared to results from $p+p$ reactions (open circles). The compilation of data is from Ref. 2. The dashed-dotted line in the figure shows the predictions of the SMES.[1]

Kaons are the lightest strange hadrons and $\langle K^+\rangle$ accounts for about half of all the anti-strange quarks produced in Pb+Pb collisions at AGS and SPS energies. K^+ and K^0 carry a dominant fraction of all produced $\bar{s}$-quarks exceeding 95% in Pb+Pb collisions at 158 A·GeV if open strangeness is considered. Because $\langle K^+\rangle \cong \langle K^0\rangle$ in approximately isospin symmetric collisions of heavy nuclei, the K^+ yield is nearly proportional to the total strangeness production and only weakly sensitive to the baryon density. As a significant fraction of s-quarks (about 50% in central Pb+Pb collisions at 158 A·GeV) is carried by hyperons, the number of produced anti-kaons, K^- and $\overline{K}^0$, is sensitive to both the strangeness yield and the baryon density. In the E_s ratio all main carriers of strange and anti-strange quarks are included. The neglected contribution of $\overline{\Lambda}$ and other hyperons and anti-hyperons is about 10% at SPS energies. Both the $\langle K^+\rangle/\langle\pi^+\rangle$ and E_s ratios are approximately, within 5% at SPS energies, proportional to the ratio of total multiplicity of s and $\bar{s}$ quarks to the multiplicity of pions. It should be noted that the $\langle K^+\rangle/\langle\pi^+\rangle$ ratio is expected to be similar (within about 10%) for $p+p$, $n+p$, and $n+n$ interactions at 158 A·GeV, whereas the E_s ratio is independent of the isospin of nucleon-nucleon interactions.

For $p+p$ interactions both ratios show monotonic increase with energy. However, very different behavior is observed for central Pb+Pb (Au+Au) collisions. The steep threshold rise of the ratios characteristic for confined matter then settles into saturations at the level expected for deconfined matter. In the transition region (at low SPS energies) a sharp maximum is observed caused by a higher strangeness to entropy ratio in the confinend matter than in the deconfined matter. As seen in Fig. 4 the measured dependence is consistent with that predicted within the SMES.[1]

4. The Step

The energy density at the early stage increases with increasing of collision energy. At low and high energies, when a pure confined or deconfined phase is produced, this leads to an increase of the initial temperature and pressure. This, in turn, results in an increase of the transverse expansion of a matter and consequently a flattening of the transverse mass spectra of final state hadrons. One may expect an 'anomaly'[4–6] in the energy dependence of the transverse hadron activity as the temperature and pressure are approximately constant in the mixed phase.

The experimental data on the transverse mass spectra are usually parameterized by a simple exponential dependence:

$$\frac{dN}{m_T dm_T} \propto \exp\left(-\frac{m_T}{T^*}\right) , \tag{3}$$

where $m_T = (m^2 + p_T^2)^{1/2}$. The inverse slope parameter T^* is sensitive to both the thermal and collective motion in the transverse direction. Hydrodynamical transverse flow with collective velocity v_T modifies the Boltzmann m_T-spectrum of hadrons. At low transverse momenta, it leads to the result (T_{kin} is a kinetic freeze-out temperature): $T^*_{low-p_T} \cong T_{kin} + \frac{1}{2} m\, v_T^2$. A linear mass dependence of T^* is supported by the data for hadron spectra at small p_T. However, for $p_T \gg m$ the hydrodynamical transverse flow leads to the mass-independent blue-shifted 'temperature': $T^*_{high-p_T} = T_{kin}\, [(1+v_T)/(1-v_T)]^{1/2}$. Note that a simple exponential fit (3) neither works for light π-mesons, $T^*_{low-p_T}(\pi) < T^*_{high-p_T}(\pi)$, nor for heavy (anti)protons and (anti)lambdas, $T^*_{low-p_T}(p, \Lambda) > T^*_{high-p_T}(p, \Lambda)$ (see e.g., Refs. 7,8).

Kaons are the best and unique particles among measured hadron species for observing the effect of the modification of the equation of state due to the onset of the deconfinement in hadron transverse momentum spectra. The arguments are the following:[6] 1) the kaon m_T–spectra are only weakly affected by the hadron re-scattering and resonance decays during the post-hydrodynamic hadron cascade at the SPS and RHIC energies; 2) a simple one parameter exponential fit (3) is quite accurate for kaons in central A+A

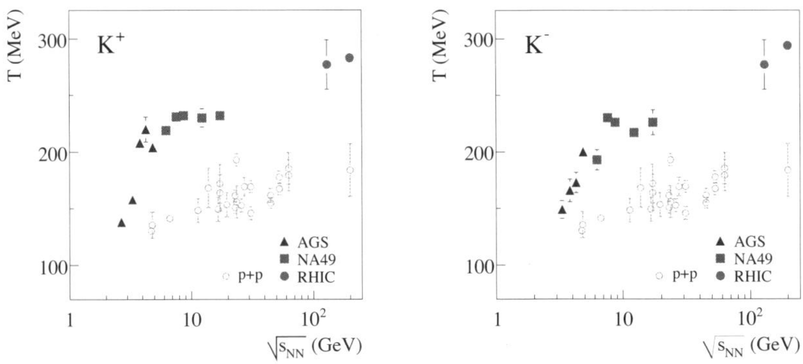

Fig. 5. Energy dependence of the inverse slope parameter T^* of the transverse mass spectra of K^+ (*left*) and K^- mesons (*right*) measured at mid-rapidity in central Pb+Pb and Au+Au collisions. The $K^\pm$ slope parameters are compared to those from $p+p$ reactions (open circles). The compilation of data is from Ref. 2.

collisions at all energies, i.e. $T^*_{low-p_T}(K) \approx T^*_{high-p_T}(K)$ for kaons; 3) the high quality data on m_T-spectra of K^+ and K^- mesons in central Pb+Pb and Au+Au collisions are available in the full range of relevant energies. Thus, one expects[6] that T^* for kaons increases when either pure confined or deconfined phase is produced at the early stage of A+A collision, but T^* remains approximately constant when the initial matter is in a mixed phase state. The energy dependence of the inverse slope parameter fitted to the K^+ and K^- transverse mass spectra for central Pb+Pb (Au+Au) collisions is shown in Fig. 5.

The striking features of the data can be summarized and interpreted as follows. The T^* parameter increases strongly with collision energy up to the SPS energy point 30 AGeV. This is an energy region where the creation of confined matter at the early stage of the collisions is expected. Increasing collision energy leads to an increase of the early stage temperature (see Fig. 1, *left*) and pressure. Consequently the transverse activity of produced hadrons, measured by the inverse slope parameter, increases with increasing energy. The T^* parameter is approximately independent of the collision energy in the SPS energy range 30 – 158 AGeV. In this energy region the transition between confined and deconfined matter is expected to be located. The resulting modification of the equation of state "suppresses" the hydrodynamical transverse expansion and leads to the observed plateau structure in the energy dependence of the T^* parameter.[6,9]

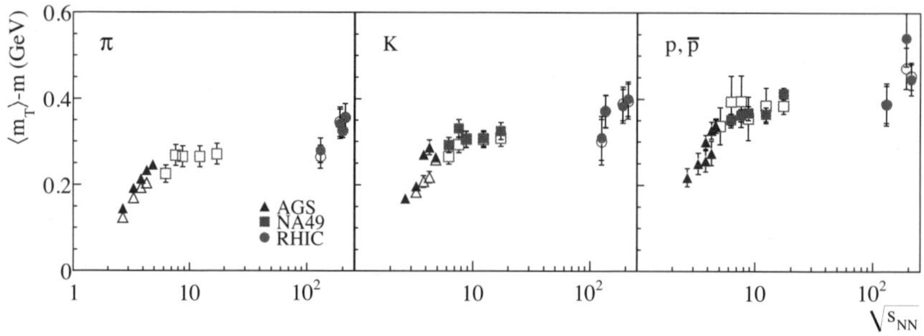

Fig. 6. Energy dependence of the mean transverse mass, $\langle m_T \rangle - m$, measured at mid-rapidity in central Pb+Pb an Au+Au collisions for $\pi^{\pm}$ (left), $K^{\pm}$ (middle), and p and $\bar{p}$ (right). In the plots, positively (negatively) charged hadrons are indicated by the full (open) symbols. The compilation of data is from Ref. 2.

At higher energies (RHIC data), T^* again increases with the collision energy. The equation of state at the early stage becomes again stiff, the early stage temperature and pressure increase with collision energy, and this results in increase of T^* too. The parameter T^* appears to increase smoothly in $p+p$ interactions as shown in the left panel of Fig.5.

For the transverse mass spectra of pions and protons the inverse slope parameter depends on the transverse mass interval used in the fit. The mean transverse mass $\langle m_T \rangle$ provides an alternative characterization of the m_T-spectra. The energy dependence of $\langle m_T \rangle - m$ for pions, kaons and (anti)protons is shown in Fig.6. The results shown in Fig.6 demonstrate that the approximate energy independence of $\langle m_T \rangle$ in the SPS energy range is a common feature for all particles investigated.

5. Conclusions

The energy scan program at the CERN SPS together with the measurements at lower (LBL, JINR, SIS, BNL AGS) and higher (BNL RHIC) energies yielded systematic data on energy dependence of hadron production in central Pb+Pb (Au+Au) collisions. The anomalies in the energy dependence of the hadron production are seen in central A+A collisions and absent in the data of $p+p$ reactions. They indicate that the onset of the deconfinement in central Pb+Pb collisions is located at about 30 AGeV.

Acknowledgments.

I am thankful to Marek Gaździcki for fruitful collaboration. I am also thankful to Walter Greiner for his permanent support and encouraging of these studies. This work was in part supported by the Program of Fundamental Research of the Department of Physics and Astronomy of NAS, Ukraine.

References

1. M. Gaździcki and M. I. Gorenstein, Acta Phys. Polon. B **30**, 2705 (1999).
2. C. Alt *et al.* [NA49 Collaboration], Phys. Rev. C **77**, 024903 (2008).
3. M. Gaździcki, Z. Phys. C **66**, 659 (1995) and J. Phys. G **23**, 1881 (1997).
4. L. Van Hove, Phys. Lett. B **118**, 138 (1982).
5. C.M. Hung and E. Shuryak, Phys. Rev. Lett. **75**, 4003 (1995); Phys. Rev. D **57**, 1891 (1997).
6. M. I. Gorenstein, M. Gaździcki, and K. Bugaev, Phys. Lett. B **567**, 175 (2003).
7. D. Teaney, J. Lauret, and E.V. Shuryak, Phys. Rev. Lett. **86**, 4783 (2001).
8. M. I. Gorenstein, K. Bugaev, and M. Gaździcki, Phys. Rev. Lett. **88**, 132301 (2002); Phys. Lett. B **544**, 127 (2002); Phys. Rev. C **68**, 01790 (2003).
9. M. Gaździcki, M.I. Gorenstein, F. Grassi, Y. Hama, T. Kodama, and O. Socolowski Jr., Braz. J. Phys. **34**, 322 (2004); Y. Hama, F. Grassi, O. Socolowski, T. Kodama, M. Gaździcki, M. Gorenstein, Acta Phys.Polon. B **35**, 179 (2004).

EXTREME NEUTRON RICH SECTOR OF THE NUCLEAR CHART: NEW HORIZONS!

K. A. GRIDNEV[1,2], V. N. TARASOV[3], D. K. GRIDNEV[2], D. V. TARASOV[3], S. SCHRAMM[2], X. VIÑAS[4] AND WALTER GREINER[2]

[1] *Saint Petersburg State University, Russia*
[2] *FIAS, Frankfurt University, Germany*
[3] *NSC, Kharkov Institute of Physics and Technology, Ukraine*
[4] *University of Barcelona, Spain*

Using Hartree-Fock + BCS approach we analyze the behavior of the neutron drip line and predict the appearance of stability peninsulas. The conditions and mechanism for appearançe of such peninsulas are analyzed and the properties of newly predicted stable isotopes are investigated.

Keywords: Hartree-Fock calculations; stability.

1. Introduction

One of the major problems in nuclear physics having also interdisciplinary importance is the positioning of the neutron drip line. The fact that many of the nuclei predicted theoretically are not found experimentally is due to shortcomings of existing experimental techniques. The technique is developing with time as can be seen from Ref. 1, where the authors report on the newly discovered neutron rich nuclei. The standard theoretical approach in locating the neutron drip line is to take a stable nucleus and load it with neutrons until it reaches the point of saturation, that is such number of neutrons, which makes the isotope go through hadronic decay thereby releasing extra neutrons. This method, however, implies a simple structure of the drip line, namely, that any straight line on the nuclear chart, which corresponds to a fixed number of protons, crosses the neutron drip line only once. Yet it might happen that the drip line has a more complicated structure. In the vicinity of magic numbers or quenched magic numbers the following scenario can take place. At some point nuclei loose their stability but then gaining more neutrons their stability becomes restored. This leads to the formation of stability peninsulas on the nuclear chart.

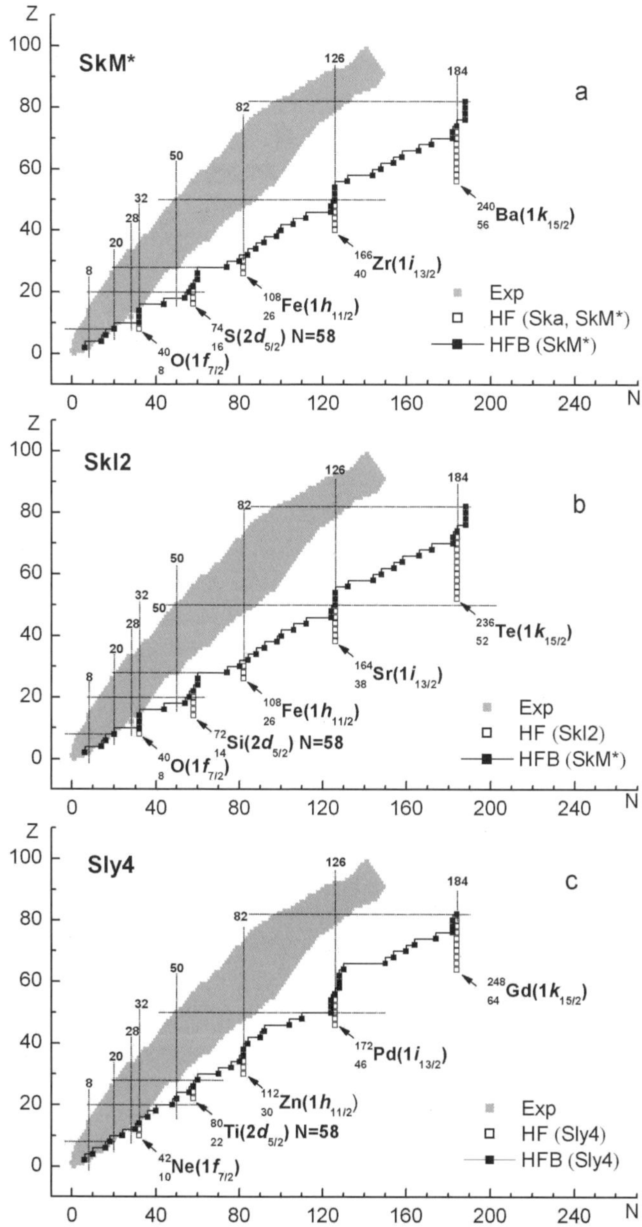

Fig. 1. Nuclear chart. Grey area shows experimentally known nuclei. Light squares show nuclei stable against 1 neutron emission for SkM*(a), SkI2(b) and Sly4(c) forces. Dotted lines correspond to magic numbers 8, 20, 28, 50, 82, 126, 184 and N=32. Solid line and black squares represent 1 neutron drip line obtained with SkM* and Sly4 forces within the HFB approach.[4]

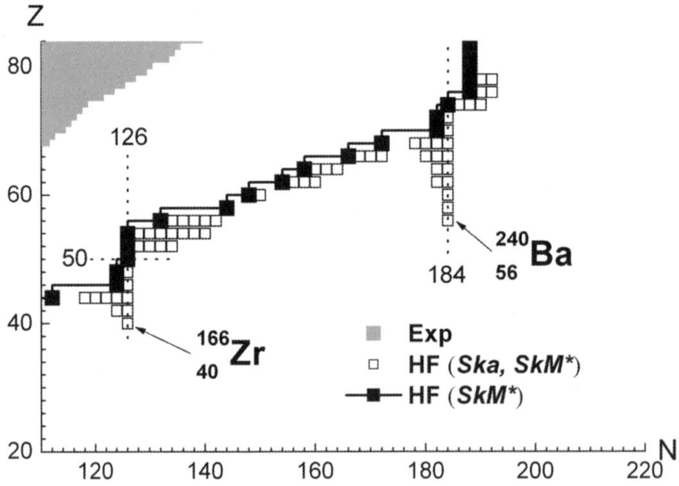

Fig. 2. Fragment of the nuclear chart around N=126 and N=184. The notations are that of Fig. 1.

The analysis of the phenomenon of stability restoration through adding neutrons has been undertaken in Refs. 2,3. We have considered the whole row of neutron rich isotopes like Pb, Zr, Ar, Kr, Rn, Gd , Ba, S as well as many other elements. In these cases we concentrated mainly on the phenomenon of stability restoration through adding neutrons that may lead to discovery of new neutron rich isotopes. Having found such isotopes we also investigated their properties like masses, deformation, root mean square radii etc. Some of the results are discussed below.

We use the non-relativistic Hartree-Fock equations with the BCS pairing scheme and various Skyrme forces. There does not exist a perfect set of forces that would make an excellent fit to the existing experimental data, not even within the stability valley. Thus it is always important to check the results with various Skyrme forces, especially when one speaks of drip lines. By definition the one or two neutron separation energies on these lines would be very small, sometimes within 0.1 Mev. Unfortunately, various Skyrme forces do not agree with each other over the whole nuclear chart to that degree of accuracy. So it would be naive to expect that different Skyrme forces would all predict a unique neutron drip line. Yet, the more forces predict a stable element the better. In the Hartree-Fock calculations we used Skyrme forces Ska, SkM*, Sly4, and SkI2, see Ref. 3 and references therein.

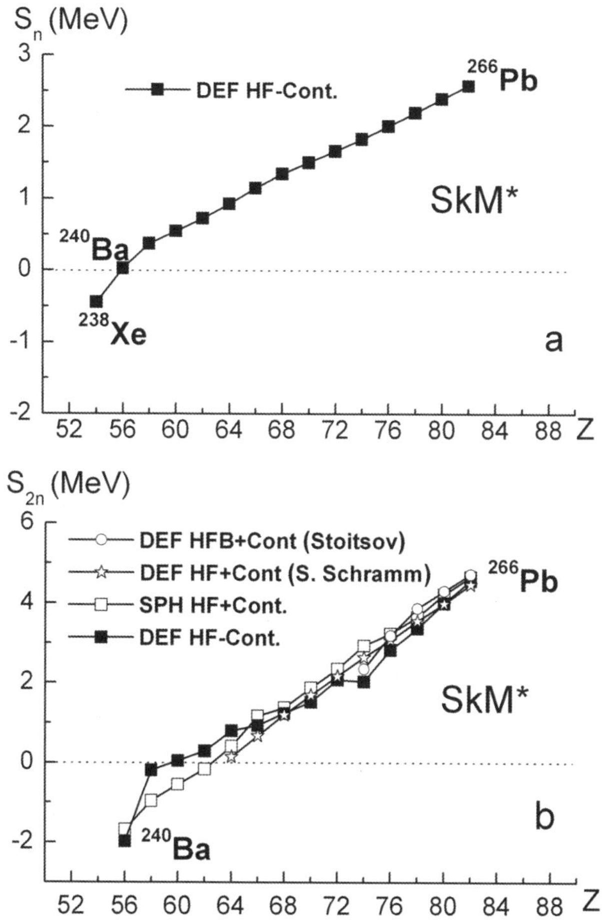

Fig. 3. One neutron (a) and two neutron (b) separation energies for isotones at N=184. Black squares are DHF calculations. Light squares are SPH calculations, which take continuum into the pairing scheme. Light circles represent HFB calculations.[4] Stars show grid calculations.[5]

2. Results and Discussion

The DHF calculations are performed with the harmonic oscillator basis, where the basis parameters are optimized on each iteration, for details see Refs. 2,3. The optimization procedure consists in choosing optimal oscillator frequencies, which minimize the total energy within a given basis dimension on each iteration. Such procedure facilitates the calculation and, more important, introduces appropriate corrections to the asymptotics of the basis functions, which should most closely match the spatially extended wave functions near the drip line. The procedure is described in detail in Ref. 2. The maximal basis dimension used in the calculations corresponds to the principal oscillator quantum number $N_0 = 24$, which gives 2925 basis functions. The results converge already at $N_0 = 18$, which makes the basis dimension equal 1330. One neutron separation energies are obtained using the Koopmans theorem.[2]

The observed stability peninsulas with respect to one neutron emission are shown in Figs 1, 2 for different Skyrme forces. It is easy to see from Figs. 1, 2 that the formation of peninsulas on the neutron drip line happens at the same for all forces N values but peninsula edges have different Z values. Fig. 3 shows the dependence of one and two neutron separation energies for the fixed neutron number $N = 184$.

In the BCS pairing scheme of DHF calculations we include only bound one particle states. In spite of ignoring the continuum states this method still provides a good agreement with the HFB (Hartree-Fock-Bogoliubov) calculations,[4] see Fig. 6 for comparison. Since most nuclei, which lie on the stability peninsulas are spherical we also use a spherical code (SPH), which solves the equations directly rather than using a particular basis. In the BCS scheme of the spherical code we implement the inclusion of quasi-bound continuum states, which are confined under the centrifugal barrier. To spot such states one introduces a fictitious "wall" forcing the wave functions vanish beyond it. The states, which remain localized when the wall is being moved at a large distance, are considered quasistable and taken into account in the pairing scheme. We compared our method to the computer program originally developed by Rutz and Bürvenich, which performs calculations on a grid. The calculations were performed by S. Schramm. A typical comparison of two methods can be seen in Fig. 3(b). The results of this comparison were quite satisfactory. The expected deviations happen due to the different treatment of pairing.

A very important point, which should be stressed, is that nuclei forming stability peninsulas are spectrally bound in the sense that there exists a well-defined ground state wave function, which minimizes the energy functional for such nuclei. At this point they become well-defined compact objects and the question about their lifetime becomes correctly formulated. And though for some nuclei it may be energetically favorable to get rid of two or more neutrons, a large centrifugal barrier of the last filled levels may serve as an indication that this lifetime would be large. For some nuclei the energetically favorable decay would be into four or more neutrons, which

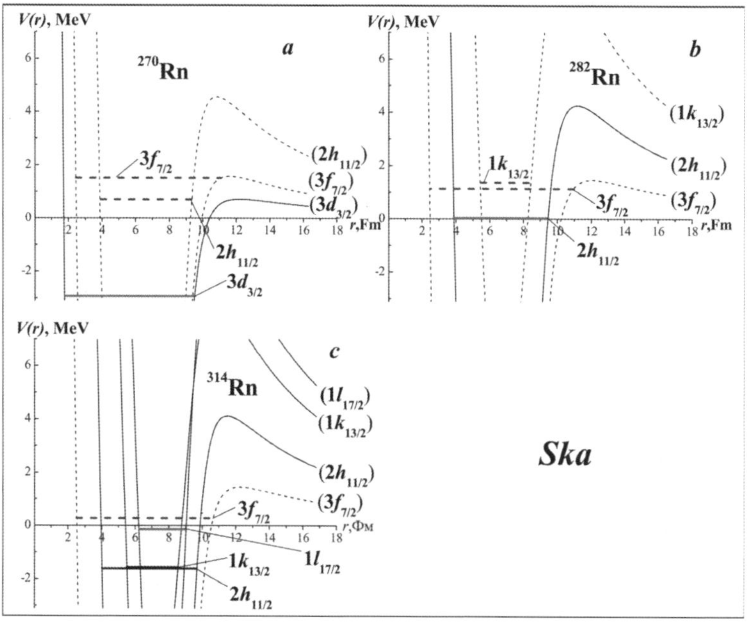

Fig. 4. Effective single–particle potentials for Rn isotopes in SHF calculations. It is shown how the single particle levels responsible for stability move towards the discrete spectrum.

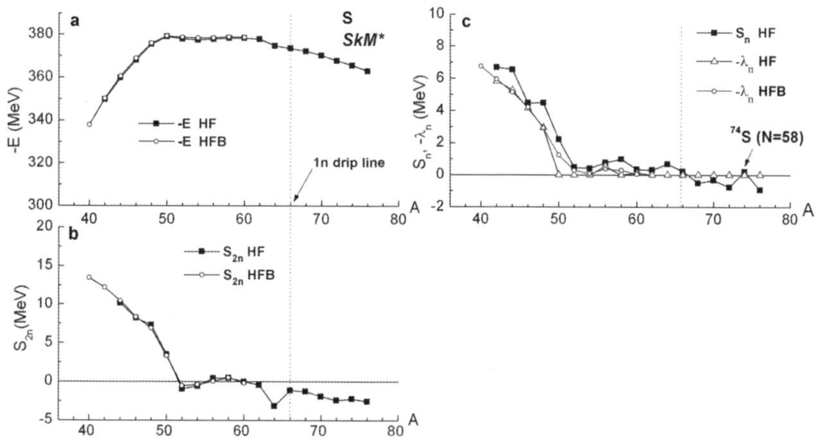

Fig. 5. One and two neutron separation energies for Sulfur isotopes.

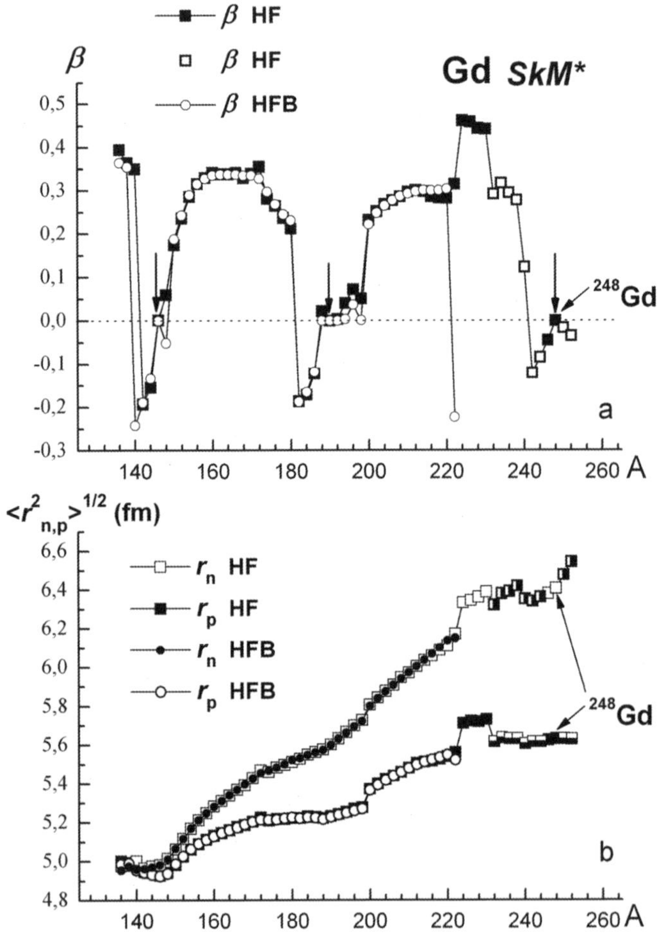

Fig. 6. Deformations and root mean square radii for Gd isotopes.

enhances the lifetime considerably (such decays were not observed experimentally so far). Let us also mention that all spectrally stable nuclei in our calculations appear as such in grid calculations.

Recently,[2,3] we have analyzed the mechanism that leads to the formation of stability peninsulas. The formation of such stability peninsulas always has the same pattern. When one adds neutrons to a nucleus lying close to the stability peninsula the last filled level gradually immerses from the continuum into the bound spectrum. For example, in Fig. 5 we show certain spherical isotopes of Radon and the levels that are responsible for their stability. We have performed a check of our results with respect to different forces. We found that Ska and SkM* forces occupy an intermediate position between SkI2 and Sly4, in the sense that more elements are stable with SkI2 and less with Sly4. The forces SkI2 are the most optimistic, at the same time the formation of stability peninsulas for Sly4 is rather rare. The difference between two of these forces is depicted in Fig. 1.

Let us take SkM forces and list for each peninsulas the isotopes and the corresponding filled subshells that are responsible for the stability of those isotopes.

- **1f7/2** ^{40}O
- **2d5/2** ^{76}Ar, ^{74}S
- **1h11/2** ^{110}Ni, ^{108}Fe
- **1i13/2** ^{174}Cd, ^{172}Pd, ^{170}Ru, ^{168}Mo, ^{166}Zr
- **1k15/2** ^{256}Hf, ^{254}Yb, ^{252}Er, ^{250}Dy, ^{248}Gd, ^{246}Sm, ^{244}Nd, ^{242}Ce, ^{240}Ba.

In Figs. 1, 2 one finds that the peninsulas occur at conventional magic numbers but also at quenched magic numbers like $N = 32$ or $N = 58$. Fig. 6 shows one and two neutron separation energies for Sulfur isotopes, which illustrates this shell quenching. As mentioned above for all isotopes we calculate their properties. For example, Fig. 7 shows root mean square radii and deformations for Gadolinium isotopes.

References

1. T. Baumann *et al.*, Nature **449**, 1022 (2007)
2. K.A. Gridnev *et al.*, Eur. Phys. J. **A 25**, S01, 353 (2005); Nucl. Phys. **69**, 3 (2006); J. Mod. Phys. **E 15**, 673 (2006).
3. V. N. Tarasov *et al.*, J. Mod. Phys. **E 17**, 1273 (2008);
4. M.V. Stoitsov *et al.*, Phys. Rev. **C 61**, 034311 (2000); Phys. Rev. **C 68**, 054312 (2003); http://www.fuw.edu.pl/ dobaczew/thodri/thodri.html
5. K. Rutz, Doctoral Thesis, University of Frankfurt (1998); T. Bürvenich, Doctoral Thesis, University of Frankfurt (2001).

SEARCH FOR SUPERHEAVY ELEMENTS IN NATURE

A. G. POPEKO

Flerov Laboratory of Nuclear Reactions,
Joint Institute for Nuclear Research,
Joliot-Curie str., 6, Dubna, 141980, Russia
E-mail: popeko@jinr.ru

Attempts to find superheavy elements in Nature started in the middle of sixties when predictions of the possible existence of longliving nuclei in the far transuranium region have been made. During following 15 years hundreds of geological samples, their processing products and probes of meteorites were studied. In all experiments only upper limits of the superheavy element concentration in the studied samples have been determined.

Keywords: Superheavy elements; $A > 220$; transactinide elements.

1. Background

The term "superheavy nuclei" was introduced by J.A. Wheeler in 1955.[1] After extrapolation of mass formulae and analysis of the data on half-lives of heavy nuclei the author concluded, that one can "expect the existence of nuclei twice as heavy as known nuclear species" (up to $Z = 147$, $N = 500$), and that "the limits to the stability of **superheavy nuclei** are set primarily by neutron escape and by spontaneous fission.

Petrzhak and Flerov, having analyzed the tendency to reduction of nuclei lifetimes,[2] have estimated the limiting nuclear charge $Z_{max} = 120$.

The possibility to exist in Nature of superheavy elements (SHE) - elements heavier than ^{238}U depends on two determinatives:

- it is necessary that one of superheavy nuclides would have a lifetime long enough to cross "half" the Galaxy (of about 2×10^4 y.) to be found in cosmic rays, or that, comparable with the age of the Earth (of about $4\text{-}5\times10^9$ y.), to be found in meteorites or terrestrial samples;
- in the Universe should exist a mechanism leading to the formation of superheavy elements.

2. Prediction of Existence of SHE

According to the liquid drop model, there should be the linear dependency between the logarithm of a spontaneous fission half-life and the fissility parameter - Z^2/A. Fig. 1 shows the dependency of a $\log T_{1/2}(s)$ on the fissility parameter x ($x = (Z^2/A)/(Z^2/A)_{crit}$ at $(Z^2/A)_{crit} = 50.883.$[3]) Apparently, the first semiquantitative explanations of deviations of $\log T_{1/2}$ from the linear dependence have been given by Swiatecki.[4]

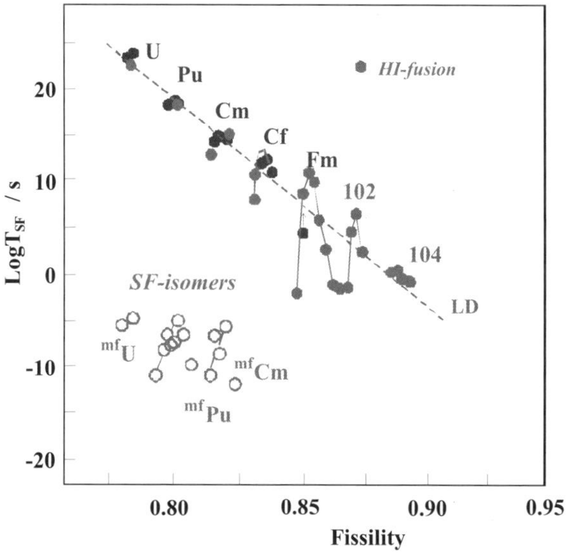

Fig. 1. Half-lives with respect to spontaneous fission as a function of the fissility parameter x.

For the first time "realistic" predictions of magic numbers beyond ^{208}Pb have been done by A. Sobiczewski, F.A. Gareev and B.N. Kalinkin[5] after analyzing single particle nucleon levels in a Woods-Saxon potential well. The possible magic numbers were found to be $Z = 114$ and $N = 184$.

Theoretical predictions on the next closed shells numbers vary strongly depending on the model. Following the well-known proton and neutron shells with $Z = 82$ and $N = 126$ (^{208}Pb), the shell correction amplitude has a maximum for the superheavy nucleus $^{298}114$ at $N = 184$ in macro-microscopic models. After calculations performed using the Hartree–Fock–Bogoliubov (HFB)-model or self-consistent relativistic mean-field models, the closed spherical proton shells are predicted at $Z = 120$ or 126.[6]

3. Accuracy of SHE Lifetimes Predictions

Fission barrier heights of the most stable nuclei in the region $Z \sim 114$ can lie in the range $(9.6 \div 13)$ MeV.[7,8] Variation of fission barriers by ± 10 % changes the spontaneous fission half-life by a factor of $\approx 10^{\pm 2}$.

The differences in inertia parameters make 30 %, that brings the uncertainty of $T_{1/2}(SF)$, as shown in,[7] by a factor of $\approx 10^{\pm 6}$. The total uncertainty of estimations of spontaneous fission half-lives of superheavy nuclei can reach, apparently, $10^{\pm 10}$.

For calculations of half-lives against α-decay, both - "strong" theory of Gamov[9] and empirical formulae[10] were applied. Uncertainty of predictions, connected only with the choice of an estimation method (at the same value of Q_α) makes $10^{\pm 2}$. Variations of Q_α value by ± 10 % can cause the change of $_{1/2}(\alpha)$ by a factor of $10^{\pm 5}$. The overall uncertainty of the α-decay half-life estimation can also reach $10^{\pm 10}$.

The position of the β-stability line can be predicted with the accuracy of 2 - 3 mass units.[11] Because parities and spins of initial and final states in nuclei are inaccessible now, β-decay properties were estimated also by empirical methods.[12] Variations of Q_β value by 1 MeV can cause the change of $_{1/2}(\beta)$ by a factor of $10^{\pm 3}$.

Modern ideas of the influence of nuclear shells on stability of nuclei and results of calculations allow to conclude that the existence of superheavy nuclei with $Z \sim 114$ and $N \sim 184$ is reliably proved theoretically. Estimations of half-lives of such nuclei cover a wide range from fractions of a second till 10^{10} years and more.

4. Spontaneous Fission Properties of SHE

Probabilities of triple fission of nuclei with $Z \sim 114$ should be $10^3 - 10^4$ times higher, than at fission of known nuclei.[13] The TKE of fission fragments of superheavy isotopes will make ~ 300 MeV at fission into two, and ~ 400 MeV at fission into three fragments.[14,15]

Average number of prompt neutrons - $\overline{\nu}$ at fission of $^{298}114$ should be ~ 11.[14] Authors[15] estimated $\overline{\nu} = 8 - 10$. Kolb[13] concluded, that at a double fission $\overline{\nu} = 8$, however, at a triple fission $\overline{\nu} = 2 - 4$. Hoffman[16] analyzing correlations of mass distribution modes and average numbers of fission neutrons concluded, that at symmetric fission $\overline{\nu} \sim 6$, whereas at asymmetric one $\overline{\nu} \sim 12$.

Uncertainties in predictions should be taken into account at design of detectors for SHE searches.

5. Possible Mechanisms of Nucleosynthesis of SHE

For more details see the report of I. Mishustin to this conference. In an explosion of supernova the prompt capture of multiple neutrons - r-process becomes possible.

Shramm and Fowler,[17] Tonder[11] and Klapdor[18] have shown that at supernova explosions formation of nuclides with $Z \sim 110$ and $A \sim 300$ is possible. The relative abundance of superheavy elements in supernova products can reach $SHE/^{238}U = 0.1 \div 1$.[17]

To the opposite conclusion Howard, Nix[19] and Mathews, Viola[20] have come. According to[19,20] nuclide with masses $A > 275$ cannot be formed in astrophysical r-process.

More recent calculations[21] revealed that the abundance of superheavy elements in the r-process can be comparable with that of uranium, but the yield of SHE depends strongly on forecasts of properties of $\beta-$delayed, neutron-induced, and spontaneous fission.

Among exotic objects neutron stars are considered. At matter density $5 \cdot 10^{14}$ g/cm^3 the equilibrium charge of nuclei will make $Z = 450$ and their mass $A = 8000$.[22] The rupture of neutron stars crust at a "star quake" should be accompanied by explosions with energy release of $\sim 10^{39}$ erg.[23] Such energy could be enough for removal to infinity of $\sim 10^{20}$ g of a super-dense substance of a neutron star. Probably, superheavy elements can be formed after decays of products of such emissions.

Modern ideas about mechanisms of nucleosynthesis do not forbid formation of superheavy elements. Due to explosions or "star-quakes" interstellar clouds could be enriched by heavy and superheavy elements. Probably such nuclei are a part of nuclear component of cosmic rays. Superheavy elements, could enter in processes of evolution of planetary bodies and be captured in some rocks or minerals.

6. Choice of Objects to Search for SHE

The choice of objects perspective for discovery of superheavy elements is based on results of predictions of their chemical and physical properties. Most predictions are based on extrapolations of properties within considered group of elements along with calculations of electronic configurations including relativistic effects.[24–26]

Elements $113 \geq Z \geq 118$ should be, correspondingly, analogues of elements from thallium to radon, and a regular filling of the 7-shell should occur. The next noble gas, or to be more exact - a noble liquid with the boiling temperature of +15 C should be the element with $Z = 118$.[26] Elements 119 and 120 should behave like alkaline metals. Starting with the element $Z = 121$ there should happen the filling of $5g$- and $6f$- electron shells, and elements $121 \geq Z \geq 154$ should form a series of superactinides.[26]

Unfortunately, due to uncertainties of predicted position of the center of stability "island" ($110 \geq Z \geq 130$) one can consider as light chemical analogue of SHE almost any element from the D.I. Mendeleev's periodic system of the elements.

7. Experimental Search for SHE in Nature

The search for heavy elements in cosmic rays, terrestrial samples, and meteorites was in the 70's and 80's one of the extensive experimental investigations (see reviews[27–30]).

7.1. *Search for SHE nuclei in cosmic rays*

Efforts of experimentalists were appreciably forced by Fowler's studies[31] of nuclear component of cosmic rays. In photoemulsion layers exposed at an upper atmosphere, two tracks have been found out, which were attributed to stops of nuclei with $Z > 105$.

More perspective method of studying heavy component of cosmic rays has been proved by Flerov, Otgonsuren and Perelygin.[32] The method is based on revealing of traces from stops of nuclei in minerals from some meteorites, which could accumulate tracks during many millions years.

In olivine crystals from the Marjalahti meteorite[33] two tracks, attributed to nuclei with $Z > 97$, were observed. In a more recent study of Marjalahti meteorite[34] among 853 tracks with $Z > 50$, detected in 27 olivine crystals, 4 tracks, which could be attributed to the $Th - U$ group, were found. No tracks of heavier elements were observed.

The charge and energy of ultra-heavy nuclei will be measured using silicon detectors, aerogel and acrylic Cherenkov counters, and a scintillating optical fiber hodoscope in the frame of the ENTICE - the Energetic Trans-Iron Cosmic-ray Experiment).[35]

7.2. *Study of isotopic anomalies*

Accumulation of fission fragments of superheavy nuclei should lead to deviations in isotope abundances of known elements. In the seventies very popular was the hypothesis of Anders, Heyman,[36] and Rao[37] about the origin of excess of heavy xenon isotopes in some meteorites - due to the spontaneous fission of superheavy elements. However, this hypothesis, despite its attraction, encountered a number of serious contradictions.

7.3. *Study of radiation damages in crystals*

Decay of superheavy nuclei can result as well in radiation damages in crystalline media. It is known that at α-decay of isotopes contained in mineral inclusions, the so-called radioactive halos - spherical zones around a central mineral grain, will be formed. "Giant" halos having unusually large diameters corresponding to $\approx$ 14 MeV, an energy predicted for nuclides with $Z \sim 126$, were observed[38] in biotite from Madagascar. In order to identify elements around $Z = 126$ by characteristic x-rays, the inclusions - monazite crystals were irradiated by protons. In the first experiments[39] $L_{\alpha 1}x$-rays of elements 116, 124 and 126 were observed. Later, a more careful analysis and advanced technique for the excitation of x-rays[40] disproved the early results.

7.4. *Mass-separation of SHE*

Several authors, e.g.[7,41] tried to separate elements with masses $A = 250 - 350$ using mass-separation technique. Because of a background of molecular complexes of uranium and thorium the sensitivity of a "simple" separation method, as a rule, does not exceed $10^{-10} - 10^{-11}$ g/g. Application of double mass-separation[42] allows to reduce considerably a background from molecular complexes, but reduces as well the overall sensitivity of this method.

Stephens[43] applied for studying of isotopic structure of platinum ore in experiments on searches for eka-*Pt* a more sophisticated technique - a tandem generator with a special beam line for ion analysis. This method is known as accelerator mass spectrometry (AMS).

In the latest experiments on the search for long-living roentgenium (Rg, $Z = 111$) isotopes in gold samples[44] using the VERA set-up, the sensitivity of 10^{-16} g/g was reported.

8. Direct Registration of SHE Decays

The most sensitive method of search for superheavy elements is connected with direct registration of their radioactive decays. At interpretation of results of such measurements one usually specify the limiting detectable dimensionless concentration C g/g of a searched element, assuming that their half-life makes $_{1/2} = 10^9$ years. The searching methods based on detection of β-particles or γ-quanta can be ruled out due to extremely high background.

8.1. *Study of natural α-activities*

Experiments on studying weak natural α-activities were carried out for many years. A history of observations of 4.4-MeV α-activity over the period 1924 -1979 in radiogenic haloes, zinc ores, monazite, thorite, huttonite, ultrabasic and other abyssal rocks, osmiridium, uranium ores, and raffinates of uranium is given in[45] (for more recent results see[44]). To search for superheavy elements in nature this method do not seems having prospects due to presence of numerous natural and man-caused α-emitters. The use of more advanced technique of detecting coinciding α-particles and characteristic x-rays[46] can give definitive results, but the overall sensitivity is insufficient for the searching experiments.

8.2. *Study of spontaneously fissioning activities*

Essentially higher sensitivity of SHE searches provide methods based on detecting of spontaneous fission. This method appears as a universal one, because superheavy nuclides or the nearest products of their α-decay should undergo spontaneous fission. The whole accumulated experimental material allows to assert, that in nature exists only one spontaneously fissioning nuclide - uranium-238. Thus the problem of search of new elements by detecting their spontaneous fission can be reduced to reliable detecting of fission events and clearing up their possible origin due to uranium, contained in the sample.

A serious problem cause interactions of active cosmic radiation components - protons, neutrons and muons with heavy nuclei in probes or with detector parts. Whereas the nucleon component can be removed with a shielding of $\approx$ 10 mwe (meter of water equivalent, $\approx$ 4 m of concrete), for suppression of muons hundreds of meters shielding are needed. The fluxes of muons were measured in[47] up to 5200 mwe in connection with searches for rare processes.

8.2.1. *Detecting of Fission Fragments*

The first searches for SHE's spontaneous fission events were performed using solid state track detectors: Pb-glasses and glasses containing Bi, Tl, and W, which could accumulate tracks during $5 \div 200$ years.[48] As it was found out later[49,50] with the use of sandwiched plastic track detectors, the additional background has been produced by cosmic rays.

Even higher sensitivity, but unfortunately also uncertainty, can be reached by etching internal fission tracks in suitable minerals, where they were accumulated over millions of years.[51,52]

Especially for the searches of rare spontaneous fission events of SHE, big proportional counters were developed.[53] The cathodes with an area of ≈ 1.6 m^2 were covered with a powdered samples layer, having the surface density ≈ 3 mg·cm^{-2}. To reduce the "cosmic" background, the counters were operated in a room with a shield of 5 mwe.

A tricky detector - spinner[54] has been used for the observations of spontaneous fission of supeheavy elements in chemical compounds, minerals, and in targets which were irradiated with 24-GeV protons.[55] The principle of operation of spinner is similar to that of bubble-chamber, but the negative pressure is produced in a working liquid by centrifugal forces. The main problem consist in the preparation of samples, which must be solvable in organic substances like ethanol.

8.3. *Detecting of prompt fission neutrons*

The method, providing the highest sensitivity at the searches for SHE, is the detection of prompt neutrons accompanying fission. Practically all samples are transparent for neutrons, thus probes of up to 100 kg can be inspected. Another advantage is that there is no need to destroy the surveyed sample.

The advanced method of detecting spontaneous fission is based on detection of multiple neutron emission.[56] The detection of 2 and more neutrons unambiguously indicates the occurrence of spontaneous fission. Several kinds of neutron multiplicity detectors based on ^3He-filled proportional counters were designed at Flerov laboratory.[57,58]

This method provide for detecting spontaneous fission events the efficiency from 15-30% ($\overline{\nu} \approx 2$) to about 50% ($\overline{\nu} \approx 4$). The expected $\overline{\nu}$-values for superheavy elements were discussed earlier.

The major part of measurements was performed in the salt mine at a depth of 1100 mwe (passive shielding) with additional suppression of cosmic muons by active shielding based on Geiger-M̀uller counters. Long term measurements with high-purity samples (metallic lead, ferric oxide, melted quartz) showed that the "cosmic" background was less than 1 event per year. Thus, the only one source of background which should be considered is the spontaneous fission of uranium.

8.4. *Search for SHE in terrestrial samples*

At the first stage of searches several dozens of various ores and their processing products, products of metallurgy, ferromanganese nodules, chemical preparations rich in platinum metals and gold, volatile metals such as Hg, Tl, Pb, Bi, rare earth elements were studied. The limits for the abundance of SHE were found to be less than 10^{-12} g/g.[59,60]

8.5. *Search for SHE in Cheleken peninsula geothermal waters*

The heavy volatile metals content (e.g. thallium, lead, etc.) of the Cheleken Penninsula (the south-eastern coast of the Caspian Sea) water[61] is nearly 100 times that of oceanic water. These metals could escape from the Earth's crust, together with other volatile components. The extraction of heavy elements was carried out[61] using a vinyl-pyridine anion-exchange resin. Some 2,000 m^3 of the water were passed through a column containing 850 kg of the resin.

During a 88-day exposure of 9 kg of the saturated resin a total of 42 spontaneous fission events (0.5 events per day) have been recorded at neutron multiplicity detectors.[62] The analysis of the uranium content of the saturated resin, carried out using different methods ($(2-3) \times 10^{-8}$ g/g), had shown that the background due to spontaneous fission of uranium did not exceed 1 event from 42. Thus it was supposed that the detected spontaneously fissioning nuclide belongs to the region of superheavy elements.

Experiments were carried out aiming at the concentration of the detected activity by extraction of various chemical elements from 170 kg of the saturated resin. The fission counting rate of the obtained hydroxides was 5 counts per day, which corresponded to about 50 % of the initial spontaneous fission activity of the resin. It should be mentioned, that 1 atom of ^{252}Cf per 5 g of saturated resin could explain the observations. Further attempts[63] to concentrate the detected activity failed.

8.6. *Search for SHE in meteorites*

Samples of meteorites "Saratov", "Efremovna" and "Allende" have been selected from the meteoric collection of Academy of Science of USSR. These meteorites belong to the class of chondrites and represent the less differentiated substance of the Solar system. The low uranium concentration in chondrites $3 \cdot 10^{-9} - 4 \cdot 10^{-8}$ g/g allowed one to realize the maximum sensitivity.

Our investigations of meteorites were performed in 1972 - 1976.[64–66] The observed counting rate was, on the average, 1 event per 5 days for 10 kg of a sample, what was $10 \div 30$ times higher than could originate from uranium contents and other background sources. Thus it was assumed that a longlived spontaneously fissioning nuclide, possibly belonging to the region of SHE, is present in the studied samples of meteorites.

The isolation of the observed unknown nuclide (together with other volatile elements) has been performed by Zvara[67] from several kilograms of the Allende meteorite in hydrogen and then oxygen flows. Authors obtained an indication of its volatility in the elemental or oxidation state, but further attempts to increase the concentration of observed activity failed.

8.7. *Discussion of early results*

Fig. 2 shows the correlation between the spontaneous fission activity observed in several samples, with the uranium concentration.

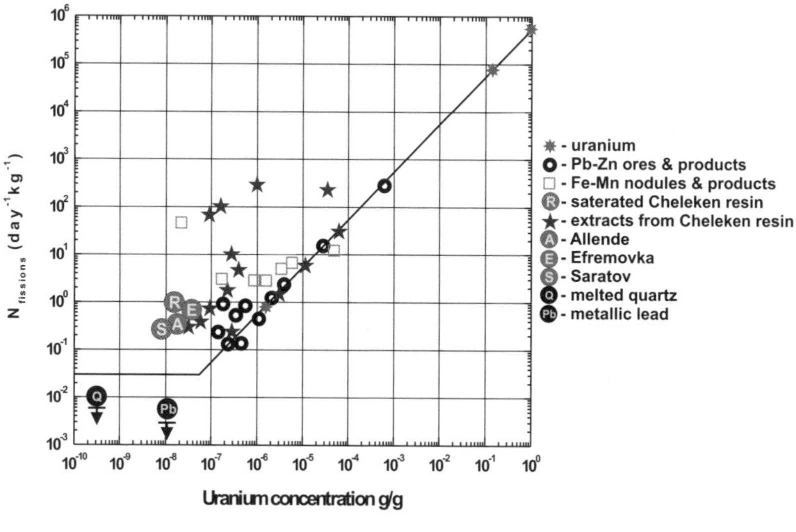

Fig. 2. Correlation between the uranium concentration and the spontaneous fission activity detected in the samples, recalculated to $events \cdot day^{-1} \cdot kg^{-1}$.

As one can see the counting rate for the majority of survived samples corresponds to the spontaneous fission activity of uranium. The activity in meteorites and products of processing of Cheleken geothermal waters exceeds that from uranium. But, also it did not show any clear correlation to known elements, and it did not follow them in attempts of further concentration.

In all experiments only upper limits of the superheavy element concentration in the studied samples have been determined. All this resulted in a pessimistic view on the possible existence of SHE in nature, and the searching experiments were practically stopped in the mid of 80-th.

9. New Samples for SHE Searches

The experimental data accumulated, and development of modern microscopic models during passed 30 years, simulated new approach to the search for perspective objects.[68] A noticeable increase in $T_{1/2}(\alpha)$ and $T_{1/2}(SF)$ may be expected for nuclei with $Z < 110$, which have not been yet looked for.

In accordance with the calculations,[69] lifetimes of the isotopes $^{290292}Hs$ and $^{290293}Ds$ fall within the range $10^{10} \div 10^{14}$ s, and according to,[70] these isotopes are β-stable (or have long lifetimes).

Considering different nuclei as objects for studies, it turns out that for element 108 – Hs, the chemical homologue of Os, the chances to be found in terrestrial samples could be favorable. The search for rare decays may be undertaken with a metallic sample of raw Os.

Perspectives for SHE search, can follow from the discovered high volatility of Cn ($Z = 112$) and element $Z = 114$. These elements can be gases (noble) at normal conditions - the boiling temperature of Cn is (360±100)K. Thus e.g. one can look for SHE in heavy fractions of Xe production.

These experiments are running now with an especially designed neutron multiplicity detector[58] in the underground laboratory in Modane (France).

10. Conclusion

The problem of the existence of superheavy elements in nature belongs to the must fundamental, because it affects nuclear and atomic physics, quantum chemistry, astrophysics, cosmology, and undoubtedly the efforts to solve it will be continued.

References

1. J.A. Wheeler, in *Proceedings of the International Conference on the Peaceful Uses of Atomic Energy, Geneva, August 1955* (United Nations, New York, 1956) Vol. **2**, p. 155.
2. K.A. Petrzhak, G.N. Flerov, *Uspekhi. Fiz. Nauk* **73**, 655, (1961).
3. Yu.Ts. Oganessian and Yu.A. Lazarev, *Treatise on Heavy-Ion Science* ed. D.A. Bromley (Plenum, New York, 1985), **4**, 3.
4. W.J. Swiatecki, *Phys. Rev.* **100**, 937(1955).
5. A. Sobiczewski, F.A. Gareev, B.N. Kalinkin, *Phys. Lett.* **22**, 500 (1966).
6. A. Sobiczewski, K. Pomorski, *Prog. Part. Nucl. Phys.*, **58**, 292 (2007).
7. S.G. Nilsson, S.G. Thompson, S.F. Tsang, *Phys. Lett.* **28B**, 458, (1969).
8. M. Bolsterli, E.O. Fiset, J.R. Nix, *Phys. Rev. C* **5**, 1050, (1972).
9. E.O. Fiset, J.R. Nix, *Nucl. Phys.* **A193**, 647, (1972).
10. D.F. Jackson et al,. *Nucl. Phys*, **A319**, 1, (1979).
11. F. Tondeur, *Z. Phys*, **A297**, 61, (1980).
12. J. Randrup, et al., *Phys. Rev. C* **13**, 229, (1976).
13. D. Kolb, *Phys. Lett*, **65B**, 319, (1976).
14. J.R. Nix, *Phys. Lett.*, **30B**, 1, (1969).
15. H. Shmitt, U. Mosel, *Nucl. Phys*, **A186**, 1, (1972).
16. D.C. Hoffman, *In Proc. Intern Symp. on superheavy elments, Lubboc, Texas, 1978*, Pergamon Press Inc. 1978, p. 89.
17. D.N. Shramm, W.A. Fowler, *Nature*, **231**, 103, (1971).
18. H.V. Klapdor, et al., *Z. Phys.*, **299**, 213, (1981).
19. W.M. Howard, J.R. Nix, *Nature*, **247**, 17, (1974).
20. G.J. Mathews, V.E. Viola, *Nature*, **261**, 382, (1976).
21. I.V. Panov, I.Yu. Korneev, and F.-K. Thielemann, *Phys. At. Nucl.*, **72**, 1026, (2009).
22. G. Baym, H. Bethe, C. Pethick, *Nucl. Phys.*, **175**, 225, (1971).
23. G.S. Bisnovaty-Kogan, V.M. Chechetkin, *Soviet Uspekhi Phys. Nauk*, **127**, 263, (1979).
24. B. Fricke, W. Greiner, J.T. Waber, *Theoret. Chim. Acta*, **21**, 235, (1971).
25. G.T. Seaborg, *Phys. Scripta*, **10A**, 5, (1974).
26. B. Fricke, *Struct. and Bond*, **21**, 89, (1975).
27. G. Herrmann, *Nature*, **280**, 543, (1979).
28. G.N. Flerov, G.M. Ter-Akopian, *Atomki Koezlemenyk*, **21**, 93 (1979).
29. G.N. Flerov, G.M. Ter-Akopian, *Rep. Prog. Phys.*, **46**, 817 (1983).
30. G. Herrmann, in *The Chemistry of Superheavy Elements*, ed M. Schädel, Kluwer Acad. Publ., 291, (2003).
31. P.H. Fowler, et al., *Proc. Roy. Soc.*, **A301**, 39, (1967).
32. G.N. Flerov, O. Otgonsuren, V.P. Perelygin, *Izvestia AN SSSR, ser. fiz*, **39**, 388, (1975).
33. V.P. Perelygin, S.G. Stetsenko, *Soviet. Lett. to J.E.T.Ph.*, **32**, 622, (1980).
34. A.B. Aleksandrov, et al., *Bull. Lebedev Phys. Inst.*, **35**, 276, (2008).
35. W.R. Binns, et al., *in Proc 31-th Int. Cosmic Ray Conf.*, Łódź, 2009.
36. E. Anders, D. Heyman, *Science*, **164**, 821, (1969).
37. M.N. Rao, *Nucl. Phys*, **A170**, 69, (1970).

38. R.V. Gentry, *Science*, **169**, 670, (1970).
39. R.V. Gentry, et al., *Phys. Rev. Lett.*, **37**, 11, (1976).
40. C.J. Sparks, et al., *Phys. Rev. Lett.*, **40**, 507, (1978).
41. J. McMinn, et al., *Nucl. Instr. and Meth*, **139**, 175, (1976).
42. C. Stephan, et al., *J. Physique*, **36**, 105, (1975).
43. W. Stephens, et al., *Phys. Rev. C*, **21**, 1664, (1980).
44. F. Dellinger, et, al., *Phys. Rev. C*, **83**, 015801, (2011).
45. R.J. Dougan, J.D. Illige, E.K. Hulet, *Preprint UCRL-52886*, University of California, Livermore, California, (1979).
46. V. Kusch, et al., *Atomnaya Energiya*, **31**, 159, (1971).
47. A.S. Malgin, O.G. Ryazhskaya, *Phys. of At. Nucl.*, **71**, 1769, (2008).
48. G.N. Flerov, V.P. Perelygin, *Sov. J. Atom. Energy*, **26**, 603, (1969).
49. G.N. Flerov, V.P. Perelygin, O. Otgonsuren, *Sov. J. Atom. Energy*, **33**, 1144, (1972).
50. F.H. Geisler, P.R. Phillips, R.M. Walker, *Nature*, **244**, 428, (1973).
51. P.B. Price, R.L. Fleischer, R.T. Woods, *Phys. Rev. C*, **1**, 1819, (1970).
52. L.L. Kahskarov, et al., *Microsymposium*, **38**, MS042, (2003).
53. G.N. Flerov, et al., *Yad. Fiz.*, **20**, 472 (1974).
54. B. Hahn, *Il Nuovo Cimento*, **22**, 650, (1961).
55. K. Behringer, et al., *Phys. Rev. C*, **(**, 48, (1974).
56. G.M. TerAkopian, A.G. Popeko, et al., *Preprint JINR R135394*, Dubna, 1970. *Soviet Patent* No 450508.
57. G.M. Ter-Akopian, A.G. Popeko, at al., *Nucl. Instr. and Meth.*, **190**, 119, (1981).
58. A. Svirikhin *et al.*, *AIP Conf. Proc.* **1175**, 297 (2009).
59. G.N. Flerov, et al., *Yad. Fiz.*, **20**, 639, (1974).
60. G.N. Flerov, et al., *Yad. Fiz..* **21**, 9, (1975).
61. Yu.T. Chuburkov, et al., *Radiochimija*, **16**, 827, (1974).
62. G.N. Flerov, et al., *Z. Physik*, **A292**, 43, (1979).
63. Yu.T. Chuburkov, A.G. Popeko, N.K. Skobelev, *Sov. Radiochem.*, **30**, 108, (1988).
64. A.G. Popeko, et al., *Phys.Lett.*, **52B**, 417, (1974).
65. A.G. Popeko, et al., *Yad. Fiz.*, **21**, 1220, (1975).
66. G.N. Flerov, et al., *Yad. Fiz.*, **26**, 449, (1977).
67. I. Zvara, et al., *Yad. Fiz.*, **26**, 455, (1977).
68. Yu.Ts. Oganessian, *J. Phys. G* **34**, R165 (2007).
69. R. Smolanczuk, *Phys. Rev. C*, **56**, 812, (1997).
70. P. Möller, J. R. Nix, and K.-L. Kratz, *At. Data Nucl. Data Tables*, *66*, 131, (1997).

THE FACILITY FOR ANTIPROTON AND ION RESEARCH FAIR

H. STÖECKER[1,2,4] AND C. STURM[1,3]

[1] *GSI Helmholtzzentrum für Schwerionenforschung GmbH, Planckstr. 1, D-64291 Darmstadt, Germany*

[2] *Institut für Theoretische Physik, Goethe Universität Frankfurt, Max-von-Laue Str. 1, D-60438 Frankfurt Main, Germany*

[3] *Institut für Kernphysik, Goethe Universität Frankfurt, Max-von-Laue Str. 1, D-60438 Frankfurt Main, Germany*

[4] *Frankfurt Institute for Advanced Studies, Ruth-Moufang-Str. 1, D-60438 Frankfurt am Main, Germany*

On October 4th, 2010, nine countries signed the international agreement on the construction of the Facility for Antiproton and Ion Research FAIR. The new facility is going to be constructed within the next eight years adjacent to the existing accelerator complex of the GSI Helmholtz Centre for Heavy Ion Research at Darmstadt/Germany, expanding the research goals and technical possibilities substantially. Providing a broad spectrum of unprecedented fore-front research at worldwide unique accelerator and experimental facilities, FAIR will open the way for a large variety of experiments in hadron, nuclear, atomic and plasma physics as well as applied sciences which will be briefly described in this article.

Keywords: GSI; FAIR; relativistic nucleus-nucleus collisions; CBM; HADES; neutron star; first order phase transition; QCD critical endpoint; open-charm; multi-strange hyperon; rare isotope beam; Super-FRS; supernova; nucleosynthesis; strong electromagnetic fields; high energy density matter; biological effectiveness; radiation damage; material research; antiproton beam; PANDA; high precision spectroscopy; charmonium; hybrid meson; glueball; XYZ state; hypernuclei.

Introduction

The GSI laboratory at Darmstadt/Germany was founded 40 years ago through the efforts of scientists from the nearby universities. The goal of the founding fathers was to combine efforts and resources, and to create a forefront research facility for nuclear physics and related areas. They proposed the construction of a novel heavy-ion accelerator that would exceed the technical capabilities of a single university. The success of the concept and the strong involvement from universities, first from Germany, but soon from most of Europe and worldwide, widened the scope of the facility. Together with a significant energy upgrade about 20 years ago, the facility has evolved into a major international research center using beams of heavy ions.

The Facility for Antiproton and Ion Research, FAIR,[1–3] builds on this tradition. It will provide an extensive range of particle beams from protons and their antimatter partners, antiprotons to ion beams of all chemical elements up to the heaviest one, uranium, with in many respects world record intensities. As a joint effort of 16 countries [a] the new facility builds, and substantially expands, on the present accelerator system at GSI, both in its research goals and its technical possibilities. Compared to the present GSI facility, an increase of a factor of 100 in primary beam intensities, and up to a factor of 10000 in secondary radioactive beam intensities, will be a technical property of the new facility. Other characteristics will be excellent beam qualities of both primary and secondary beams. This will be achieved through innovative beam handling techniques, many aspects of which have been developed at GSI over recent years with the present system. This includes in particular electron-beam cooling of high-energy, high-charge state ion beams in storage rings and bunch compression techniques.

After the official launch of the project on November 7th, 2007, on October 4th, 2010, nine countries[b] signed the international agreement on the construction of FAIR. Because of the long lead times for civil-construction planning, civil work for the first buildings of FAIR will start in 2012 and first beams will be delivered in 2017/2018. The start version of FAIR, the so-called *Modularized Start Version*,[4,5] includes a basic accelerator SIS100 (module 0) as well as three experimental modules (module 1-3) as it is illustrated in Fig. 1. The superconducting synchrotron SIS100 with a circumference of 1100 meters and a magnetic rigidity of 100 Tm is at the heart of the FAIR accelerator facility. Following an upgrade for high intensities, the existing GSI accelerators UNILAC and SIS18 will serve as an injector. Adjacent to the SIS100 synchro tron are two storage-cooler rings and ex-

[a]In alphabetical order: Austria, China, Finland, France, Germany, Greece, India, Italy, Poland, Romania, Russia, Slovenia, Slovakia, Spain, Sweden, and the United Kingdom.
[b]In alphabetical order: Finland, France, Germany, India, Poland, Romania, Russia, Slovenia and Sweden.

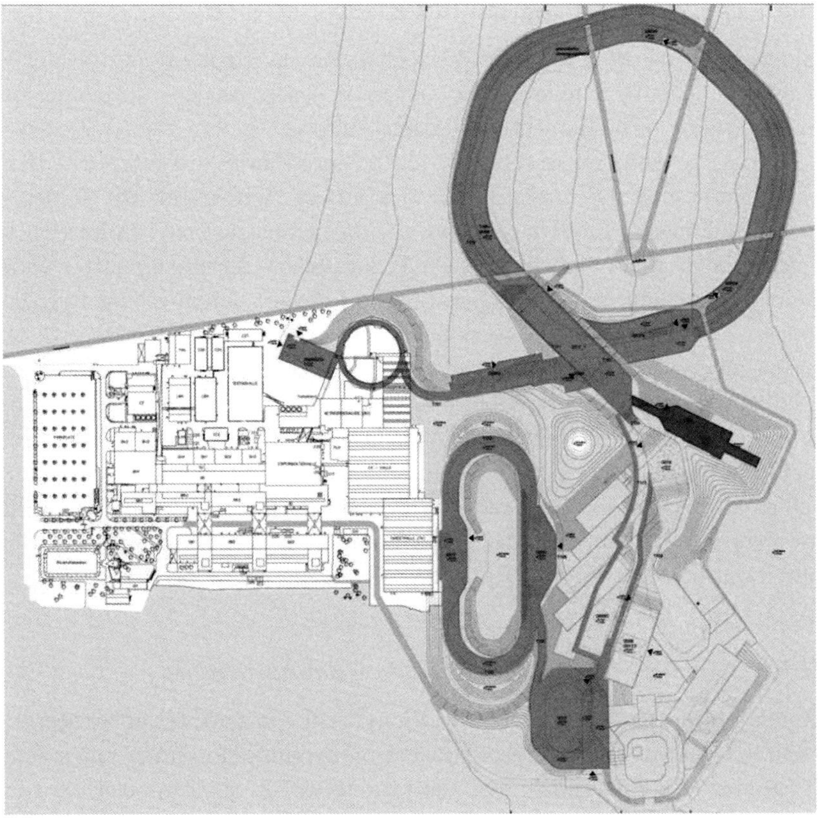

Fig. 1. (color online) On the left the existing GSI facility is shown. Displayed in color is the so called Modularized Start Version of FAIR including module 0, 1, 2 and 3. Coloring: the 100 Tm super conducting synchrotron SIS100 (module 0) - green; the experimental area for CBM/HADES (module 1) - red; the NuSTAR facility including the Super-FRS (module 2) - yellow; The Antiproton facility including the PANDA experiment (module 3) - orange. Not shown is the additional experimental area above ground for the APPA community (module 1). These colored parts are financed within the presently available firm funding commitments of the 16 partner countries.

periment stations, including a superconducting nuclear fragment separator (Super-FRS) and an antiproton production target. The Modularized Start Version secures a swift start of FAIR with outstanding science potential for all scientific pillars of FAIR within the current funding commitments. Moreover, after the start phase and as additional funds become available the facility will be upgraded by experimental storage rings enhancing capabilities of secondary beams and upgraded by SIS300 providing particle energies 20-fold higher compared to those achieved so far at GSI.

The Experimental Program of FAIR

The main thrust of FAIR research focuses on the structure and evolution of matter on both a microscopic and on a cosmic scale – deepening our understanding of fundamental questions such as :
How does the complex structure of matter at all levels arise from the basic constituents and the fundamental interactions ? How can the structure of hadronic matter be deduced from the strong interaction ? In particular, what is the origin of hadron masses ? What is the structure of matter under the extreme conditions of temperature and density found in astrophysical objects ? What was the evolution and the composition of matter in the early Universe ? What is the origin of the elements in the Universe ?

The approved FAIR research program embraces 14 experiments. These form the four scientific pillars of FAIR and offer a large variety of unprecedented forefront research in hadron, nuclear, atomic and plasma physics as well as applied sciences. Already today, over 2500 scientists and engineers are involved in the design and preparation of the FAIR experiments. The four scientific pillars are APPA, CBM, NuSTAR, and PANDA.

APPA – Atomic Physics, Plasma Physics and Applications

Atomic physics with highly charged ions[6] will concentrate on two central research themes: (i) the correlated electron dynamics in strong, ultra-short electromagnetic fields including the production of electron-positron pairs and (ii) fundamental interactions between electrons and heavy nuclei, in particular the interactions described by Quantum Electrodynamics, QED. Here bound-state QED in critical and supercritical fields is the focus of the research programme. In addition, atomic physics techniques will be used to determine properties of stable and unstable nuclei and to perform tests of predictions of fundamental theories besides QED.

For Plasma physics the availability of high-energy, high-intensity ion-beams enables the investigation of High Energy Density Matter in regimes of temperature, density and pressure not accessible so far.[7] It will allow probing new areas in the phase diagram and long-standing open questions of basic equation-of-state research can be addressed. FAIR will provide beams of high energy and very high intensity, whose biological effectiveness was never studied in the past, but whose contribution to the dose equivalent in space is very significant. It will be possible to investigate the radiation damage induced by cosmic rays and protection issues for Moon and Mars missions. Furthermore, the intense ion-matter interactions with projectiles of energies above 1 GeV/u will enable systematic studies of material modifications.

CBM/HADES – Compressed Baryonic Matter

Violent collisions between heavy nuclei promise insight into an unusual state in nature, that of highly compressed nuclear matter. In addition to its relevance for understanding fundamental aspects of the strong interaction, this form of matter may exist in various so far unexplored phases in the interior of neutron stars and in the core of supernovae. The mission of high-energy nucleus-nucleus collision experiments worldwide is to investigate the properties of strongly interacting matter under these extreme conditions. At very high collision energies, as available at RHIC and LHC, the measurements concentrate on the study of the properties of deconfined QCD matter at very high temperatures and almost zero net baryon densities. Results from lattice QCD indicate that the transition from confined to deconfined matter at vanishing net baryon density is a smooth crossover, whereas in the region of high baryon densities, accessible with heavy-ion reactions at lower beam energies, a first-order phase transition is expected.[8] Its experimental confirmation would be a substantial progress in the understanding of the properties of strongly interacting matter.

Complementarily to high-energy nucleus-nucleus collision experiments at RHIC and LHC, the CBM experiment[9,10] as well as HADES[11,12] at SIS100/300 will explore the QCD phase diagram in the region of very high baryon densities and moderate temperatures by investigating heavy-ion collision in the beam energy range 2 - 35 AGeV. This approach includes the study of the nuclear matter equation-of-state, the search for new forms of matter, the search for the predicted first order phase transition to the deconfinement phase at high baryon densities, the QCD critical endpoint, and the chiral phase transition, which is related to the origin of hadron masses. In the case of the predicted first order phase transition, basically one has to search for non-monotonic behaviour of observables as function of collision energy and system size. The CBM experiment at FAIR is being designed to perform this search with a large range of observables, including very rare pro bes like charmed hadrons. Produced near threshold, their measurement is well suited to discriminate hadronic from partonic production scenarios. The former requires pairwise creation of charmed hadrons, the latter the recombination of c-quarks created in first chance collisions of the nucleus-nucleus reaction. Ratios of hadrons containing charm quarks as a function of the available energy may provide direct evidence for a deconfinement phase.

The properties of hadrons are expected to be modified in a dense hadronic environment which is eventually linked to the onset of chiral symmetry restoration at high baryon densities and/or high temperatures. The experimental verification of this theoretical prediction is one of the most challenging questions in modern strongly interacting matter physics. The dileptonic decays of the light vector mesons (ρ,ω,ϕ) provide the tool to study such modifications since the lepton daughters do not undergo strong interactions and can therefore leave the dense hadronic medium essentially undistorted by final-state interaction. For these investigations the ρ meson plays an important role since it has a short lifetime and through this a large probability to decay inside the reaction zone when created in a nucleus-nucleus collision. As a detector system dedicated to high-precision di-electron spectroscopy at beam energies of 1 – 2 AGeV, the modified HADES detector at SIS100 will measure e^+e^- decay channels as well as hadrons[13,14] up to 10 AGeV beam energy. Complementarily, the CBM experiment will cover the complete FAIR energy range by measuring both the e^+e^- and the $\mu^+\mu^-$ decay channels.

Most of the rare probes like lepton pairs, multi-strange hyperons and charm will be measured for the first time in the FAIR energy range. The goal of the CBM experiment as well as HADES is to study rare and bulk particles including their phase-space distributions, correlations and fluctuations with unprecedented precision and statistics. These measurements will be performed in nucleus–nucleus, proton–nucleus, and proton–proton collisions at various beam energies. The unprecedented beam intensities will allow studying extremely rare probes with high precision which have not been accessible by previous heavy-ion experiments at the AGS and the SPS.

NuSTAR – Nuclear Structure, Astrophysics and Reactions

The main scientific thrusts in the study of nuclei far from stability are aimed at three areas of research: (i) the structure of nuclei, the quantal many-body systems built by protons and neutrons and governed by the strong force, towards the limits of stability, where nuclei become unbound, (ii) nuclear astrophysics delineating the detailed paths of element formation in stars and explosive nucleosynthesis that involve short-lived nuclei, (iii) and the study of fundamental interactions and symmetries exploiting the properties of specific radioactive nuclei.

The central part of the NuSTAR programme at FAIR[15,16] is the high acceptance Super-FRS with its multi-stage separation that will provide high intensity mono-isotopic radioactive ion beams of bare and highly-ionized exotic nuclei at and close to the driplines. This separator, in conjunction with high intensity primary beams with energies up to 1.5 AGeV, is the keystone for a competitive NuSTAR physics programme. This opens the unique opportunity to study the evolution of nuclear structure into the yet unexplored territory of the nuclear chart and to determine the properties of many short-lived nuclei which are produced in explosive astrophysical events and crucially influence their dynamics and associated nucleosynthesis processes.

PANDA – AntiProton ANnihilation in DArmstadt

The big challenge in hadron physics is to achieve a quantitative understanding of strongly interacting complex systems at the level of quarks and gluons. In $p\bar{p}$-annihilation, particles with gluonic degrees of freedom as well as particle-antiparticle pairs are copiously produced, allowing spectroscopic studies with unprecedented statistics and precision. The PANDA experiment at FAIR[17–19] will bring new fundamental knowledge in hadron physics by pushing the precision barrier towards new limits. The charmonium ($c\bar{c}$) spectroscopy will take advantage of high-precision measurements of mass, width and decay branches. Particular emphasis is placed on mesons with open and hidden charm, which extends ongoing studies in the light quark sector to heavy quarks, and adds information on contributions of the gluon dynamics to hadron masses. The search for exotic hadronic matter such as hybrid mesons or heavy glueballs gains enormously by precise scanning of resonance curves of narrow states as well. Recently, this field has attracted much attention with the surprise observation at electron-positron colliders of the new X, Y and Z states with masses around 4 GeV. These heavy particles show very unusual properties, whose theoretical interpretation is entirely open. Additionally, the precision gamma-ray spectroscopy of single and double hypernuclei will allow extracting information on their structure and on the hyperon-nucleon and hyperon-hyperon interaction.

Conclusion

On October 4th, 2010, and after about ten years of negotiations, R&D and writing reports, nine countries signed the international agreement on the construction of the Facility for Antiproton and Ion Research FAIR.

Construction of the first FAIR buildings will start in 2012, so that the first beams will be delivered in 2017/2018. The start version of FAIR, the so-called *Modularized Start Version*, includes the superconducting synchrotron SIS100 as well as three experimental modules to perform experiments for all research pillars. It will allow to carry out an outstanding and world-leading research programme in hadron, nuclear, atomic and plasma physics as well as applied sciences. Due to the high luminosity which exceeds current facilities by orders of magnitude, experiments will be feasible that could not be done elsewhere. FAIR will expand the knowledge in various scientific fields beyond current frontiers. Moreover, the exploitation of exiting strong cross-topical syn ergies promise novel insights.

References

1. H.H. Gutbrod et al. (Eds.), *FAIR Baseline Technical Report*, ISBN 3-9811298-0-6 Nov. 2006.
2. W. F. Henning, *FAIR: Recent developments and status*, Nucl. Phys. A 805 (2008) 502.
3. H. Stöcker, *FAIR: Challenges Overcome and Still to be Met*, published in Conf.Proc.C0806233:moycgm01, 2008.
4. C. Sturm, B. Sharkov and H. Stöcker, *1,2,3 ... FAIR !*, Nucl. Phys. A 834 (2010) 682c - 687c .
5. Green Paper of FAIR: *The Modularized Start Version* , https://www.gsi.de/documents/DOC-2009-Nov-124-1.pdf , October 2009
6. Th. Stöhlker et al. , *Atomic physics with highly-charged ions at the future FAIR facility: A status report* , Nuclear Instruments and Methods in Physics Research B 261 (2007) 234 - 238.
7. I.V. Lomonosov and N.A. Tahir, *Prospects of High-Energy Density Matter Research at the Future FAIR Facility at Darmstadt* , Nucl. Phys. News 16 1 (2006) 29 - 35.
8. B. Friman, C. Höhne, J. Knoll, S. Leupold, J. Randrup, R. Rapp, P. Senger (Editors) , *The CBM Physics Book*, Lecture Notes in Physics 814 (2011) 1 - 994
9. P. Senger et al. [CBM Collaboration] , *Compressed baryonic matter: Experiments at GSI and FAIR*, Phys. Part. Nucl. 39 (2008) 1055.
10. P. Senger et al. [CBM Collaboration] , *Probing dense baryonic matter*, Prog. Part. Nucl. Phys. 62 (2009) 375.
11. G. Agakishiev et al. [HADES Collaboration] , *The High-Acceptance Dielectron Spectrometer HADES* , Eur. Phys. J. A 41 (2009) 243
12. J. Stroth et al. [HADES Collaboration] , *Di-electron emission from resonance matter*, Prog. Part. Nucl. Phys. 62 (2009) 481.
13. I. Fröhlich for the HADES Collaboration , *Future perspectives at SIS-100 with HADES-at-FAIR*, arXiv:0906.0091 [nucl-ex].

14. P. Tlusty et al. [HADES Collaboration] , *Future experiments with HADES at FAIR*,
AIP Conf.Proc. 1322 (2010) 116-124
15. R. Krücken et al. [NuSTAR Collaboration] , *The Nustar Facility At FAIR*, J. Phys. G 31 (2005) S1807.
16. B. Rubio and T. Nilsson, *NuSTAR*, Nucl. Phys. News 16 (2006) 9.
17. Physics Performance Report for PANDA: *Strong Interaction Studies with Antiprotons*,
arXiv:0903.3905v1 [hep-ex]
18. K. Fohl et al. [PANDA Collaboration] , *The Panda Detector At The Future FAIR Laboratory*,
Eur. Phys. J. ST 162 (2008) 213 .
19. J. S. Lange for the PANDA Collaboration , *The Panda Experiment: Hadron Physics With Antiprotons At FAIR*, Int. J. Mod. Phys. A 24 (2009) 369.

K^- AND $\bar{p}$ SPECTRA FOR Au+Au COLLISIONS AT $\sqrt{s} = 200$ GeV FROM STAR, PHENIX AND BRAHMS IN COMPARISON TO CORE-CORONA MODEL PREDICTIONS

C. SCHREIBER, K. WERNER AND J. AICHELIN*

SUBATECH, Université de Nantes, EMN, IN2P3/CNRS
4 rue Alfred Kastler, 44307 Nantes cedex 3, France

Based on results obtained with event generators we have launched the core-corona model. It describes in a simplified way but quite successfully the centrality dependence of multiplicity and $< p_t >$ of identified particles observed in heavy ion reaction at beam energies between $\sqrt{s} = 17$ GeV and 200 GeV. Also the centrality dependence of the elliptic flow, v_2, for all charged and identified particles could be explained in this model. Here we extend this analysis and study the centrality dependence of single particle spectra of K^- and $\bar{p}$ measured by the PHENIX, STAR and BRAHMS collaborations. We find that also for these particles the analysis of the spectra in the core-corona model suffers from differences in the data published by the different experimental groups, notably for the pp collisions. As for protons and K^+ for each experience the data agree well with the prediction of the core-corona model but the value of the two necessary parameters depends on the experiments. We show as well that the average momentum as a function of the centrality depends in a very sensitive way on the particle species and may be quite different for particle which have about the same mass. Therefore the idea to interpret this centrality dependence as a consequence of a collective expansion of the system, as done in blast way fits may be premature.

1. Motivations

There is ample evidence by now that in heavy ions collisions at beam energies which can be reached at the colliders at CERN and in Brookhaven a plasma of quarks and gluons is created. Such a state is predicted by lattice gauge calculations at high density and/or temperature. This plasma is a very short living state - it lasts less than 10^{-23} seconds - but it is assumed that this time is sufficiently long for reaching equilibrium. Assuming such an early equilibrium, whose origin is still debated, hydrodynamical calcu-

*Invited speaker.

lations describe many details of the observables. It was surprising that the multiplicity of identified stable particles in the most central collisions agrees almost perfectly with that expected for a statistical distribution at a freeze out temperature of around 170 MeV and a small baryon chemical potential.

For symmetric systems the number of projectile participants equals that of target participants, independent of the centrality. If each particpant contributes the same energy in the center of mass system and if the system come to equilibrium one does not expect that the multiplicity per participating nucleon varies with centrality. In the experiments such a variation has been observed, however. In addition the centrality dependence depends strongly on the particle species. Whereas for π this ratio is almost constant, for multi strange baryons this ratio varies by a large factor, a phenomenon which has been dubbed strangeness enhancement.

The basic aussumption in statistical model calculations is that geometry does not play a role and *all* nucleons come to statistical equilibrium, means that all phase space configurations compatibel with the overall quantum numbers become equally probable. In simulations of the heavy ion reactions on an event by event basis,[1] however, it has been observed that this not not the case. Nucleons closee to the surface of the interaction region suffer from less collisions than thos in the center of the reaction and there is a nonnegligibla fraction of nucleons which scatter only once and therefere they will not come to an equilibrium with their environment. The relative fraction of these surface nucleons decreases with centrality.

This observation has motivated the core-corona model in which is it assumed that nucleons which scatter initially only once (corona particles) are not part of the equilibrated source but produce particles as in pp collisions whereas all the other come to statistical equilibrium (core particles). Of course this fast transition between core and corona particles is a crude approximation but it allows to define from experimental pp and central AA data the centrality dependence of the different observables. Studies have shown that the present quality of data does not allow for a more refined definition of the transition between core and corona particles. It has been further verified that the core-corona model describes quantitatively the results of the much more involved EPOS simulation program.

In a series of papers[2–4] it has been shown that the core-corona model describes quite nicely the centrality dependence of the multiplicity of identified particles, $< p_t >$ of identified particles, spectra of protons and K^{+}[5] and even of v_2 observed in AuAu and PbPb collisions. The latter has been considered as a test ground for the shear viscosity needed to describe heavy ion data in viscous hydrodynamical calculations. The core-corona model describes this data without any reference to a viscosity. The prediction for the CuCu data are completely determined by the AuAu data and agree with data as far as data have been published.

In this contribution we study the observed single particle p_t spectra for antiprotons, and K^- at midrapidity. These two particles are interesting because:

a) Antiprotons are created in the reaction whereas a part of the protons are just shifted towards midrapidity. It is therefore interesting to see whether the spectra show differences.

b) To search for domains in p_t where the deviations from the core-corona prediction is different for particles and antiparticles and to try to interpret those deviations, if they exist, in physical terms.

2. Centrality and Core-Corona Fraction

In order to determine the centrality dependence of the specta we have first of all to know the relative contribution of core and corona particles as a function of the centrality. The core-corona model relies on a single parameter: $f(N_{core})$, the fraction of core nucleons as a function of the centrality. Along with the number of participans, N_{part}, it is calculated by a Monte-Carlo simulation based on a Glauber model for hadrons in the nucleus. The parameters of the Glauber distribution are the only freedom of this model. We apply here the EPOS approach. The results are presented in Ref. 5.

The STAR[7] and PHENIX[9] collaboration found other and mutually different values of N_{part} for the same centrality bins. The values of the two collaborations agree within error bars but since this is a completely theoretical quantity it is not clear why the difference cannot be avoided. Because different N_{part} yield different f_{core} it becomes difficult to compare the different experiments with the same model parameters.

3. Results

3.1. *Centrality dependence of $< p_t >$ of identified particles*

In this contribution we investigate the centrality dependence of $< p_t >$ for different particles as measured by the STAR,[8] collaboration. We observe that if we plot $< p_t >$ as a function of the centrality for p, K^+ and π^+ we observe indeed a mass dependence 1(a) which can be well described by a blast wave fit in which the increase of $< p_t >$ depends on the mass. That the situation is more complicated is displayed in Fig. 1 which shows

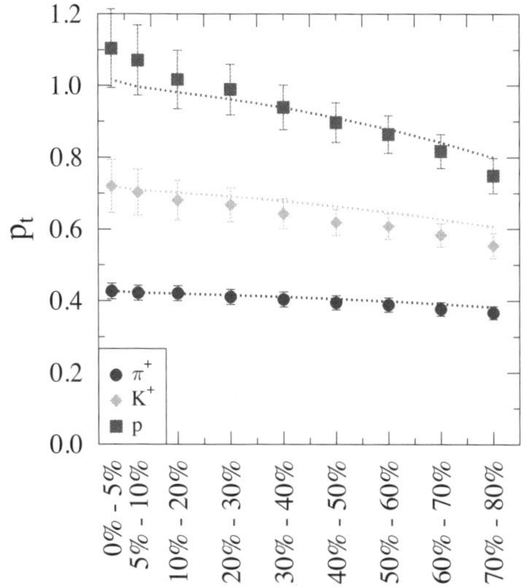

(a) Core - Corona N_{part}

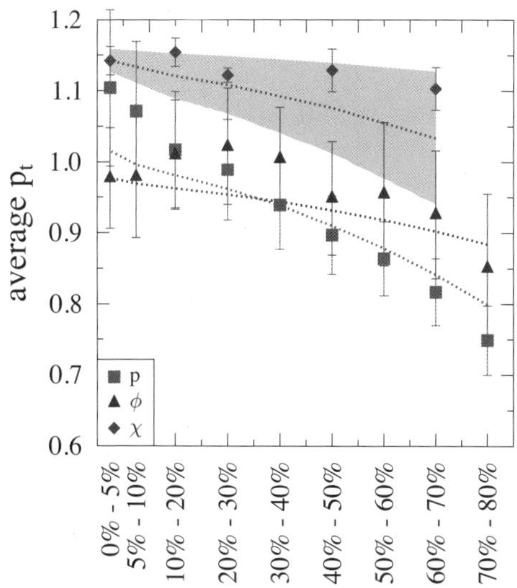

Fig. 1. $< p_t >$ as a function of the centrality for different identified particles. On the left hand side we display the centrality dependence $< p_t >$ for p, K^+ and π^+ as compared with the results of the core-corona model. On the right hand side we display the centrality dependence $< p_t >$ for different particles having about the same mass.

that participles with about the same mass have a quite different centrality dependence of $< p_t >$. This is well described in the core corona model (lines). For the Ξ we display as well the theoretical uncertainty by the shaded area.

In the core-corona model the centrality dependence of the multiplicity of a given particle species i in a centrality bin containing N_{part} participants is given by :

$$M^i(N_{part}) = N_{part}. \left[f_{core}(N_{part}).M^i_{core} + (1 - f_{core}(N_{part})).M^i_{corona}\right] \quad (1)$$

where M^i_{core} is the multiplicity per core participant and M^i_{corona} the multiplicity per corona participant. There are several ways to determinate these two values: one can either calculate them from integrated fits or one can use directly the published values, which are the results from a fit of a specific form (blast wave model) to the experimental spectra. We chose the latter in the present study: we use the multiplicity measured in pp and divided by a factor of two for M^i_{corona}. Then we extract M^i_{core} from the most central multiplicity using eq. (1). If there was no suitable pp data like for PHENIX, we use instead the most peripheral AA bin and eq. (3) to determine M^i_{corona}. The result is displayed in fig. 1 as straight line. We observe that for all experiments the centrality dependence is that which is expected in the core-corona model but, as already said, the large difference between the experiments does not allow for an unique value of M^i_{corona} and of M^i_{core}.

3.2. *Spectra*

After being formed during the confinement phase transition, the hadrons interact on the way to the detector. The results of EPOS[6] demonstrates that this rescattering is present but changes the spectra only at low p_t, where almost no experimental data are available. EPOS succeeds to reproduce the measured spectra in between a factor of two and, in consequence, to reproduce the change of the spectral form from central to peripheral collisions. The origin of this success, however, is little transparent due to the complexity of the approach. Therefore we decided to study the spectra also in the core-corona model.[2–4] In this model we can calculate which spectra would be expected if no final state interactions among hadrons take place and we can use the difference to the data to learn something about the final state interaction. Assuming no final state interactions, in the core-corona model the spectra are superpositions of two contributions: the core contribution and the corona contribution. The corona distribution $\frac{d^2N_i^{corona}}{2\pi.p_t.dp_t.dy}$ is the measured pp spectra divided by two, the core contribution $\frac{d^2N_i^{core}}{2\pi.p_t.dp_t.dy}$ is obtained from the experimental spectra for the most central AA collisions corrected for the corona contribution (see table I) and divided by the number of core participants. Then in the core-corona model the spectra for a given centrality is given by:

$$\frac{d^2M_i}{2\pi.p_t.dp_t.dy} = N_{part}[(1 - f_{core}(N_{part}))\frac{d^2N_i^{corona}}{2\pi.p_t.dp_t.dy} + f_{core}(N_{part})\frac{d^2N_i^{core}}{2\pi.p_t.dp_t.dy}]. \tag{2}$$

Fig. 2 presents the results for K^- mesons. We see in the top row the spectra measured by the different experimental groups in comparison with the predictions of the core corona model. The left hand side presents the STAR data, the middle panel the PHENIX data and the right hand side the BRAHMS data. The middle row displays the difference between theoretical predictions and the data. The shaded region marks the error bars (which are taken as the averaged error bar over the centrality bins). We observe that for almost all STAR data points the theoretical predictions are in the experimental error bars. In peripheral reactions there is a tendency that at large p_t the core-corona model is above the data. For semi central reactions model and data are in agreement for all p_t values. This is not trivial at all as the bottom row shows. There we display the inverse slope parameter obtained by fitting the experimental and theoretical spectra by a thermal spectra

$$\frac{d^2N}{2\pi.m_t.dm_t.dy} = C.m_t.e^{\frac{-m_t}{T}} \tag{3}$$

with m_t being the *transverse mass* of the considered particle:

$$m_t = \sqrt{{p_t}^2 + m^2} \tag{4}$$

where p_t is the *transverse momentum* of the particle (with respect to the beam axis) and m its free mass. Even if the curves are not exactly exponential and therefore the value of the inverse slope parameters depends on the fit range the bottom panel shows clearly that the slope varies considerably from central to peripheral reactions (which is in the core corona approach a consequence of the different invariant slope parameters in pp and central AA collisions). Also the PHENIX data are compatible with the core-corona model besides the second last and third last centrality bin where the deviations, expected from fig. 1, show up. We note in passing that for those centrality bins the core-corona models agrees well with the STAR data.

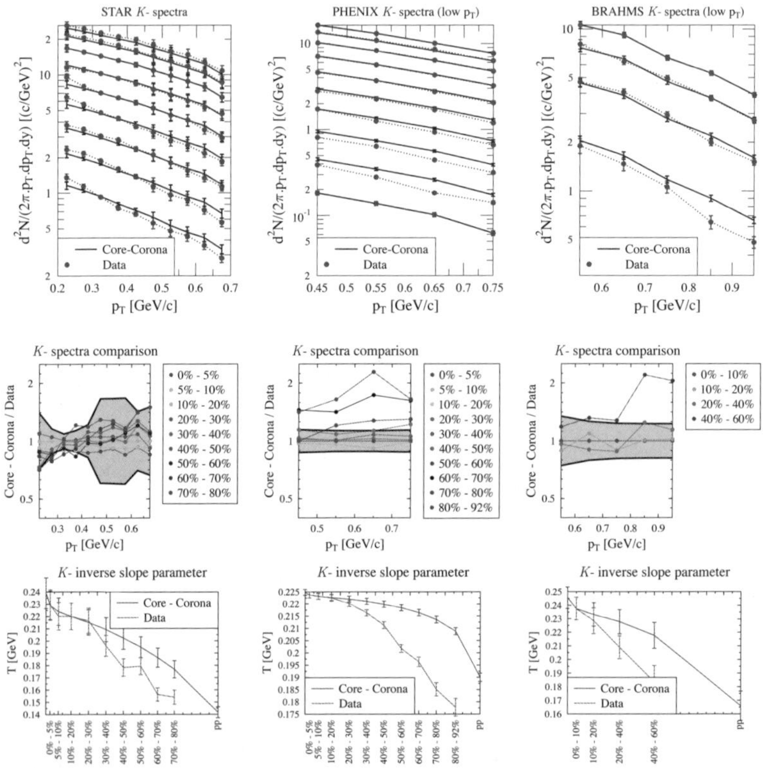

Fig. 2. Centrality dependence of the K^- spectra measured by the STAR (left), PHENIX (middle) and BRAHMS (right) collaborations in comparison with the predictions of the core-corona model. Top: The experimental spectra in comparison with the model prediciton. Middle: Ratio of the prediction of the core-corona model and the experimental data. The shaded area mark the experimental errors. Bottom: Inverse slope parameter T obtained by fitting data and theory by eq. 3.

The central BRAHMS data are also compatible with the model but we see deviations for the most peripheral bin. It is remarkable that the peripheral STAR data are almost exponential whereas those of PHENIX are not and even less those of the BRAHMS collaboration. Comparing the three experimental spectra with the model we can conclude that for each experiment the majority of data is well described by the model. Deviations are specific for the experiment. There are no systematic deviations.

Fig. 3 show the same quantities for the antiprotons. For the STAR data see an almost perfect agreement between data and model predictions. The

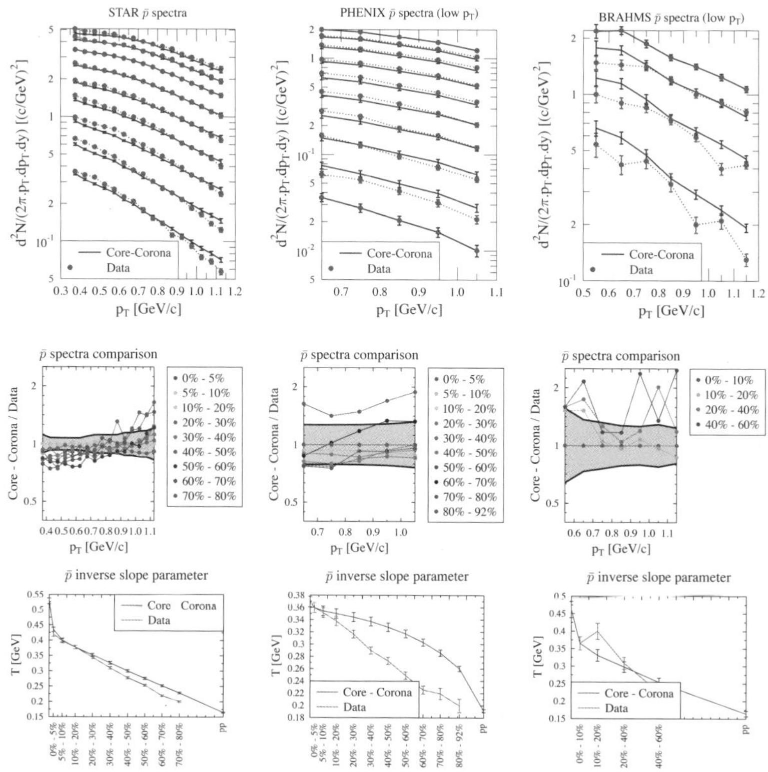

Fig. 3. Centrality dependence of the antiproton spectra measured by the STAR (left), PHENIX (middle) and BRAHMS (right) collaborations in comparison with the predictions of the core-corona model. Top: The experimental spectra in comparison with the model prediciton. Middle: Ratio of the prediction of the core-corona model and the experimental data. The shaded area mark the experimental errors. Bottom: Inverse slope parameter T obtained by fitting data and theory by eq. 3.

only exceptions are, as for the K^-, peripheral data at large p_t and the data at p_t around .4-.5 GeV. The middle left figure demonstrates this in detail. We see that besides the high p_t points in peripheral collisions the spectral form is reproduced by the core-corona model in between the error bars. This is far from being trivial. The inverse slope parameter of the spectrum varies by a factor of 2 between central and peripheral reactions and so does the inverse slope parameter of the model due to the large difference between in the inverse slope parameters in pp and central AA. Almost the same is true for the PHENIX data. Here the model overpredicts the data of the second to last centrality bin by an almost constant factor. In the BRAHMS proton data the form varies from central to peripheral reaction, what is not seen in the STAR and PHENIX data. This can also not be reproduced in the core-corona model

Thus the K^- and $\bar{p}$ spectra can be described in the core-corona model as well as the K^+ and p spectra.[5] The spectra of particle and antiparticles are rather similar and the systematic deviations between core-corona prediction and data almost identical. It is, however, impossible to describe the spectra of the different collaborations by a common model. This is a consequence of the fact that the data of the different collaborations for peripheral reactions are not compatible.

In conclusion, we have shown that all three available data sets for p_t spectra of $\bar{p}$ and K^- can be well described in the core-corona model. There are deviation but where they occur varies from experiment to experiment. Unfortunately no common parameter set can be found which describes the three experiments simultaneously. This is due to the strong differences between the experimental results for more peripheral events and due to the differences of the spectra measured in pp where the multiplicities differ up to a factor of two between the three experiments.

Acknowledgment:

We would like to thanks Dr. Roehrich and Dr. Yang for making us the Brahms pp data available and for an interesting discussion.

References

1. K. Werner, Phys. Rev. Lett. **98**, 152301 (2007) [arXiv:0704.1270 [nucl-th]].
2. J. Aichelin and K. Werner, Phys. Rev. C **79** (2009) 064907 [Erratum-ibid. C **81** (2010) 029902] [arXiv:0810.4465 [nucl-th]].
3. J. Aichelin and K. Werner, arXiv:1001.1545 [nucl-th].
4. J. Aichelin and K. Werner, J. Phys. G: Nucl. Part. Phys. 37 (2010) 094006. arXiv:1008.5351 [nucl-th].
5. C. Schreiber, K. Werner and J. Aichelin, arXiv:1012.2066 [nucl-th].
6. K. Werner, I. Karpenko, T. Pierog, M. Bleicher and K. Mikhailov, arXiv:1004.0805 [nucl-th].
7. B. I. Abelev *et al.* [STAR Collaboration], Phys. Lett. B **673**, 183 (2009) [arXiv:0810.4979 [nucl-ex]].
8. B. I. Abelev *et al.* [STAR Collaboration], Phys. Rev. C **79** (2009) 034909 [arXiv:0808.2041 [nucl-ex]].
9. S. S. Adler *et al.* [PHENIX Collaboration], Phys. Rev. C **69**, 034909 (2004) [arXiv:nucl-ex/0307022].

10. I. Arsene *et al.* [BRAHMS Collaboration], Phys. Rev. C **72**, 014908 (2005) [arXiv:nucls-ex/0503010].
11. H. Yang, Particle Production in p+p and Au+Au collisions at $\sqrt{s_{NN}} = 200$ GeV, PhD Thesis, Univeristy of Bergen (Norway) , 2007

MULTISCALE APPROACH TO THE PHYSICS OF ION-BEAM THERAPY: THERMO-MECHANICAL DAMAGE

A. V. SOLOV'YOV*, A. V. YAKUBOVICH AND E. SURDUTOVICH

Frankfurt Institute for Advanced Studies
Ruth-Moufang-Str. 1, 60438 Frankfurt am Main, Germany
**E-mail: solovyov@fias.uni-frankfurt.de*

We present a brief overview of the multiscale approach towards the understanding of processes responsible for the radiation damage caused by energetic ions. This knowledge is important because it can be utilized in the ion-beam cancer therapy, which is one of the most advanced modern techniques to cure certain types of cancer. The central element of the multiscale approach is the theoretical evaluation and quantification of DNA damage within cell environment. We consider different pathways of DNA damage and focus on the the illustration of the thermo-mechanical effects caused by the propagation of ions through the biological environment and in particular on the possibility of the creation of the shock waves in the vicinity of the ion tracks. We demonstrate that at the initial stages after ion's passage the shock wave is so strong that it can contribute to the DNA damage due to large pressure gradients developed at the distances of a few nanometers from the ionic tracks. This novel mechanism of the DNA damage provides an important contribution to the cumulative biodamage caused by low-energy secondary electrons, holes and free radicals.

The biological effects of irradiation with heavy ions are known for many years and they have been employed for both medical and radio-protection purposes, see Ref. 1 and references therein.

When energetic ions are incident on tissue, they lose energy mostly by ionizing molecules of tissue and thus produce secondary electrons and free radicals. These products of ionization are largely responsible for the DNA damage, which causes cell death, see Ref. 1 and references therein. The cell death is the object of evaluation whether it may be desirable in the course of therapy or not desirable as a result of exposure to radiation. In any case, the ultimate biological result of irradiation with ions is different from that of photons.

This difference can be quantified using the so called Relative Biological Effectiveness (RBE). The RBE is a ratio of the dose (energy deposited to the tissue) due to photons to that due to the projectiles used in a given therapy bringing about the same biological effect. For instance, the larger is this ratio in particular conditions (location of a tumor, presence or absence of oxygen on site, etc.) the better is the therapy suited; e.g., protons may be better than heavier ions for some parts of the tumor and vice versa. Apparently, the RBE is the most significant value for making a decision about the desired therapy or for estimating danger of exposure to a particular radiation.

The problem is, however, that the RBE, cannot be easily calculated. The approaches used to calculate the RBE, such as the Local Effect Model and Microdosimetric Kinetic Model[2–4] derive it as a function of the dose distribution after the calibration is done at a particular dose. These models are currently used for the dose optimization and treatment planning, but they are not directly tied to physical, chemical, or biological effects. We think that a phenomenological approach built on a variety of these effects is scientifically attractive and one day may become practical.

The main obstacle to understanding radiation damage to DNA on the microscopic level is that the scenario includes events on many spatial, temporal, and energetic scales; e.g., time scales for relevant processes vary from 10^{-22} s to minutes, hours, or even longer times. Indeed, 10^{-22} s is the characteristic time of nuclear reactions, which take place when an incident ion collides with nuclei of the medium; 10^{-17} s is that of ionization and excitation of molecules of the medium, which are the leading processes of energy loss by the projectile, 10^{-12} s is that of the transport of secondary electrons formed as a result of the above ionization, 10^{-5} s is that of DNA damage, and longer times correspond to DNA repair by different mechanisms.

In recent years, however, it has been shown that a detailed understanding of these processes is indeed possible due to new advances in the theoretical and experimental tools developed in atomic and molecular physics (AMP). An impressive example, among others, has been a discovery of a resonant mechanism to be at the origin of the damage carried out by very low energy electrons,[5] via the occurrence of a dissociative electron attachment. This understanding on a quantum mechanical level paved the way to new research on possible low energy damage effects.[6] Thus, the atomic and molecular physics community has become more involved in this field of research and has initiated an increasingly fruitful collaboration with the radiation research community all over the world, especially within Europe. The development of new therapeutic tools, such as hadron therapy,[7] brachytherapy,[8] laser-driven electron beams,[9] etc., new challenges for solving radiation protection problems, such as those connected with space exploration, the increasing use of radio diagnostics, and the acknowledged broad exposure to radiation, which exists in every day life, have further stimulated research in the field.

These developments require that the AMP community provides data, such as cross sections, as well as models for understanding of these new fields. These up-to-date tools of AMP make the advanced research in the field of radiation damage possible in order to provide further improvement of radiotherapies as well as consistent protocols of radiation protection. For example, the AMP expertise is required if we are to develop new radiosensitisers, i.e., molecules which can emphasize radiation effects on biological targets. These molecules act in combination with the radiation (e.g., by binding to a specific site of the target molecule) and their usage can lead to the exploitation of radio-chemotherapy, i.e., a concurrent radio- and chemotherapy protocol, more effective against radioresistent tumors.

The growing attention to this multidisciplinary research has successfully generated, among other events, a series of RADAM, i.e., radiation damage in biomolecular systems, conferences, initiated by the COST action (2004-2007) and, due to its success, still continues on an annual basis after the end of the action.

A quest for fundamental physical understanding on qualitative and quantitative levels is present and many groups, both theoretical and experimental, that are involved in research in the field of radiotherapy and ion-beam therapy in particular. This activity has recently been boosted by the development of collaborative projects between the fundamental and applied (clinical) communities, for example by the approval of the ESF COST Action Nano-IBCT *"Nanoscale insights into ion-beam cancer therapy"*.

The phenomenological multiscale approach to radiation damage, developed in a series of works,[10–15] is aimed at the fundamental understanding of the key effects involved in ion-beam therapy on different scales. A wide spectrum of aspects, such as ion propagation in a medium, details of ionization of the dense medium, heat transfer from ion tracks, different pathways of DNA damage, etc., was addressed by this approach.

However, there is one aspect that has been mentioned in various papers on DNA radiation damage but has never been thoroughly developed. This is an aspect of heat transfer initiated by the energy loss of projectiles, in particular heat transfer during the earliest stage after the ion's passage. The characteristic times of this stage are $10^{-15} - 10^{-9}$ s. These times are longer than the characteristic time of primary ionization (10^{-15} s) and shorter than typical times of conformational changes in DNA such as unwinding, which are measured in μs or even longer times (See Ref. 1 and references therein). Nonetheless, the events, which happen on this intermediate time scale set the initial conditions for the next scales and may be important for the subsequent dynamics of the medium.

1. Hydrodynamic Expansion on a Nanometer Scale

Consider a single ion passing through tissue, represented by liquid water at 300 K. This ion enters the medium with an energy of $\sim$ 200 .. 400 MeV/u and having relativistic velocity ($\approx 0.7c$), but then loses its energy colliding with water molecules. The main channel of energy loss is ionization, although excitation, nuclear fragmentation, and other channels of interaction are also present. The energy loss per unit length ($|dE/dx|$) is called the linear energy transfer (LET) and it is an important quantity, since many properties of radiation in relation to radiation damage are based on it. For ions, the LET is characterized by a peak close to the end of their penetration range in the tissue referred to as the Bragg peak. For a single ion, this peak is very sharp, e.g., for carbon ions, the peak's maximum value is about 1000 $keV/\mu m$ and its width is about 1 mm. The plateau in front of the peak is 10 .. 20 $keV/\mu m$.[13] This feature of the LET dependence is the major advantage of ions as projectiles compared to other particles. However, this remarkable sharpness of the Bragg peak for a single ion is lost in the beam because of the initial energy width of the beam and especially because of the ion's multiple scattering. This effect, known as energy straggling, reduces the maximum height of the peak to about 120 $keV/\mu m$ and increases its width to almost 3. However, since each ion's damaging effect is independent of the other ions,[14] in this paper, we will be interested in the LET of a single ion.

A single ion produces secondary electrons, whose average energy (at the Bragg peak) is about 45 eV.[13] Higher energy δ-electrons propagate further away from the ion track and may not fully contribute to the effects discussed in this paper, but the low-energy electrons transfer their energy to water molecules.[13] According to the results of Ref 14, the maximum temperature is reached at 0.1 ps after the ion's passage. By this time, an average 45 eV secondary electron, which has a mean free path of about 0.3 nm, will collide about 1000 times, which makes thermalization possible.

We will assume that the energy is transferred from electrons to the water molecules and the latter are contained within a small volume by the ion track. This volume is approximately cylindrical, since during this time, the

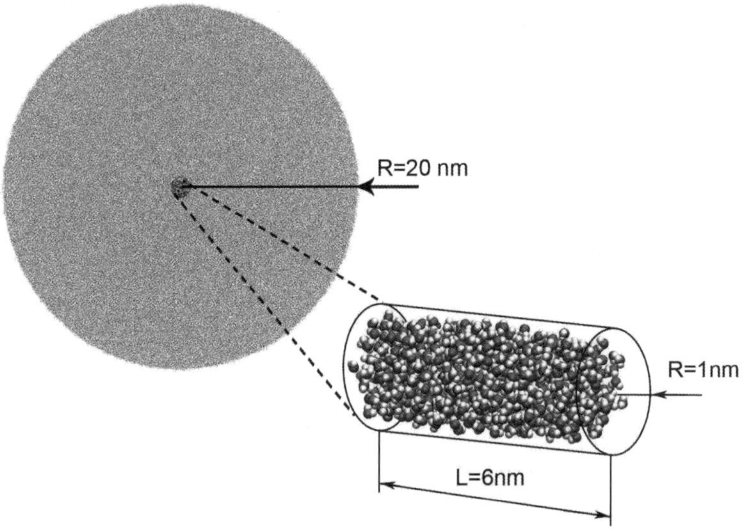

Fig. 1. The water cylinder of $20nm$ in radius and $6nm$ in length. By red and gray colors are shown oxygen and hydrogen atoms of water molecules, respectively. The energy of the heavy ion at the initial moment of time is deposited to the kinetic energy of water molecules being at the distance of $1nm$ from the ion's track. The total number of atoms in the system exceeds 700.000.

(on the average 0.3 MeV/u) ion passes $0.8\mu m$ along the track, while the radius within which the electron-water energy transfer had occurred is well within 1 nm. The pressure inside this cylinder is very high, $>$ 103 MPa, according to Ref. 16, and this prompts the following cylindrical shock wave, which we intend to explore.

2. MD Simulation of Strong Explosion

In this section we present the results of molecular dynamics (MD) simulation of the temperature spike in the water medium caused by the propagation of the heavy ion. We consider the final part of carbon ion's trajectory in the vicinity of the Bragg peak. In the vicinity of the Bragg peak LET of carbon ion can be up to 1 keV/nm of ion's trajectory.[14]

The MD simulations were conducted using the following procedure. We have performed the MD simulation of water molecules arranged in a cylinder-like shape around the ion's track (see Fig. 1). The cylinder was chosen to be 20 nm in diameter and 6 nm in length. The trajectory of the ion was assumed to pass through the cylinder's axis (z-axis). We have used the periodic boundary conditions along the axis of the water cylinder and repulsive boundary conditions in the xy plain. Water was simulated using the TIP3P[17] parameterization for water molecules. The equilibration of the system for 100ps at 300 K was performed in the NVT ensemble using NAMD software package[18] and the CHARMM22[17] forcefield, and using a time step of 0.5 fs. The temperature control was maintained using a Langevin thermostat with a damping coefficient of 5 ps. The cut-off of Van-der-Waals and Coulomb interactions were chosen to be equal to 6 and 10 Å, respectively.

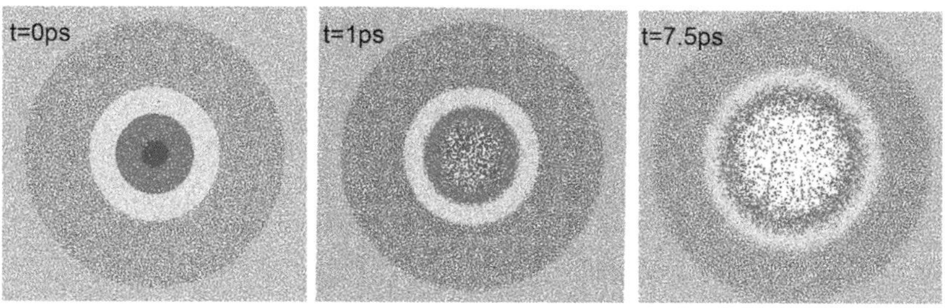

Fig. 2. The snapshots of molecular dynamics simulation of the heat spike. Only oxygen atoms of water molecules are shown. Figures on the left, center and on the right correspond to 0, 1 and 7.5 ps after ion's passage, respectively. By violet, yellow, orange and red colors are shown water molecules initially located 10nm, 5nm, 3nm and 1nm away from the ion's track, respectively. From the left figure it is seen the formation of a strongly vacuated cylinder with radius of $\sim 3 - 5nm$ at $7.5ps$ after ion's passage.

After equilibration of the system the coordinates and velocities of all particles at the final time step of the simulation were recorded. In the current work we have assumed that the deposition of the ion's energy into the kinetic energy of water molecules occurs instantaneously. We have also assumed that ion's energy is deposited evenly between water molecules located not farther than 1 nm from ion's trajectory. Further, for the simplicity of notations we will call the coaxial with ion's trajectory cylinder of the radius 1nm as *small* cylinder. Therefore, in order to simulate the heating of the medium by the ion we have increased the velocities of the water molecules in the vicinity of the ion's track (see Fig. 1). In particular, the absolute value of the velocities of water molecules in the small cylinder was multiplied by a certain factor α. This factor was calculated from the following equation:

$$\sum_{i=1}^{N} \frac{(\alpha v_i)^2}{2m_i} = \frac{3}{2}NkT + l\frac{|dE|}{dx}, \tag{1}$$

where N is the total number of atoms in the small cylinder, v_i, m_i, k and T are the velocity and mass of $i-th$ atom, Boltzmann constant and absolute temperature, respectively. l and $\frac{|dE|}{dx}$ are the length of the cylinder and linear energy transfer of the ion. In the current paper $l = 6nm$ and $\frac{|dE|}{dx} = 1keV/nm$.

The molecular dynamics simulation of the heat spike were started using the recorded after equilibration coordinates and velocities of the atoms. The velocities of the atoms in the small cylinder were increased by a factor α derived in Eq. (1). The molecular dynamics simulations were carried out in NVE microcanonical ensemble using periodic boundary conditions along z-axis of the cylinder. The MD smulations were performed for the first 10 ps after the ion's passage. In Fig. 2 is presented the structure of the system at 0, 1 and 7.5 ps after ion's passage. From Fig. 2 it is seen that the increase of the thermal energy of the water molecules in the small cylinder leads to the formation of the supersonic shock wave. At 7.5 ps after ion's passage it is clearly seen the formation of a strongly vacuated wake of the wave front. The radius of the space with the substantially decreased pressure can be estimated as being equal to $\sim 3-5$nm. From the snapshots of MD simulation it is evident that the properties of the water medium at the very first stages after the ion propagation are substantially different form that of the properties of the water at equilibrium conditions. The change of the properties of water environment should result in change of all other physical characteristics of the medium that are important for DNA damage, such as the diffusion rate of thermal electrons and holes, water ionization cross sections, etc.

Another plausible consequence of the formation of the shock wave in the vicinity of the ion's track can be the mechanical damage of DNA, proteins, lipids, etc. Indeed, the gradients of pressure on the front of the shock wave can be extremely high and can lead to the brake of the covalent bonds in the system due to mechanical stretching. In this paper we do not perform the detailed analysis of the influence of the shock waves on the structure of biomolecules and leave this question open for further investigations.

One of the important characteristics of the propagating shock wave is the radial velocity of the particles in the vicinity of the wave front. In Fig. 3 is presented the dependence of the radial velocity of water molecules on time after ion's passage and on distance from ion's track. From Fig. 3

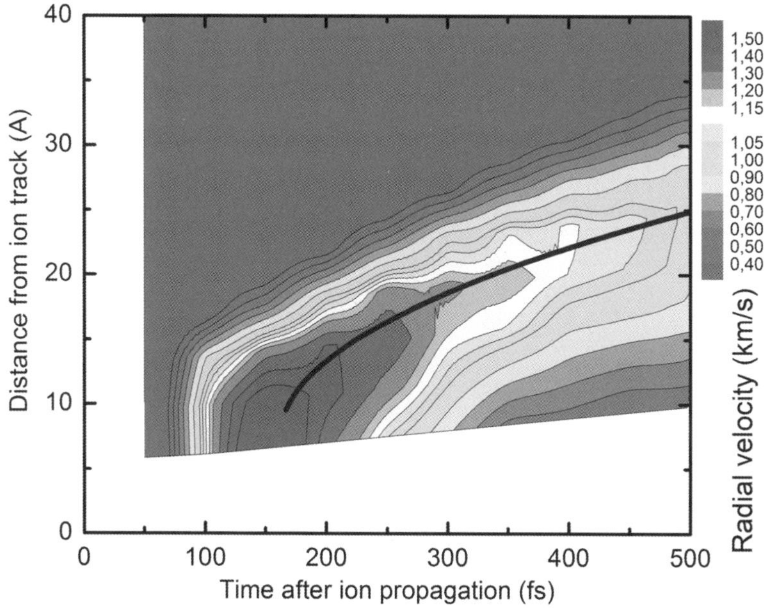

Fig. 3. Dependence of the radial velocity of water molecules on time after ion's passage and on distance from ion's track. By bold solid line are shown the predictions of the theoretical model for the shock wave propagation. The theoretical model is developed in Ref. 16 and considers the scenario of the so-called strong explosion in cylindrical geometry.

it is seen that the radial velocities of water molecules can be higher than 1 km/s at the initial stage after ion's passage. However, with the increase of time the speed of the wave front itself and the speed of water molecules in the wave front decrease. In Ref. 16 was developed a theoretical model of the strong explosion in cylindrical coordinates. In 16 was shown that speed of the wave front is inversely proportional to the square root of time after the ion's passage. However, in that work the water was considered as a continuous medium and the deposition of ion's energy was assumed to occur on the line of ion's trajectory. In this work we consider the situation when the kinetic energy of the ion is deposited in the cylinder of the radius of 1 nm. Therefore, the formation of the wave front occurs at a certain distance from ion's trajectory. From Fig. 3 it is seen that the formation of a wave front requires $\approx$ 150 fs. Note that this time is larger than 100 fs, the characteristic time that is necessary for the deposition of the kinetic energy of the secondary electrons into the kinetic energy of water molecules.

This observation supports the assumption that we can consider the process of energy deposition into water molecules as instant. Indeed, at timescales smaller than 100 fs no substantial rearrangement of the geometry of the system can be observed.

In Fig. 3 by bold solid line is shown the predictions of the theoretical model from 16. In 16 it was shown that the radius of the wave front R can be calculated as follows:

$$R = \beta\sqrt{t}\left[\frac{\epsilon}{\rho_1}\right]^{1/4}, \tag{2}$$

where β is the coefficient depending on the properties of the medium (0.86 for water), t is time after ion's passage, ϵ is LET and ρ_1 is the density of water in front of the wave front.

In the results of theoretical model shown in Fig. 3 we have shifted the origin of the explosion by 150 fs in time and by 0.75 Å in space in order to account for the non-point nature of the shock wave formation region. From Fig. 3 it is seen that the predictions of the theoretical model are in a good agreement with the results of the molecular dynamics simulations. This fact supports both the theoretical model and the approach based on molecular dynamics simulations.

3. Conclusions

In this work, we have performed a brief overview of the multiscale approach,[11,19] which forms the basis for the understanding of the complex cascade of processes leading to the DNA damage and the cell death. This theoretical approach and modeling based on it requires a large amount of atomic and molecular data on the cross sections and probabilities of various processes, atomic and molecular structure. These data partially exist or they can be obtained in the gas phase using the traditional methods of atomic and molecular physics and then corrected for the medium effects. Alternatively, direct measurements of various processes in the condense phase can be performed.

The special emphasis of this paper is on the discussion of thermo-mechanical effects in the water medium caused by the heavy ion propagation. It is demonstrated that within a few nanometer distances from the ion track there arises a shock wave of the strength sufficient to produce the direct mechanical damage of biomolecules. This thermo-mechanical effect can result in damaging not only the DNA molecules, but also other components of the cell architecture, such as membranes, reducing their chance for survival and thus increasing the relative biological effectiveness.

References

1. I. Baccarelli, F. Gianturco, E. Scifoni, A. Solov'yov and E. Surdutovich, *Eur. Phys. J. D* **60**, p. 1 (2010).
2. T. Elsaesser, M. Kraemer and M. Scholz, *Int. J. Radiat. Oncol. Biol. Phys.* **71**, 866 (2008).
3. T. Elsaesser and M. Scholz, *Radiat. Res.* **167**, 319 (2007).
4. M. Waligorski, R. Hamm and R. Katz, *Nucl. Tracks Radiat. Meas.* **11**, p. 309 (1986).
5. B. Boudaïffa, P. Cloutier, D. Hunting, M. A. Huels and L. Sanche, *Science* **287**, p. 1658 (2000).
6. B. Michael and P. O'Neill, *Science* **287**, p. 1603 (2000).
7. U. Amaldi and G. Kraft, *Rep. Prog. Phys.* **68**, p. 1861 (2005).
8. B. Thomadsen, M. Rivard and W. Butler, *Brachytherapy Physics* (Medical Physics Publishing, Madison, WI, 2005).
9. C. Geddes, C. Toth, J. van Tilborg, E. Esarey, C. Schroeder, D. Bruhwiler, C. Nieter, J. Cary and W. Leemans, *Nature* **431**, p. 538 (2004).
10. E. Surdutovich and A. Solov'yov, *Europhys. News* **40/2**, p. 21 (2009).
11. A. Solov'yov, E. Surdutovich, E. Scifoni, I. Mishustin and W. Greiner, *Phys. Rev.* **E79**, p. 011909 (2009).
12. E. Scifoni, E. Surdutovich and A. Solovyov, *Phys Rev. E* **81**, p. 021903 (2010).
13. E. Surdutovich, O. Obolensky, E. Scifoni, I. Pshenichnov, I. Mishustin, A. Solov'yov and W. Greiner, *Eur. Phys. J. D* **51**, 63 (2009).
14. M. Toulemonde, E. Surdutovich and A. Solov'yov, *Phys. Rev. E* **80**, p. 031913 (2009).
15. E. Surdutovich, E. Scifoni, and A. Solov'yov, *Mutat. Res.* **704**, 206 (2010).
16. E. Surdutovich and A. Solov'yov, *Phys. Rev. E* **82**, p. 051915 (2010).
17. A. MacKerell, D. Bashford, M. Bellott, R. Dunbrack, J. Evanseck, M. Field, S. Fischer, J. Gao, H. Guo, S. Ha *et al.*, *J. Phys. Chem. B* **102**, p. 3586 (1998).
18. J. C. Phillips, R. Braun, W. Wang *et al.*, *J. Comp. Chem.* **26**, p. 1781 (2005).
19. E. Surdutovich, A. Yakubovich and A. Solov'yov, *Eur. Phys. J. D* **60**, p. 101 (2010).

NON-EXTENSIVE THERMODYNAMICS OF LATTICE GAUGE FIELDS

T. S. BIRÓ

MTA KFKI RMKI,
P.O.Box 49, Budapest 114, H-1525, Hungary
e-mail: tsbiro@rmki.kfki.hu

Z. SCHRAM*

Department of Theoretical Physics, University of Debrecen,
P.O.Box 5, Debrecen, H-4010 Hungary
**e-mail: schram@phys.unideb.hu*

We implement a non-conventional statistical distribution, the canonical Tsallis distribution for lattice field theoretical calculations. We use a so called super-statistical approach: here the application of the Tsallis distribution is realized via introducing a fluctuating temperature. The fluctuations of the inverse temperature follow a Gamma distribution. We study the possibilities to evaluate expectation values in the superstatistical system and generalize the usual direct numerical method using a Metropolis algorithm. Our method is suitable for performing numerical calculations in order to investigate the thermodynamical properties of the strongly interacting superstatistical matter, or to derive the equation of state in this non-extensive approach.

Keywords: Tsallis distribution; superstatistics; SU(2); equation of state.

1. Introduction

Lattice field theory, a systematic non-perturbative method has been successfully applied for studying the special features of strong interactions and, in general, the thermodynamics of strongly interacting matter.[1–5] Usually, lattice simulations are based on a conventional canonical ensemble approach where configurations follow an exponential distribution. Experimental particle spectra, on the other hand, are not purely exponential. In heavy-ion collisions both exponential and power-law regions have been observed in the measured pion, kaon, etc. spectra.[6–14]

Experimentally measured hadron spectra may reflect statistical properties of the observed system, i.e. the strongly interacting matter in extreme conditions may show non-conventional thermodynamical behaviour characterized by non-conventional distributions. Such distribution with a power-law like tail is the so-called Tsallis distribution which may be derived from a non-extensive energy addition rule.[15] This distribution has been widely used recently in many areas where statistical models have been applied. For instance, it leads naturally to the mass spectrum of hadrons when they are combined from massless partons with Tsallis-like energy distributions, and results in a Hagedorn-type limiting temperature.[16] Fits using the Tsallis distribution to the observed particle spectra in heavy-ion reactions have also been remarkably successful.[17,18]

Non-conventional distributions are based in general on non-additive composition rules,[19] or on a non-conventional entropy formula (which replaces the Boltzmann entropy). Such an entropy formula is the Tsallis entropy,[15] which is not an extensive quantity. One can find a monotonic function of it, however, which is proven to be extensive.

Non-extensive thermodynamics can be viewed as an effective theory for non-equilibrium phenomena and for systems where long-range interactions and long-range correlations are present. Applying it to describe the thermodinamical properties of the strongly interacting matter under extreme conditions is particularly interesting.

If E denotes the energy of a state of the system, the Tsallis distribution is given with the probability distribution

$$w(E) = \frac{1}{Z_{TS}} \left(1 + \frac{\beta E}{c}\right)^{-c} \tag{1}$$

(β is the inverse temperature). In the $c \to \infty$ limit it restores the Gibbs factor:

$$\lim_{c\to\infty} w(E) = \frac{1}{Z_G} \exp(-\beta E). \tag{2}$$

The quantity $q = 1 + 1/c$ is the Tsallis index. It is easy to show that the following identity holds:

$$\left(1 + \frac{\beta E}{c}\right)^{-c} = \frac{c^c}{\Gamma(c)} \int_0^\infty dt\, t^{c-1}\, e^{-ct} e^{-t\beta E}. \tag{3}$$

In (3) one can recognize the Gamma distribution:

$$w_c(t) = \frac{c^c}{\Gamma(c)}\, t^{c-1}\, e^{-ct}. \tag{4}$$

The identity (3) is remarkable: it shows that averages with the Tsallis distribution can be evaluated as averaging over different β valued Gibbs expectation values. This is a specific example of the superstatistical approach,[20] where the inverse temperature is not a simple quantity: it follows a Gamma distribution.

2. Superstatistical Approach in Lattice Field Theory

Now we apply the above equations to implement lattice field theory simulations with Tsallis statistics instead of a Gibbs one. The temperature on the lattice is determined by the length in the Euclidean time direction, $\beta = N_t a_t$ (N_t is the lattice extension in time direction, a_t is the timelike lattice spacing), the lattice fields satisfying periodic boundary conditions. In a usual lattice simulation the timelike and spacelike lattice spacings equal, and the volume of the system is $V = {N_s}^3 {a_s}^3$ (N_s is the lattice size in space direction, a_s is the spacelike lattice spacing). In our approach, so as to realize a fluctuating temperature, we use an anisotropic lattice where the fluctuation of the inverse temperature is simulated by a Gamma distributed anisotropy parameter, $t = a_t/a_s$. The mean value of t is 1, the finite width, $1/\sqrt{c}$, is a parameter in this approach which can be taken from the observed particle spectra. In the following we apply the method for SU(2) gauge fields.
The partition function in the superstatistical approach is given by

$$Z = \int dt\, w_c(t) \int \mathcal{D}U\, e^{-S[U,t]} \tag{5}$$

where the SU(2) lattice action is

$$S = \frac{4}{g^2} \sum \left(\frac{a_t}{a_s} P_s + \frac{a_s}{a_t} P_t \right) = \frac{4}{g^2} \sum \left(t P_s + P_t/t \right). \tag{6}$$

The contribution of spacelike and timelike plaquettes, P_s and P_t, in a generic form is $P = 1 - \frac{1}{2}\mathrm{Tr}(U_{plaq})$ with U_{plaq} being the ordered product of the SU(2) link variables around the elementary spacelike or timelike plaquette. The expectation value of an operator which includes a timelike link with power ν, $\hat{A} = t^\nu A$ can be calculated as

$$\langle A \rangle = \frac{\int \mathcal{D}U \, \int dt\, w_c(t)\, e^{-S(t,U)} t^\nu A}{\int \mathcal{D}U \, \int dt\, w_c(t)\, e^{-S(t,U)}}. \tag{7}$$

The evaluation of the above integral can be achieved in several ways. By integrating over the asymmetry parameter t we can derive an effective action which turns out to be a logarithm of a Bessel-function. In this report however, instead of using the effective action, we calculate the expectation values via a direct numerical simulation. We outline the method in the next section.

3. Numerical Simulations

In the direct numerical approach we start with the usual canonical Gibbs simulation, in our instance using a Metropolis update algorithm. We have to modify the algorithm according to our previous discussion. Now there is an additional variable, the asymmetry, which has to be generated according to a Gamma distribution. The characteristic values of the distribution are determined via the Tsallis parameter c. We performed simulations for different c values: for $c = 5.5, 13.5, 32.0$ and for $c = 1024.0$. The lower values correspond to parameters derived from high-energy collision experiments, $c = 1024$ is meant to approximate the $c \to \infty$ (Gibbs) limit, and $c = 32$ is chosen as an intermediate value to show the transition. At a given c we generated numerically random asymmetry values of t. We investigated the distributions of the realized asymmetry parameter values and we found that for a really smooth reconstruction of the Euler-Gamma distribution one should generate random values in the order of 10^5. Our actual numerical simulations were performed for 1000 random values of the asymmetry parameter at each c value and at each coupling. Obviously about 1000 random values cannot follow precisely the required Gamma distributions. In fact, there are apparent fluctuations. However, in average the values cover the Gamma distributions quite well and Euler-Gamma fits to the 1000 generated numbers are rather good, reproducing the initial c parameters of the distributions. Nevertheless, longer simulation runs would be preferable, but these should be limited to reasonable lengths of computing time.

In general, once an asymmetry parameter is determined we can apply the standard Metropolis sweep to update the gauge fields. Then a new asymmetry parameter is thrown and so on. The ensemble averages are calculated with averaging the gauge configurations over the generated asymmetry values, just as in the canonical Monte-Carlo simulations. In this numerical approach several technical questions may arise, partly connected to the sequential organization of the algorithm. In the following we summarize our choices.

We perform simulations for the pure SU(2) gauge theory in 3+1 dimensions on $10^3 \cdot 2$ and 10^4 lattices. We use the action given in equation (6). After heating up the initial configurations (performing 1000 Monte Carlo sweeps through the lattice) we throw a new asymmetry parameter for the whole lattice before each update sweeps to measure the observables.

4. Equation of State

To determine the equation of state for the superstatistical SU(2) Yang-Mills lattice gauge theory we calculate the energy density and pressure similar way as it is done in the conventional case. There also an asymmetry parameter is used to perform the volume and temperature differentiations independently. However, in our calculation the asymmetry parameter $t = a_t/a_s$ cannot be set to unity after obtaining the derivatives. Rather, we have to write it in the expressions explicitly.

The pressure and energy density is obtained from the partition function (5) according to standard thermodynamical rules with $\beta = 1/T$ inverse temperature:[2,4]

$$p = \frac{1}{\beta}\frac{\partial}{\partial V} \ln Z, \quad e = -\frac{1}{V}\frac{\partial}{\partial \beta} \ln Z. \tag{8}$$

It is well known that for an ideal gas $p = e/3$. Therefore it is particularly interesting to consider the interaction measure,

$$\Delta = e - 3p = -\frac{1}{\beta V}\frac{d \ln Z}{d \ln a}. \tag{9}$$

From the above equations we obtain

$$ea^4 = \frac{4}{g^2}\langle tP_s - P_t/t\rangle + a\frac{\partial}{\partial a}\left(\frac{4}{g^2}\right)\langle tP_s + P_t/t\rangle \tag{10}$$

$$pa^4 = \frac{1}{3}\frac{4}{g^2}\langle tP_s - P_t/t\rangle - a\frac{1}{3}\frac{\partial}{\partial a}\left(\frac{4}{g^2}\right)\langle tP_s + P_t/t\rangle. \tag{11}$$

These formulae are extensions of the standard expressions used so far in lattice gauge theory in Gibbs-Boltzmann thermodynamics.[22]

In our calculations here, for the energy density (and pressure) we consider the first term only, so we can directly compare the values to the results of the pioneering work of Engels et. al.[23] For the interaction measure, as the first terms cancel, we use the perturbative result in leading order for the derivative of the coupling.

There is one more important issue regarding the above equations. Namely, (10) and (11) give significantly non-zero values even if we consider a system at zero temperature, i.e. if we use a symmetric, N^4 lattice at small inverse couplings (10^4 lattice in our case). This is a consequence of the fluctuating asymmetry parameter which causes that $P_s t \neq P_t/t$ even in the space-time symmetric case. Naturally, this effect is remarkable for the physically interesting c values and disappears as $c \to \infty$ (see Fig. 1). For this reason we use a simple renormalization method where we calculate the quantities

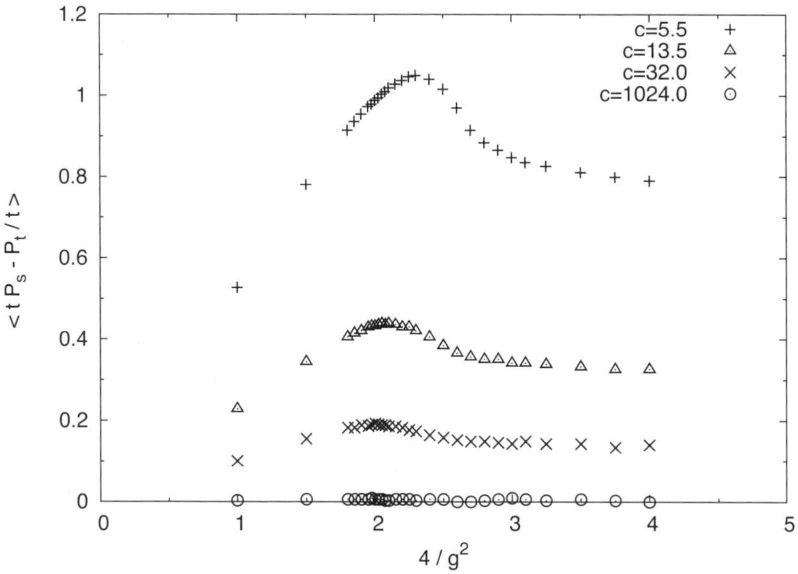

Fig. 1. The action difference on a symmetric, 10^4 lattice as a function of the coupling for different c values. For $c = 1024$ the values are practically zero as expected.

on an asymmetric lattice ($10^3 \cdot 2$ lattice) first and subtract from them the corresponding quantities obtained on the symmetric, 10^4 lattice. This is the usual, simple renormalization procedure for the interaction measure, which is given by the following expression:

$$\Delta = 2a \frac{\partial}{\partial a} \left(\frac{4}{g^2} \right) \left(\langle tP_s + P_t/t \rangle_{asym} - \langle tP_s + P_t/t \rangle_{sym} \right). \tag{12}$$

In the following we will display our results as a function of T/T_c. In usual SU(2) lattice gauge theory the critical couplings are known quite precisely for different N_t lattice sizes.[24] This allows us to translate the $4/g^2$ values on our $N_t = 2$ lattice to that of T/T_c, where T_c is the critical temperature for a conventional ($c = \infty$) lattice system in the thermodinamical limit.

Our numerical results for the energy density and for the interaction measure in the approximation mentioned above are shown in Fig. 2. One can see a slight shift of the curves to higher temperatures as c is getting smaller. Of course it is not easy to draw a precise conclusion about the consequences of that shift at this stage of the simulations. One needs at least more statistics and it would also be desirable to investigate the neglected terms in the expressions for the thermodynamical observables. Values at high temperatures should also be taken with caution: at temperatures, let say larger than about $4T_c$, the symmetric lattice does not represent any more a system at zero temperature. Here the lattice spacing becomes so small that the lattice ceases to represent a real thermodynamical system. However, it is clear that even after the subtraction of the symmetric contributions there is an apparent effect due to the applied nonconventional distribution.

In the high temperature limit the energy density e should approach the value for the free gluon gas, $e \sim T^4$. This is the well known Stefan-Boltzmann relation. Therefore it is usual (and interesting) to consider values of e/T^4 in lattice simulations. As we evaluate expectation values on the lattice where the temperature fluctuates, we might consider possibilities like $e/\langle T\rangle^4$, $e/\langle T^4\rangle$ or $\langle e/T^4\rangle$. These expressions give significantly different results in our cases because the temperature is given via the asymmetry parameter t as $\langle T\rangle/t$ and the expectation values of different powers of t depend strongly on the Tsallis parameter c.
Estimations for free massless Tsallis gas show that

$$\left(\frac{e}{T^4}\right)_c = \left(\frac{e}{T^4}\right)_\infty \cdot \frac{\langle t^{-4}\rangle_c}{\langle t^{-1}\rangle_c} = \left(\frac{e}{T^4}\right)_\infty \cdot \langle t^{-3}\rangle_{c-1}. \tag{13}$$

We plot our numerical results for $\langle(e-3p)/T^4\rangle$ and for $\langle e/T^4\rangle$ in Figs. 3 and 4. The $\langle t^{-3}\rangle_{c-1}$ values for $c = 5.5, 13.5, 32.0$ and 1024.0 are $6.996, 1.705, 1.225$ and 1.006, respectively. Similar ratios (at least approximately) can be deduced from the data given in Fig. 4. Nevertheless, besides the lack of sufficient statistics (which we mentioned earlier), simulations on larger lattices are needed for exploring the weak coupling, high temperature limit and the Stefan-Boltzmann relation for the SU(2) Tsallis gas. Work on this and other problems such as the question of thermalization, as well as on improving the statistics of our calculations is in progress.

Acknowledgments

This work has been supported by the TÁMOP 4.2.1./B-09/1/KONV-2010-0007 project. The project is implemented through the New Hungary Development Plan, co-financed by the European Social Fund and the European Regional Development Fund. Support by the Hungarian National Research Fund OTKA (K68108) is also acknowledged. The work was partly supported by the Helmholtz International Center for FAIR within the framework of the LOEWE program (LandesOffensive zur Entwicklung

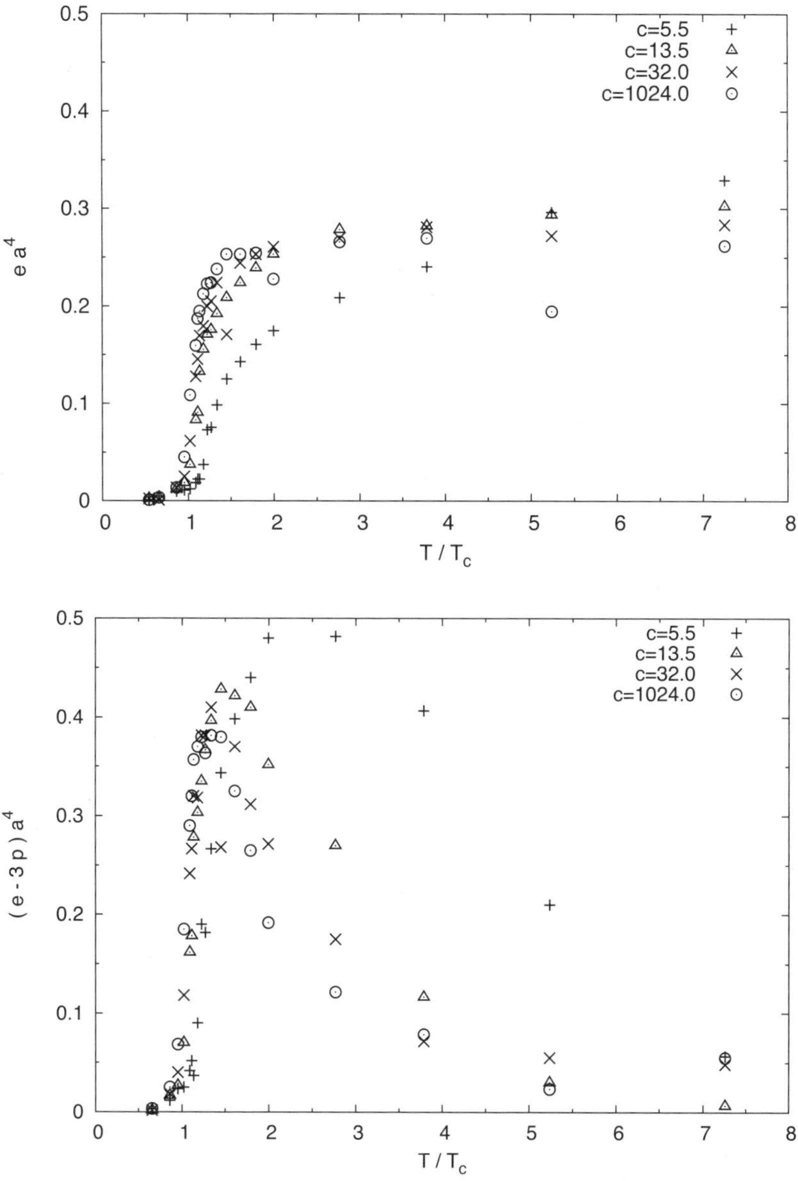

Fig. 2. The energy density and the interaction measure in lattice units vs. temperature for the Tsallis parameters $c = 5.5$, 13.5, 32.0, and 1024.0. T_c is the critical temperature for a conventional ($c = \infty$) lattice system in the thermodinamical limit.

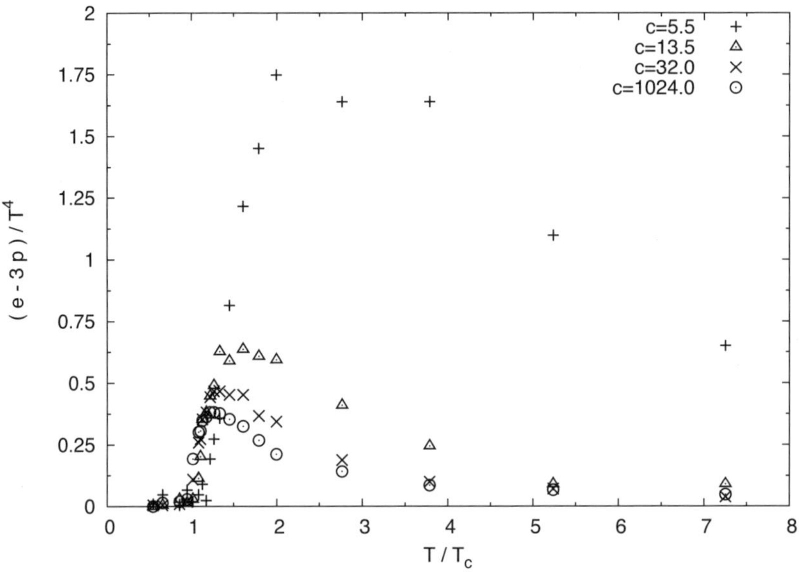

Fig. 3. The interaction measure divided by T^4 vs. temperature for the Tsallis parameters $c = 5.5$, 13.5, 32.0, and 1024.0. T_c is the critical temperature for a conventional ($c = \infty$) lattice system in the thermodinamical limit.

Wissenschaftlich-ökonomischer Exzellenz) launched by the State of Hesse.

References

1. M. Creutz, *Quarks, gluons and lattices* (Cambridge University Press, 1983).
2. I. Montvay and G. Münster, *Quantum Fields on a Lattice* (Cambridge University Press, 1994).
3. J. Smit, *Introduction to Quantum Fields on a Lattice* (Cambridge University Press, 2002).
4. H. Rothe, *Lattice Gauge Theories: An Introduction(3rd ed.)* (World Scientific, 2005).
5. C. Gattringer and C.B. Lang, *Quantum Chromodynamics on the Lattice* (Springer, 2010).
6. ALEPH Collaboration (R. Barate et al.), *Phys. Rep.* **294**, 1 (1998).
7. NA49 Collaboration (S.V. Afanasjev et al.), *Phys. Rev. C* **66**, 054902 (2002).
8. PHOBOS Collaboration (B.B. Beck et al.), *Phys. Rev. Lett.* **91**, 052303 (2003).
9. UA1 Collaboration (C. Albajar et al.), *Nucl. Phys. B* bf 335, 261 (1990).
10. NA35 Collaboration (T. Alber et al.), *Eur. Phys. J.* **C2**, 643 (1998).
11. STAR Collaboration (B.I. Abelev et al.), *Phys. Rev. C* **81**, 054907 (2010).

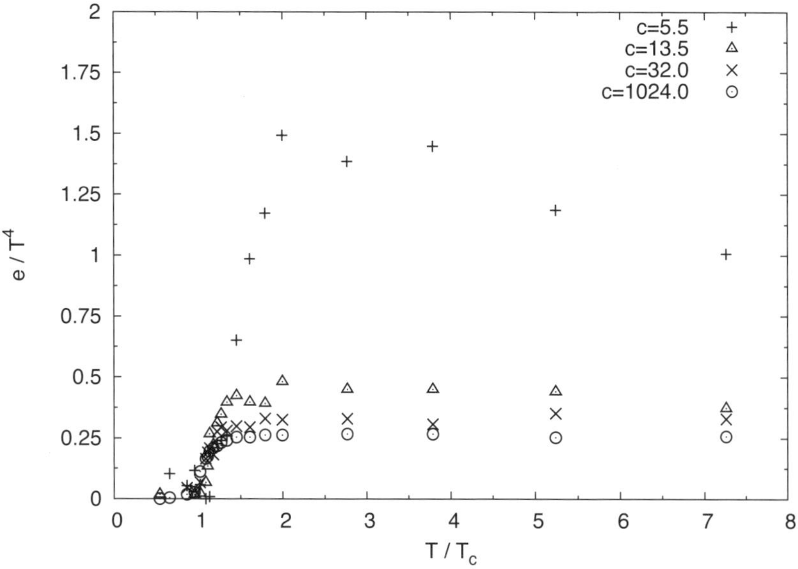

Fig. 4. The energy density divided by T^4 vs. temperature for the Tsallis parameters $c = 5.5$, 13.5, 32.0, and 1024.0. T_c is the critical temperature for a conventional ($c = \infty$) lattice system in the thermodinamical limit.

12. STAR Collaboration (Yichun Xu et al.), *Nucl. Phys. A* **830**, 701C (2009).
13. ALICE Collaboration (K. Aamodt et al.), arXiv: 1007.0719, 2010
14. CMS Collaboration (V. Khachatryan et al.), *Phys. Rev. Lett.* **105**, 022002 (2010).
15. C. Tsallis, *Introduction to Nonextensive Statistical Mechanics* (Springer Science+Business Media, New York, LLC 2009).
16. T.S. Biró and A. Peshier, *Phys. Lett. B* **632**, 247 (2006).
17. K. Ürmössy et al., *EPJ WoC*, Proc. of HCBM2010.
18. F. Siklér et al., *EPJ WoC*, Proc. of HCBM2010.
19. T.S. Biró, *EPL* **84**, 56003 (2008).
20. C. Beck, E.D.G. Cohen, H.L. Swinney, *Phys. Rev. E* **72**, 056133 (2005); C. Beck, *Phys. Rev. Lett.* **98**, 064502 (2007).
21. T.S. Biró and Z. Schram, *Particles and Nuclei Letters A: Heavy Ions*, (2010) (in press).
22. F. Karsch, *Nucl. Phys. B* **205[FS5]**, 285 1982.
23. J. Engels, F. Karsch, H. Satz and I. Montvay, *Phys. Lett. B* **101**, 89 (1981).
24. J. Fingberg, U. Heller and F. Karsch, *Nucl. Phys. B* **392**, 493 (1993).

DESCRIPTION OF *pp* COLLISIONS AT ULTRARELATIVISTIC ENERGIES

L. BRAVINA*, L. V. MALININA†, M. S. NILSSON, E. ZABRODIN†

Department of Physics, University of Oslo,
N-0316 Oslo, Norway
**E-mail: larissa.bravina@fys.uio.no*
† also at Institute of Nuclear Physics, Moscow State University,
RU-119899 Moscow, Russia

J. BLEIBEL

Institut für Physik, WA 331, Johannes Gutenberg Universität Mainz,
D-55099 Mainz, Germany
and Max-Planck-Institut für Metallforschung,
D-70569 Stuttgart, Germany

Both bulk features of particle production and various correlations of particles in proton-proton collisions at energies 200 GeV $\leq \sqrt{s} \leq$ 14 TeV are studied within the Monte Carlo quark-gluon string model. The model relies on Reggeon Field theory accomplished by string phenomenology. The available experimental data are reproduced quite well. Predictions are made for collisions at $\sqrt{s} = 14$ TeV.

Keywords: ultrarelativistic hadronic collisions; Reggeon Calculus; Monte Carlo simulations; short-range; long-range and femtoscopy correlations

1. Introduction

Proton-proton collisions at ultrarelativistic energies accessible for Large Hadron Collider (LHC) at CERN are extremely interesting because of several reasons. One objective is searching for new particles, such as eagerly awaiting Higgs gauge boson. Another goal is exploration of possible collective properties of hot and dense partonic matter, most likely the quark-gluon plasma (QGP). Up to now no such signals have been observed and elementary *pp* interactions are used as reference ones for the comparison with ultrarelativistic heavy-ion collisions. Then, the multiparticle production in *pp* collisions at energies about 1 TeV is not fully understood yet, despite the serious progress achieved in last few decades.

The main difficulty here is as follows. Quantum chromodynamics (QCD) is a good candidate for a theory of strong interactions. Experimental studies of the processes with large momentum transfer Q^2 have shown that all the predictions of perturbative QCD are in good agreement with the data. However, the dominant contribution to hadronic interactions even at high energies comes from the "soft" processes with small momentum transfer Q^2, where the coupling constant $\alpha_S(Q^2)$ is not small. Therefore, non-perturbative methods should be employed. Such an opportunity is offered to us by the quark-gluon string model (QGSM)[1] and similar to it dual parton model.[2]

2. Quark Gluon String Model

The QGSM is based on the so-called $1/N$ expansion of the amplitude for processes in quantum chromodynamics accomplished by a string-type phenomenological model that describes fragmentation of quarks into hadrons. Here N can be either the number of colors or flavors in the theory. The various graphs arising in the $1/N$ expansion have certain topologies. It appears that these diagrams correspond to processes with exchange of Regge singularities in the t-channel. For instance, the planar diagrams, representing the subprocesses with quark annihilation and exchange, can be treated as the exchange of secondary Reggeons, whereas the cylinder diagrams, describing the processes with color exchange, correspond to the Pomeranchuk poles, or exchange of Pomerons, in elastic scattering. Thus, the perturbative Reggeon Field theory (RFT) is linked to QCD.

The Monte Carlo version of the QGSM[3–6] employs statistical weights, hadron structure functions and leading quark fragmentation functions obtained from the Regge approach in[1] to choose subprocesses of string production, to compute mass and momentum of strings and to simulate string decays, respectively. The hadron-hadron collision part of the model includes single and double diffraction subprocesses, antibaryon-baryon annihilations and elastic scattering. Then, the hadron inelastic interaction cross section $\sigma_{inel}(s) = \sigma_{tot}(s) - \sigma_{el}(s)$ is split into the cross section for single diffractive interactions $\sigma_{SD}(s)$ and the cross section for non-diffractive reactions $\sigma_{ND}(s)$, as it is usually done in analysis of experimental data. By means of the Abramovskii-Gribov-Kancheli (AGK) cutting rules[7] the inelastic non-diffractive interaction cross section $\sigma_{ND}(s)$ can be expressed via the sum of the cross sections for the production of $n = 1, 2, \ldots$ pairs of quark-gluon strings, or cut Pomerons, and the cross section of double diffractive process

$$\sigma_{ND}(s) = \sum_{n=1}^{\infty} \sigma_n(s) + \sigma_{DD}(s) \ . \tag{1}$$

To find $\sigma_n(s)$ one can rely on the quasi-eikonal model[8,9] which states that

$$\sigma_{tot}(s) = \sum_{n=0}^{\infty} \sigma_n(s) = \sigma_P \, f\left(\frac{z}{2}\right) \ , \tag{2}$$

$$\sigma_n(s) = \frac{\sigma_P}{nz}\left(1 - \exp(-z)\sum_{k=0}^{n-1}\frac{z^k}{k!}\right) \ , \ k \geq 1 \tag{3}$$

$$\sigma_0 = \sigma_P\left(f\left(\frac{z}{2}\right) - f(z)\right) \ , \tag{4}$$

$$f(z) = \sum_{\nu=1}^{\infty}\frac{(-z)^{\nu-1}}{\nu\nu!} \ . \tag{5}$$

Here

$$\sigma_P = 8\pi\gamma_P \exp(\Delta\xi) \ , \tag{6}$$

$$z = \frac{2C\gamma_P}{(R_P^2 + \alpha'_P\xi)} \exp(\Delta\xi) \ . \tag{7}$$

The cross section σ_0 corresponds to diffraction contribution. The parameters γ_P and R_P are Pomeron-nucleon vertex parameters, quantity $\Delta \equiv \alpha_P(0) - 1$, and $\alpha_P(0)$ and α'_P is the intercept and the slope of the Pomeron trajectory, respectively. The quantity C takes into account the deviation from the pure eikonal approximation ($C = 1$) due to intermediate inelastic diffractive states, $\xi = \ln(s/s_0)$ and s_0 is a scale parameter.

The hard gluon-gluon scattering and semi-hard processes with quark and gluon interactions are also incorporated in the model, i.e. the eikonal $u(s, b)$ consists of two terms

$$u(s, b) = u_{soft}(s, b) + u_{hard}(s, b) \ . \tag{8}$$

Here b is the impact parameter of the collision. For the modeling of string fragmentation the Field-Feynman algorithm[10] is employed. It enables one to consider emission of hadrons from both ends of the string with equal probabilities. The break-up procedure invokes the energy-momentum conservation and the preservation of the quark numbers. Due to uncertainty principle it takes time to create a hadron from constituent quarks, e.g., fast particles are created the last. In string models two definitions of formation time are accepted:[11] the time when string is broken and all constituents of the hadron are created (constituent) or the time when the trajectories of hadron constituents (quarks) cross ("yo-yo"). In this version of QGSM we are using the constituent formation time. The formation time t_i^* and coordinate z_i^* of i-th hadron in the string center of mass can be expressed through its energy E_i^*, its longitudinal momentum p_{zi}^* and the longitudinal momenta/energies of all hadrons produced by the decay of this string as

$$t_i^* = \frac{1}{2\kappa}\left(M_s - 2\sum_{j=1}^{i-1} p_{zj}^*\right), \quad z_i^* = \frac{1}{2\kappa}\left(M_s - 2\sum_{j=1}^{i-1} E_j^*\right). \tag{9}$$

Then we calculate t_i in the laboratory frame and make the propagation of the coordinates to this point (x_i, y_i, z_i, t_i): $a_i = a_{0i} + t_i p_{ai}/E_i$, $a = x, y, z$. Note that κ acts as a scaling parameter of the particle formation time.

3. Comparison with Data

3.1. *Bulk observables*

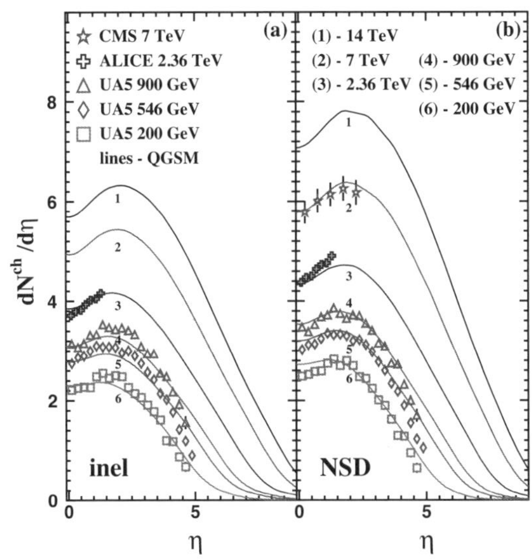

Fig. 1. The pseudorapidity spectra of charged particles in inelastic (a) and non-single-diffractive (b) events calculated in QGSM for pp collisions at $\sqrt{s} = 14\,\text{TeV}$ (1), 7 TeV (2), 2.36 TeV (3), 900 GeV (4), 546 GeV (5), and 200 GeV (6). Data are from.[12,15,16]

For the comparison with model calculations the experimental data obtained in pp and $\bar{p}p$ collisions at energies $\sqrt{s} = 200,\ 546,\ 900\ 1800,\ 2360$ and 7000 GeV[12–16] are used. To check the validity of a MC model one has to reproduce bulk features of the reaction, i.e., multiplicity, (pseudo)rapidity and transverse momentum distributions of charged particles. Pseudorapidity distributions of charged particles are shown in Fig. 1. One can clearly see the increase of charged particle density at midrapidity with rising $\sqrt{s}$, in contrast to Feynman scaling hypothesis that claims the independence of the $dN^{ch}/d\eta$ on s in midrapidity window after a certain energy threshold. Moreover, experiments at LHC have demonstrated non-linear rise of the $dN^{ch}/d\eta\,|_{\eta=0}$ with $\ln s$, whereas at energies below $\sqrt{s}$ =900 GeV this rise can be well approximated by linear dependence $dN^{ch}/d\eta\,|_{\eta=0} = a + b\,\ln s$. Note that appearance of quadratic term $\propto \ln^2 s$ is suggested by the saturation of the Froissart bound. The scheme with only soft Pomeron exchanges fails to reproduce such a non-linear behavior, and inclusion of the hard Pomeron in the MC model is absolutely necessary to describe the data at LHC. For pp collisions at $\sqrt{s} = 14\,\mathrm{TeV}$ the MC QGSM predicts

$$\frac{dN_{inel}}{d\eta}\,|_{\eta=0} = 6.1 \quad , \quad \frac{dN_{NSD}}{d\eta}\,|_{\eta=0} = 7.0 \quad . \tag{10}$$

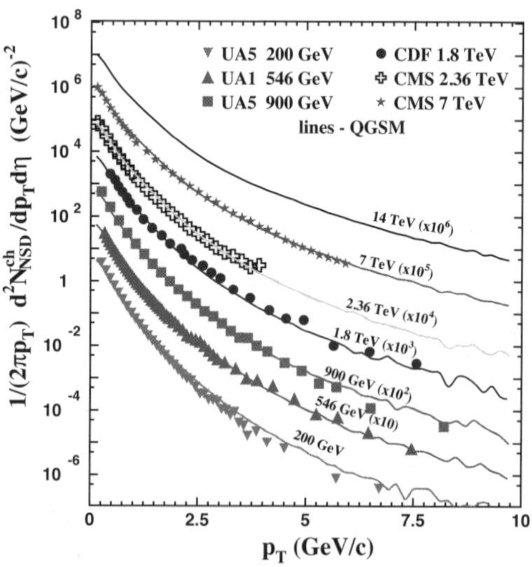

Fig. 2. P_T-distributions of the invariant cross section of charged particles in NSD pp collisions obtained in QGSM at $|y| \leq 2.5$ for all energies in question. Data are from.[12–14,16]

Figure 2 displays the invariant cross sections of charged particles in non-single diffractive pp collisions as functions of transverse momentum. The soft processes contribute mainly at low and intermediate transverse momenta, whereas at $p_T \geq 3\,\mathrm{GeV}/c$ the minijets dominate. Since the contribution of hard Pomeron exchanges to the total cross section rises with increasing collision energy, the p_T spectra of charged particles are getting harder. This is distinctly seen for the tails of the distributions at $p_T \geq 2.5\,\mathrm{GeV}/c$. Thus the average transverse momentum of the produced hadrons increases as well. For the NSD pp collisions at $\sqrt{s} = 14\,\mathrm{TeV}$ the model gives $\langle p_T \rangle = 0.56 \pm 0.03\,\mathrm{GeV}/c$.

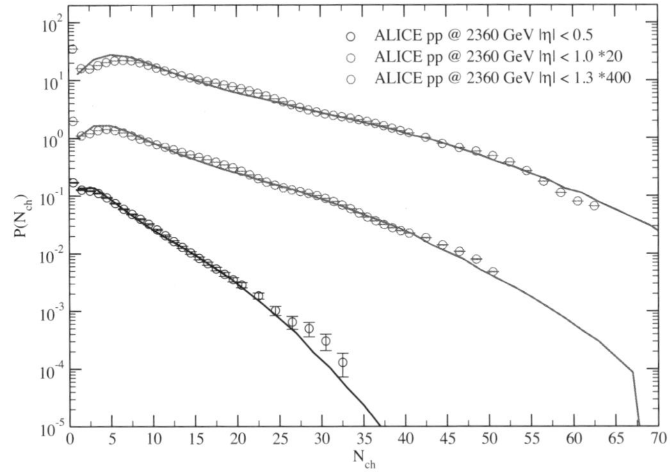

Fig. 3. Charged-particle multiplicity distributions in $|\eta| < 0.5$, $|\eta| < 1.0$ and $|\eta| < 1.3$ intervals, obtained in QGSM for pp collisions at $\sqrt{s} = 2360\,\mathrm{GeV}$. Open symbols present the ALICE data.[15]

The multiplicity distributions of charged particles calculated for pp collisions at center-of-mass energy $\sqrt{s} = 2.36\,\mathrm{GeV}$ in three pseudorapidity intervals are compared in Fig. 3 with the ALICE data. Again, the agreement between the calculations and the data is good. Both in the data and in the model results we observe wavy fluctuations at multiplicities above 30 in the two larger pseudorapidity bins. One of the possible explanations is that these fluctuations are related to fluctuations in the raw data.[15] In the QGSM, however, such wavy structure is assigned to many-Pomeron exchanges.[6,8] At lower energies the maxima of multiplicity distributions for multi-Pomeron processes are close to that for single-Pomeron exchange. With rising center-of-mass energy of the reaction these maxima are moved towards higher multiplicities. To see the distinct wavy structure one has to go to the full phase space.[8]

3.2. *Short-range, long-range and femtoscopy correlations*

Particles produced in string fragmentation are correlated in rapidity space. However, their correlation function drop exponentially with rising rapidity gap between the particles. Thus, hadrons far from each other in the rapidity space are uncorrelated, and short-range correlations emerge. For the inclusive process $a + b \to c + X$ the single particle inclusive cross section

$$f_c \equiv E\,\frac{d^3\sigma_c}{d^3p} = \frac{d^2\sigma(y_a - y_c, y_c - y_b, p^2_{T,(c)})}{dy_c d^2 p_{T,(c)}} \tag{11}$$

is independent on $y_c - y_b$ at sufficiently high energy $\sqrt{s}$ in the fragmentation region of particle a, where $y_a - y_c \sim 1$ and $y_c - y_b \gg 1$. The inclusive density f_c/σ_{el} becomes a function of only two variables, $y_a - y_c$ and $p^2_{T,(c)}$. This

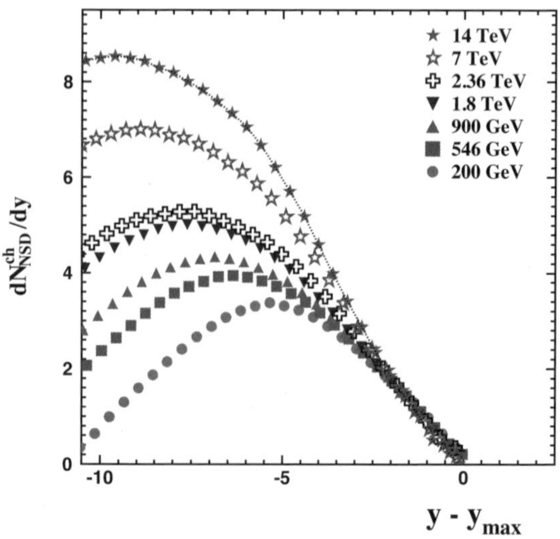

Fig. 4. The distributions $\dfrac{1}{\sigma_{NSD}}\,\dfrac{d\sigma_{NSD}}{dy}$ of charged particles as functions of rapidity difference $y - y_{max}$ obtained in QGSM for pp collisions at all energies in question.

phenomenon is called the extended longitudinal scaling.[12] It is equivalent to Feynman scaling in the fragmentation region. The fulfillment of the extended longitudinal scaling in the QGSM for pp collisions at energies up to top LHC energy is presented in Fig. 4. Here the distributions $\frac{1}{\sigma_{NSD}} \frac{d\sigma_{NSD}}{dy}$ are expressed as functions of $y - y_{max}$. The scaling holds to 10% in a broad energy range. In contrast, estimations based on extrapolations within the framework of the thermal model[17] predict much narrow distribution at $\sqrt{s} = 14$ TeV and, therefore, violation of the extended longitudinal scaling at LHC energies.

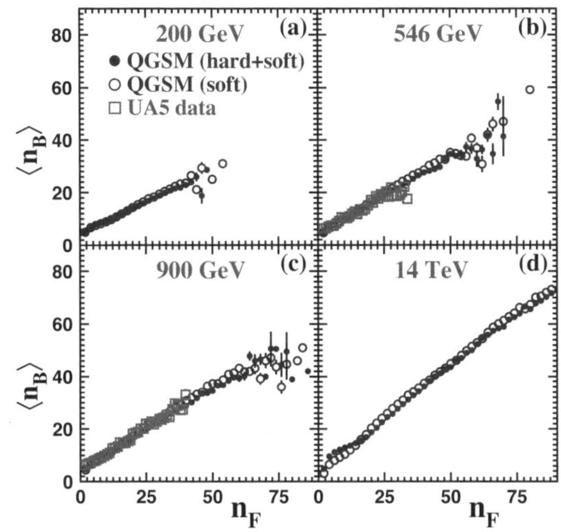

Fig. 5. Backward-forward multiplicity correlations $\langle n_B(n_F)\rangle$ for $0 \le |\eta| \le 4$ in NSD pp interactions at $\sqrt{s} = 200$ GeV (a), 546 GeV (b), 900 GeV (c) and 14 TeV (d). Open circles denote contributions of soft processes, full symbols are for all processes. Data are from.[12]

Long-range correlations between charged particles emitted in forward (F) and backward (B) hemispheres were first observed in.[18] The strength of the correlations is defined as

$$b = \frac{\langle (n_F - \langle n_F \rangle)(n_B - \langle n_B \rangle)\rangle}{\left[\langle (n_F - \langle n_F \rangle)^2\rangle \langle (n_B - \langle n_B \rangle)^2\rangle\right]^{1/2}}\,, \tag{12}$$

where n_F and n_B represent multiplicities of charged particles in forward and backward hemispheres, respectively. In Fig. 5 we show the dependence of the mean charged-particle multiplicity in the backward hemisphere $\langle n_B \rangle$, measured in the range $-4 \leq \eta \leq 0$, on the multiplicity in the forward hemisphere n_F for the symmetric range $0 \leq \eta \leq 4$ at four energies in question. Comparison with experimental data at $\sqrt{s} = 546\,\mathrm{GeV}$ and $900\,\mathrm{GeV}$ shows a good agreement between the model results and the data. The dependence looks pretty linear

$$\langle n_B(n_F) \rangle = a + b\, n_F \ , \tag{13}$$

and its slope b increases with energy. In the QGSM the rise of the strength of correlations is linked to increase of number of Pomerons, i.e. strings, with energy in the aforementioned pseudorapidity range. As one can see from Fig. 5, the correlations between $\langle n_B \rangle$ and n_F are fully determined by soft processes.

The momentum correlations are usually studied by means of two-particle correlation functions defined as a ratio of the two-particle distributions from the same event to the reference ones. Generally, the correlations are measured as a function of pair relative momenta four vector q or rather its invariant form $q_{inv} = \sqrt{q_0^2 - |q|^2}$.

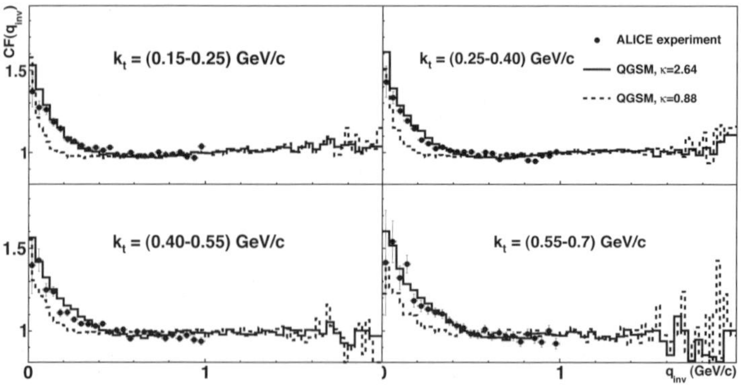

Fig. 6. The $\pi^+\pi^+$ correlation functions for pp at $\sqrt{s} = 900\,\mathrm{GeV}$ in four k_T bins, calculated in QGSM with $\kappa = \kappa_0$ (dashed line) and $3\,\kappa_0$ (full line). Cuts are $|\eta| < 0.8$ and $0.15\ GeV/c < p_T < 1.0\ GeV/c$. Data are from.[19]

The reference distribution is constructed by mixing the particles from different events of a given class. In both the STAR and the ALICE experiments the correlation function (CF) is fitted to a single-Gaussian

$$CF_{single}(q_{inv}) = \left(1 + \lambda exp(-R_{inv}^2 q_{inv}^2)\right) D(q_{inv}) \ . \tag{14}$$

The one-dimensional CFs fitted to Eq. (14) in different k_T-ranges of pp collisions at $\sqrt{s} = 900\,\text{GeV}$ are presented in Fig. 6 together with the ALICE experimental data. Because the model CFs are narrower, the extracted correlation radii are larger than those from the experiment. This result implies that the formation time implemented in the QGSM in a standard way should be diminished somehow. The best agreement with the data is obtained when the "scaling factor" κ equals $3\,\kappa_0$, as shown in Fig. 6. The mechanisms responsible for the decrease of the formation time can be related to interactions of strings, such as string fusion or percolation.

Acknowledgments

This work was supported by the Norwegian Research Council (NFR) under contract No. 185664/V30.

References

1. A. B. Kaidalov, *Phys. Lett. B* **116**, 459 (1982);
   A. B. Kaidalov, K. A. Ter-Martirosyan, *Phys. Lett. B* **117**, 247 (1982).
2. A. Capella *et al.*, *Phys. Rep.* **236**, 225 (1994).
3. N. S. Amelin, L. V. Bravina, *Sov. J. Nucl. Phys.* **51**, 133 (1990).
4. L. V. Bravina *et al.*, Phys. Rev. C **60**, 044905 (1999);
   E. E. Zabrodin *et al.*, Phys. Lett. **B508**, 184 (2001).
5. G. Burau *et al.*, Phys. Rev. C **71**, 054905 (2005);
   J. Bleibel *et al.*, Phys. Lett. **B659**, 520 (2008).
6. J. Bleibel *et al.*, *preprint* arXiv:1011.2703 [hep-ph].
7. V. Abramovskii, V. Gribov, O. Kancheli, *Sov. J. Nucl. Phys.* **18**, 308 (1974).
8. A. B. Kaidalov, *Surveys in High Energy Phys.* **13**, 265 (1999).
9. M. Baker, K. A. Ter-Martirosyan, *Phys. Rep.* **28C**, 1 (1976).
10. R. D. Field, R. P. Feynman, *Nucl. Phys. B* **136**, 1 (1978).
11. A. Bialas, M. Gyulassy, *Nucl. Phys. B* **291**, 793 (1987).
12. G. J. Alner *et al.* (UA5 Collab.), *Phys. Rep.* **154**, 247 (1987).
13. G. Arnison *et al.* (UA1 Collab.), *Phys. Lett. B* **118**, 167 (1982);
    C. Albajar *et al.* (UA1 Collab.), *Nucl. Phys. B* **335**, 261 (1990).
14. F. Abe *et al.* (CDF Collab.), *Phys. Rev. Lett.* **61**, 1819 (1988); *Phys. Rev. D* **41**, R2330 (1990).
15. ALICE Collaboration (K. Aamodt *et al.*), *Eur. Phys. J. C* **68**, 89 (2010); *Eur. Phys. J. C* **68**, 345 (2010); *Phys. Lett. B* **693**, 53 (2010).
16. CMS Collaboration (K. Khachatryan *et al.*), *JHEP* **1002**, 041 (2010); *Phys. Rev. Lett.* **105**, 022002 (2010).
17. J. Cleymans, J. Strümpfer, L. Turko, *Phys. Rev. C* **78**, 017901 (2008).
18. G. J. Alner *et al.* (UA5 Collab.), *Nucl. Phys. B* **291**, 445 (1987).
19. ALICE Collaboration (K. Aamodt *et al.*), *Phys. Rev. D* **82**, 052001 (2010).

REFLECTIONS

WALTER GREINER*

Frankfurt Institute for Advanced Studies, J. W. Goethe Universität,
Ruth-Moufang-Strasse 1, 60438 Frankfurt am Main, Germany
**E-mail: greiner@fias.uni-frankfurt.de*

This symposium was very special. It was topical: Some of the most outstanding problems in Nuclear Physics were discussed: Superheavy elements; extremely neutron rich elements, as well as nuclei with strangeness and their possible creation in the cosmos and on earth; the nuclear equation of state has to be identified within strongly compressed and hot nuclear matter as it appears in nucleus-nucleus encounters; giant nuclear systems which are short lived ($\sim 10^{-19} - 10^{-20}$ seconds) and extremely important for identifying the vacuum decay in overcritical electric fields (this is a very fundamental process - the most fundamental one in Quantum Electrodynamics!); astrophysical centers of extreme high density around which magnificent sun-like objects are Kepler-orbiting are discovered in our Galaxy by R. Genzel and colleagues (these centers are no black holes those don't exist at all because repulsive gravitational forces may play an important role - the pseudocomplex general relativity eliminates the Schwarzschild singularity); network physics for distributing energy (nuclear, wind, sun, tides,...) all over Europe (and over the world) is basic for energy consumption now and even more so in future. We heard wonderful talks and I am grateful to all the friends and speakers (from Russia, America, Europe and India) for coming to Goa. It was a great symposium! Particular thanks go to Professor Bikash Sinha and especially to Professor Debades Bandyopadhyay from Calcutta who had the idea for and organized this Goa-symposium.

My friends from Dubna deserve special embracing: In particular, Yuri Oganessian, who came here with his wife Irina and had to go home alone. The sudden death of Irina Oganessian was tragic and saddened all of us. She was a great lady, a superb violinist and a human in the true sense of the word. We will never forget her!

This symposium shortly after my 75th birthday gives me the opportunity to reflect somewhat about my scientific life.

It was in May 1964 that I returned with my wife Barbara and our first America-born son Martin from the University of Maryland to my alma mater at Freiburg/Breisgau, Germany. I had done some important work at Washington with Michael Danos: The coupling of giant resonances (proton-against neutron vibrations in nuclei) to the dynamics of the nuclear surface. It introduced a lot of structure into the photoelectric giant dipole and quadrupole motion, which was very soon experimentally discovered. A revival of photo-nuclear physics occurred. Also the width of the 1p-1h giant resonance doorway states due to their spreading into 2p-2h states was important to calculate. It received recognition. Invitations for colloquia and seminars at various places followed, in particular to Frankfurt and Darmstadt. They wanted me both as a professor and I received the offer in July 1964 and accepted the position as ordinarius at the Johann Wolfgang Goethe University in Frankfurt am Main where I also became director of the Institute for Theoretical Physics. It was the position which was held first by Max von Laue (1914-1919), Max Born (1919-1921), Erwin Madelung (1921-1949), Friedrich Hund (1951-1956), Helmut Steinwedel (1961-1963). My colleagues were Heinz Bilz (solid state theory) and Georg Süssmann (nuclear theory). A few years later when H. Bilz left for the MPI Stuttgart we had appointed Peter Fulde (solid state theory) whom I knew very well from the University of Maryland where he did his PhD. with R.A. Ferrell on the theory of superconductivity. Later, ($\sim$1969) after G. Süssmann left for München and Peter Fulde gave up his chair in Frankfurt because the radical left wing students revolted tremendously (the Egon Becker uproar) Reiner Dreizler (Atomic and Nuclear Theory), Harry Thomas (Solid State Theory - he left a couple of years later for Basel), Rainer Jelitto (Solid State Theory) and Hartmuth Haug (Solid State Theory : He succeeded Harry Thomas) became my colleagues as full professors at our institute. During my first three years as the Director of the Institut für Theoretische Physik also five C3-Professors (C3 is the administrative short name for Extraordinarius equivalent to Associate Professor with tenure) were newly established. Three of these I personally got when I negotiated with the ministry (Kultusministerium) in Wiesbaden, because I had an offer for a full professorship to the University of Virginia at Charlottesville, Va. Judah Eisenberg, M.E. Rose and Dexter Whitehead wanted me there to create a Center for Advanced Theoretical Studies. In addition to this my idea and very strong engagement for building a Joint Heavy Ion Laboratory for the Universities around Frankfurt (Darmstadt, Marburg, Gieβen, Mainz, Heidelberg) was honored. These 3 C3-professorships for Theoretical Physics should strengthen Theoretical Heavy Ion Physics and - therefore - the emerging new Joint Laboratory which became GSI (**G**esellschaft für **S**chwer-**I**onenforschung). One of the three C3-professorships we occupied with Bernhard Mrowka who, being ill with multiple sclerosis for many years and who had taken care of teaching elementary courses in theoretical physics for a long time, deserved a permanent position. This was a humanitarian act. Dieter Drechsel, Christian Toepfer, Werner Scheid, Berndt Müller, Johann Rafelski, Joachim Maruhn, Horst Stöcker and others held these Extraordinariats (C3-Professorships) for some time until they became full professors at distinguished universities all over the world. A school, the Frankfurt School of Theoretical Physics, developed. Basic in the educational program was that I introduced together with my colleagues H. Bilz and G. Süssman the teaching of courses for theoretical physics (Mechanics I, Mechanics II, Classical Electrodynamics, Quantum Mechanics) from the first semester on. Mathematical methods had to be woven in. This was quite new, because at all German universities Theoretical Physics has been taught to students after they had completed their second or third semester, i.e. from the 3rd or even 4th semester onwards.

This has been a most essential pedagogical step. The young students were fascinated by bringing them into contact with theoretical physics right away after entering the university. The exercises were done in small groups (not larger than 10 students) under supervision of an assistant. We called them Theoretica. Students could ask questions concerning the subjects discussed in the general course and had to solve the exercise problems. If they had difficulties they could ask the supervising assistant for help. The idea for organizing the education in Theoretical Physics this way stems from my distinguished teachers Prof. Hans Marschall and Prof. Theodor Schmidt at Freiburg/Breisgau University. It can be considered a revolution in the education of university physics teaching. Our experimental colleagues Werner Martienssen, Hermann Dänzer and Erwin Schopper at Frankfurt deserve thanks for their support!

When I came to Frankfurt it was clear to me that Experimental Nuclear Physics had to undergo a tremendous change. A 6 MeV van-de-Graff-accelerator or a 10-12 MeV Tandem, even of Emporer size, was out of date. Bigger accelerators were necessary, but such could not be carried by a single university. My colleague Erwin Schopper agreed to this and we came up with the idea for a joint great accelerator for the universities in the state of Hessen. Professor Peter Brix (Darmstadt) was contacted. I invited him over to Frankfurt for a discussion of this problem. He was immediately supportive of the idea and we agreed to invite also Profs. Wilhelm Walcher and Rudolf Bock from Marburg University. My dream was heavy ion physics. The study of nuclear molecules, Coulomb excitation of rotational and vibrational states to extremely high angular momenta, shape isomeric structures, and in particular, the creation of new, superheavy elements together with new atomic physics [electronic quasi molecules with very high central (two-center-) charges set up in the collision of very heavy ions] were the motivating ideas.

Rudolf Bock was extremely essential, because he was the only one who had already experimental experience with heavy ions (^{12}C was the heaviest heavy ion at those days - today we would call it a light ion). Also other ideas came up. In particular Friedrich Beck (Darmstadt) suggested to consider a several-hundred-MeV-proton cyclotron of the Indiana-type in order to study (short range) correlations between nucleons in nuclei. In a small, highly stimulating symposium of the participating professors from Frankfurt, Darmstadt and Marburg and their close collaborators both ideas were presented (F. Beck for the correlation physics and I pleaded for heavy ion physics). Over-whelmingly we decided to go for heavy ion physics.

The question was now which type of accelerator should be built.

We founded the "Arbeitsgemeinschaft Hessischer Kernphysiker" and wrote a first proposal for a "Forschungs- und Ausbildungs-Zentrum (1966)". The leading ideas, in particular the possibility to eventually create superheavy elements can already be found in this proposal. The next magic numbers for protons (Z =114) and neutrons (N =184,196) - depending on the underlying Mayer-Jensen shell model parameters - were presented. These calculations were done by my students Ulrich Mosel and Bernd Fink. Parallel to this Heiner Melder, being a student of Georg Süssmann, did a study of the single particle spectra around lead (Pb). His shell model parameters also indicated Z=114 as the next magic proton number. He presented his results at a conference at Lysekil (Sweden). Since that time the Z=114 number has become known to a broader community (Copenhagen, Berkeley). It is clear, however, that already in our first proposal of the Arbeitsgemeinschaft Hessischer Kernphysiker we presented Z =114 and also Z =126 as possible next magic proton numbers. In this context it is interesting to remark that Meldners thesis was objected by the Mathematicians of the Frankfurt faculty, because it was for them too soft work. Georg Süssmann (first referee) and I (second referee) pushed on for finally accepting Meldners thesis. Soon we did stability calculations for superheavy nuclei against fission, α-decay and electron capture. For nuclei in the middle of the stable valley rather long life-times (up to 10^{12} years) resulted (U. Mosel, J. Grumann, ...).

From Hermann Dänzers group, particularly from his collaborator Horst Klein (who became years later his successor) came the idea for a helix-accelerator (Wendellinearbeschleuniger). This group was novel and at that time lesser known. Peter Brix suggested to invite Christoph Schmelzer from Heidelberg to join. He had year-long experience in accelerator physics due to his work at CERN and had thought about a heavy ion linear accelerator. His thoughts were quite general and from a physics point of view qualitatively: the physics with complex nuclei could be a very interesting subject of research. Rudolf Bock described Christoph Schmelzer's joining of the Arbeitsgemeinschaft Hessischer Kernphysiker and documented the political maneuvers very well [see ISHIP 2006 "International Symposium on Heavy Ion Physics 2006"].

Already in 1967 we created a "Heavy ion colloquium": Every Tuesday afternoon we carried our collaborators by buses to Darmstadt, Marburg, Frankfurt and presented/discussed the new ideas and possibilities in heavy ion physics. Great enthusiasm resulted. Meanwhile the colleagues from Heidelberg woke up and wanted to erect a heavy ion accelerator at Karlsruhe, were they expected more direct access. However, when the Ministerpräsident of the state of Hessen in those days, Georg August Zinn, pleaded with Gerhard Stoltenberg (then Bundesforschungsminister) for the approval for creating the Heavy Ion Laboratory in the woods near Wixhausen between Frankfurt and Darmstadt (this place was selected by Peter Brix from the Technische Hochschule at Darmstadt) all the efforts of the Arbeitsgemeinschaft Hessischer Kernphysiker finally culminated: The new emerging laboratory was named GSI (Gesellschaft für Schwer-Ionenforschung) and Christoph Schmelzer became its first director. Together with Dieter Böhne and Norbert Angert he built the UNILAC-accelerator, which could accelerate all nuclei, also the most heavy ones like Pb and U over the Coulomb barrier, i.e. with energies higher than 6 MeV per nucleon. I did strongly plead for the heaviest ions because I saw great possibilities for the fusion of superheavy elements and - especially for the search for vacuum decay in supercritical electric fields. The UNILAC was (and still is) a marvellous accelerator!

During my first years in Frankfurt I studied the coupling of giant dipole and quadrupole resonances to the surface motion of nuclei. The proton-neutron-vibrations were described within the Steinwedel-Jensen-model and also in the particle-hole picture and for the quadrupole and octupole motions of the nuclear surface we used the collective coordinate description ($\alpha_{2\mu}$,$\alpha_{3\mu}$,..). The excitation of the 1^- and 2^+ - states by photons and electrons could be investigated by the electron acceleratros at Darmstadt and at Mainz. I got Dieter Drechsel interested in these problems.

Also the Eigenchannel Theory of Nuclear reactions has been practically put into action (H.G. Wahsweiler). With M. Danos I had developed this idea of calculating the eigenchannels, $V_c^{(l)}$ and eigenphases, δ_l of the S-matix directly in a R-matrix type way and then constructing the S-matix by $S_{cc'} = \sum_l V_c^{(l)} e^{2i\delta_l} V_{c'}^{(l)}$. Christian Toepfer applied it very successfully for calculating the dispersion effects in electron scattering.

As a young Professor in Frankfurt I assembled a few outstanding collaborators and students around me. As already mentioned, together with my colleagues H. Bilz , G. Süssmann and later R. Dreizler, Harry Thomas and R. Jelitto we decided to teach Theoretical Physics from the first semester onward (and not as up to then customary in Germany after the 3rd or 4th semester). Supplemented by Theoretica, i.e. small exercise groups of 10-12 students who had to solve given problems under the leadership of an assistant, the education in Theoretical Physics has received a new strong impetus. The very best mathematically minded students became interested in Theoretical Physics. It gave me great pleasure to work with young, brilliant minds. In this way a school of Theoretical Physics developed. I mention a few steps which come in my mind. With H. J. Weber and Max Huber I extended the Dynamic Collective Model (coupling the giant dipole resonances to the dynamical nuclear surface) to vibrational nuclei. Rich structure in the photonuclear giant resonance lead to great success, which was confirmed by experiments. This is also true for the giant quadrupole resonances. After receiving an offer for a full professorship from the University of Virginia, I went with my family to Charlottesville, Va. Werner Scheid, Christian Toepffer, Rainer Ligensa and Wilhelm Pieper accompanied me. There I wrote the 3 volumes on Nuclear Theory with Judah Eisenberg. North Holland published these books. They were - and still are - widely used and, more importantly, Judah and I became close friends. This friendship was even more exploited when Judah went to Tel Aviv/Israel and we established a close Frankfurt - Tel Aviv axis. Frequent mutual visits resulted.

The ideas of creating superheavy nuclei (U. Mosel, B. Fink, B. Fricke,...), studying nuclear molecules (Werner Scheid, J. Park,...) invoking the two -center-shell model (Holzer, Mosel, Maruhn,...) were essential steps in developing heavy ion physics. Interpreting A. Bromley's resonances found in

^{12}C-^{12}C, ^{12}C-^{16}O etc. scattering led to a close collaboration with Yale nuclear physics.

Fundamental problems like the decay of the vacuum in strong, supercritical electric fields as do appear in dynamical giant molecular systems like U+Th, U+U, U+Cm, (B. Müller, J. Rafelski, G. Soff, J. Reinhardt,...) let to great excitement. While some of the predictions and expectations, in particular those concerning quasi-molecular x-rays and δ-electrons were nicely confirmed experimentally (W. Meyerhof, J. Greenberg, Hartmut Backe), it turned out, unfortunately, that the line-structures in the positron production experiments were not reproducible (D. Schwalm, J. Greenberg, P. Kienle,...). A sort of atomic clock for the lifetime of the giant nuclear system was missing. Only 10 years ago the idea to use the mass-transfer or the energy-loss as a signal (pointer) for the atomic clock concerning the measurement of the sticking time (lifetime of the giant nuclear system) came up and has been quantified (V. Zagrebaev, M. Itkis,...). Now, utilizing this type fo atomic clock, there is great hope that the fundamental problem of vacuum decay may experimentally come to a conclusion (Sophie Heinz,...).

Around 1973 - I suggested with W. Scheid that nuclear shock waves should appear in high energy nucleus-nucleus collisions. Even though contradicted at first by an American theoretician, this paper and all the follow-ups (with H. Stöcker,...) paved the way to High Energy Heavy Ion Physics, making the study of the Quark-Gluon-Plasma (or Quark-Gluon-Meson-Baryon-Plasma) and all the very interesting and fundamental problems connected with that possible. With H. Stöcker we developed Quantum Molecular Dynamics and with Stöcker and Sorge the Relativistic Quantum Molecular Dynamics. Parallel to this the Navier-Stokes hydrodynamics (W. Scheid) was relativistically generalized and applied (J. Maruhn,..., D. Rischke).

The fragmentation theory has been developed in the 1970ies- 1980ies (H.-J. Fink, W. Scheid,...). The valleys in the potential energy surfaces (calculated within the Two-Center Shell Model utilizing the Strutinsky renormalization procedure) led on the one hand straightforwardly to an explanation of symmetric and asymmetric fission and on the other hand to the prediction of superasymmetric fission. This was first discovered by Mikhail Itkis 1982 and more recently by Friedrich Goennenwein (2010). The fragmentation theory led us also to the prediction and confirmation of cluster decay of nuclei, i.e. to cluster radioactivity (A. Sandulescu, D. Poenaru,...) which has been widely confirmed. These new decay modes entered even Encyclopedia Britannica.

I lectured regularly a few-hour-per week course with two hours-per week exercises for more that 40 years. Out of these lectures emerged my 14 books [Mechanik, Teil 1; Mechanik, Teil 2; Hydrodynamik; Klassische Elektrodynamik; Spezielle Relativitätstheorie; Quantenmechanik, Teil 1, Einführung; Quantentheorie, Spezielle Kapitel; Quantenmechanik, Teil 2, Symmetrien; Relativistische Quantenmechanik, Wellengleichungen; Quantenelektrodynamik; Eichtheorie der schwachen Wechselwirkung; Thermodynamik und Statistische Mechanik; Quantum Chromodynamics; Quantum Mechanics, An Introduction; Quantum Mechanics, Symmetries; Nuclear Theory Vol. 1: Nuclear Models; Nuclear Theory Vol. 2: Excitation mechanisms of the nucleus; Nuclear Theory Vol. 3: Microscopic theory of the nucleus; Quantum Electrodynamics of Strong Fields; Quantum Electrodynamics; Gauge Theory of Weak Interactions; Experimental Techniques in Nuclear Physics; Handbook of Nuclear Properties; Heavy Elements and Related New Phenomena; Relativistic Quantum Mechanics;] edited by H. Deutsch and Springer Companies. These books were translated into English, French and some even into Japanese and Chinese. They have found world wide recognition, some of them appeared already in their eighth edition. Nearly every week I get notes from students and physicists all over the world asking to supplement these courses by solid state physics, general relativity,... .

It gives me the pride to have had so many brilliant students and collaborators over nearly 50 years. More than 50 of them became professors at universities world-wide. My own sons, Martin and Carsten, followed their father in theoretical physics but with very independent new research fields: Martin, professor at Aarhus University, studies networks and applies them in different areas (one of them is the creation of energy networks). Carsten became my successor as professor at Frankfurt University. He engages in the depth of the very difficult questions arising in the plasma physics and the dynamics of quarks and gluons in hot and dense matter arising in high energy nucleus nucleus collisions.

In 2004 I was asked together with Wolf Singer by the President of the Johann Wolfgang Goethe-University, Rudolf Steinberg, to build up an interdisciplinary institute for Theoretical Sciences. I immediately retreated as Dean of the physics faculty and retired as a professor. Wolf Singer and I created the FIAS (Frankfurt Institute for Advanced Studies) were we study extraordinary problems in heavy ion physics (fusion of neutron-rich superheavy elements in underground atomic explosions (I. Mishustin), very neutron-rich nuclei at and outside the neutron drip line (Stefan

Schramm, Konstantin and Dima Gridnev, ...); in biology the sixth sense via magneto-reception of birds, fishes,... (Ilia Solov'yov, Klaus Schulten,...); phase transitions in nanostructures and complex biological molecules (Andrey Solov'yov, Alexander Yakubovich, Andrey Korol,...); the possibility of Gamma-lasers via channelling of charged particles in periodically bent crystals (Andrey Solov'yov, Andrey Korol,..., Hartmut Backe, E. Uggerhoj,...); the destruction of the DNA in heavy ion tumor therapy (Igor Mishustin, Andrey Solov'yov,...). Wolf Singer and his associates engage in brain research on a very broad level. Particular pleasure gives me recent work with Peter Hess (a former student of mine who is now professor at UNAM, Mexico City) on Pseudo-Complex General Relativity. It is a generalization of Einsteins General Relativity not allowing singularities like Black Holes, leading to Antigravitation at high mass densities and to a pulsating universe. I find it fascinating and have great hopes that it paves the way to a new understanding of gravitational problems and to the structure of our universe. Indeed, a widely spread and truly interdisciplinary research field of theoretical physicists together with biologists, medical doctors, chemists, experimental and technical physicists has developed.

I have great hope that FIAS evolves further into a first-class, internationally widely recognized and bright research center.